AF544887

Membrane Alterations as Basis of Liver Injury

FALK SYMPOSIUM 22

Membrane Alterations as Basis of Liver Injury

EDITED BY

Hans Popper,
Mount Sinai School of Medicine of the City University of New York,
New York,
USA

Leonardo Bianchi,
Department of Pathology,
University of Basel,
Switzerland

Werner Reutter,
Biochemisches Institut der Universität,
Freiburg/Br.,
West Germany

Proceedings of the First Part of the Basel Liver Week 1976,
held at the Hilton Hotel, Basel, Switzerland
October 6–7, 1976

Published by
MTP Press Limited
St. Leonard's House
St. Leonardgate
Lancaster, England

Copyright © 1977 MTP Press Limited

No part of this book may be
reproduced in any form without
permission from the publishers,
except for the quotation of
brief passages for the purpose
of review.

ISBN 0 85200 170 3

Printed in Great Britain at The Spottiswoode Ballantyne Press by
William Clowes and Sons Limited, London, Colchester and Beccles

Contents

List of Contributors

B. AGOSTINI
Max-Planck-Institut für
Medizinische Forschung,
Jahnstrasse 29,
D-6900 Heidelberg,
West Germany

A. C. ALLISON
Clinical Research Centre,
Watford Road,
Harrow,
Middlesex, HA1 3UJ,
England

E. BARAONA
Mount Sinai School of Medicine (CUNY),
Bronx Veterans Administration Hospital,
Bronx,
New York 10468,
USA

Ch. BAUER
Biochemisches Institut der Universität,
Hermann-Herder-Strasse 7,
D-7800 Freiburg/Br.,
West Germany

W. P. van BEEK
Division of Cell Biology,
Antoni van Leeuwenhoek-Huis,
Netherlands Cancer Institute,
Amsterdam,
The Netherlands

L. BIANCHI
Department of Pathology,
University of Basel,
Schoenbeinstr. 40,
CH-4056 Basel,
Switzerland

S. A. BOROWSKY
Medical Research Service,
Bronx Veterans Administration Hospital,
Bronx,
New York 10468,
USA

J. L. BOYER
Liver Study Unit,
Delartment of Medicine, Box 400,
University of Chicago, 950 E.59th St,
Chicago,
Illinois 60637,
USA

M. M. BURGER
Department of Biochemistry,
University of Basel,
Klingelbergstrasse, 70,
CH-4056 Basel,
Switzerland

M. G. CALDWELL
University of California, Davis
School of Medicine,
Veterans Administration Hospital,
150, Muir Road,
Martinez,
California 94553,
USA

R. COLEMAN
Department of Biochemistry,
University of Birmingham,
P.O. Box 363,
Birmingham B15 2TT,
England

G. DALLNER
Department of Pathology,
Huddinge Hospital,
Karolinske Institutet,
Stockholm,
Sweden

K. DECKER
Biochemisches Institut der Universität,
Hermann-Herder-Strasse 7,
D-7800 Freiburg/Br.,
West Germany

D. DOYLE
Department of Molecular Biology,
Roswell Park Memorial Institute,
Buffalo,
New York 14263,
USA

LIST OF CONTRIBUTORS

N. EHLER
Institut für Systematische Botanik und Pflanzengeographic der Universität, Im Neuenheimer Feld 280, D-6900 Heidelberg, West Germany

P. EMMELOT
Division of Cell Biology, Antoni van Leeuwenhoek-Huis, Netherlands Cancer Institute, Amsterdam, The Netherlands

Barbara ENGLAND
Department of Molecular Biology, Roswell Park Memorial Institute, Buffalo, New York 14263, USA

L. C. ERIKSSON
Department of Pathology, Huddinge Hospital, Karolinska Institutet, Stockholm, Sweden

W. H. EVANS
National Institute for Medical Research, Mill Hill, London, NW7, England

H. FAULSTICH
Max-Planck-Institut für Medizinische Forschung, Jahnstrasse 29, D-6900 Heidelberg, West Germany

G. FELDMANN
Unité de Recherches de Physiopathologie Hépatique, Hôpital Beaujon, F-92110 Clichy, France

M. M. FISHER
Department of Pathology, University of Toronto, Toronto, Canada

Becca FLEISCHER
Department of Molecular Biology, Vanderbilt University, Nashville, Tennessee 37235, USA

S. FLEISCHER
Department of Molecular Biology, Vanderbilt University, Nashville, Tennessee 37235, USA

W. FRANKE
Dt. Krebsforschungszentrum, D-6900 Heidelberg, West Germany

S. W. FRENCH
University of California, Davis School of Medicine, Veterans Administration Hospital, 150, Muir Road, Martinez, California 94553, USA

P. FREYCHET
Groupe de Recherches sur les Hormones Polypeptidiques et la Physiopathologie Endocrinienne, U. 145, Faculté de Médecine (Pasteur), Chemin de Vallombreuse, F-06034 Nice, France

Else FRIEDMAN
Department of Molecular Biology, Roswell Park Memorial Hospital, Buffalo, New York 14263, USA

M. FRIMMER
Institut für Pharmakologie und Toxikologie, Justus Liebig-Universität, Frankfurter Strasse 107, D-6300 Giessen, West Germany

K. FUNATSU
Department of Pathology, University of Toronto, Toronto, Canada

LIST OF CONTRIBUTORS

F. GUDAT
Department of Pathology,
University of Basel,
Schoenbeinstr., 40,
CH-4056 Basel,
Switzerland

M. J. HARDONK
Department of Pathology,
University of Groningen,
The Netherlands

F. W. HEMMING
Department of Biochemistry,
University of Nottingham Medical School,
Nottingham,
England

W. HOFMANN
Pathologisches Institut der Universität,
D-6900 Heidelberg,
West Germany

E. J. HOLBOROW
Medical Research Council Rheumatism
Unit,
Taplow,
England

G. HOLDSWORTH
Bone and Joint Research Unit,
The London Hospital Medical College,
Turner St,
London E1,
England

Esther HOU
Department of Molecular Biology,
Roswell Park Memorial Institute,
Buffalo,
New York 14263,
USA

Z. HRUBAN
Liver Study Unit,
Department of Medicine and Pathology,
University of Chicago,
Chicago,
Illinois 60637,
USA

W. A. R. HUIJBERS
Centre for Medical Electron Microscopy,
University of Groningen,
Groningen,
The Netherlands

H. E. HUXLEY
Medical Research Council Laboratory of
Molecular Biology,
Hills Road,
Cambridge, CB2 2QH,
England

W. JAHN
Max-Planck-Institut für Medizinische
Forschung,
Jahnstrasse 29,
D-6900 Heidelberg,
West Germany

B. JEANRENAUD
Laboratoires de Recherches Médicales,
64, Avenue de la Roseraie,
CH-1211 Geneva 4,
Switzerland

K. N. JEEJEEBHOY
Department of Pathology,
University of Toronto,
Toronto,
Canada

J. KARTENBECK
Dt. Krebsforschungszentrum
im Neuenheimer Feld 280,
Heidelberg-1

T. J. LAYDEN
Liver Study Unit,
Department of Medicine and Pathology,
University of Chicago,
Chicago,
Illinois 60637,
USA

M. A. LEO
Section of Liver Disease, Nutrition and
Alcoholism,
Mount Sinai School of Medicine (CUNY),
Bronx Veterans Administration Hospital,
Bronx,
New York 10468,
USA

C. S. LIEBER
Mount Sinai School of Medicine (CUNY),
Chief, Section of Liver Disease, Nutrition and Alcoholism,
Bronx Veterans Administration Hospital,
Bronx,
New York 10468,
USA

Y. le MARCHAND
Groupe de Recherches sur les Hormones Polypeptidiques et la Physiopathologie Endocrienne, U.145,
Faculté de Médecine (Pasteur),
Chemin de Vallombreuse,
F-06034 Nice,
France

M. MAURICE
Unité de Recherches de Physiopathologie Hépatique,
Hôpital Beaujon,
F-92110 Clichy,
France

I. MOLENAAR
Centre for Medical Electron Microscopy,
University of Groningen,
Groningen,
The Netherlands

D. J. MORRÉ
Department of Medical Chemistry and Pharmacology,
Purdue University,
West Lafayette,
Indiana 47907,
USA

H. P. MORRIS
Department of Biochemistry,
Howard University College of Medicine,
Washington, DC 20012,
USA

M. ODA
Department of Pathology,
University of Toronto,
Canada

C. PATZELT
Department of Biochemistry,
University of Chicago,
920 E.58th St,
Chicago,
Illinois 60637,
USA

E. PETZINGER
Institut für Pharmakologie und Toxikologie,
Justus Liebig-Universität,
Frankfurter Str. 107,
D-6300 Giessen,
West Germany

M. J. PHILLIPS
Department of Pathology,
University of Toronto,
Toronto,
Canada

HANS POPPER
at the City University of New York,
Mount Sinai School of Medicine,
New York,
New York 10029,
USA

W. REUTTER
Biochemisches Institut der Universität,
Hermann-Herder-Strasse 7,
D-7800 Freiburg/Br.,
West Germany

H. SCHIMASSEK
Institut für Biochemie I der Universität,
Im Neuenheimer Feld 328,
D-6900 Heidelberg,
West Germany

J. S. SIM
University of California, Davis School of Medicine,
Veterans Administration Hospital,
150, Muir Road,
Martinez,
California 94553,
USA

L. A. SMETS
Division of Cell Biology,
Antoni van Leeuwenhoek-Huis,
Netherlands Cancer Institute,
Amsterdam,
The Netherlands

Olga STEIN
Department of Experimental Medicine and Cancer Research,
Hebrew University Hadassah Medical School,
Jerusalem,
Israel

Y. STEIN
Lipid Research Laboratory,
Department of Medicine B,
Hadassah University Hospital,
Jerusalem,
Israel

I. STERNLIEB
Albert Einstein College of Medicine,
1300, Morris Park Avenue,
Bronx,
New York 10461,
USA

J. TWETO
P.O. Box B,
Frederick,
Maryland 21701,
USA

P. VISCHER
Biochemisches Institut der Universität,
Hermann-Herder-Strasse 7,
D-7800 Freiburg/Br.,
West Germany

O. S. VYVODA
Department of Biochemistry,
University of Birmingham,
P.O. Box 363,
Birmingham, B15 2TT,
England

A. K. WALLI
Institut für Biochemie I der
Universität,
Im Neuenheimer Feld 328,
D-6900 Heidelberg,
West Germany

Th. WIELAND
Max-Planck-Institut für Medizinische
Forschung,
Jahnstrasse 29,
D-6900 Heidelberg,
West Germany

M. H. WISHER
Department of Medicine,
St. Thomas's Hospital Medical School,
London, SE1,
England

I. YOUSEF
Department of Pathology,
University of Toronto,
Toronto,
Canada

List of Moderators

K. DECKER
Biochemisches Institut der Universität,
Herman-Herder-Strasse 7,
D-7800 Freiburg/Br.,
West Germany

R. SCHMID
Medicine Center,
University of California,
San Francisco,
California 94122,
USA

E. FARBER
Department of Pathology,
University of Toronto,
Banting Institute,
Toronto,
Canada

D. W. KING
Department of Pathology,
Columbia-Presbyterian Hospital,
New York,
New York 10032,
USA

Foreword

The investigation of biological membranes, particularly of the cell membranes, has blossomed in the last decade; information exists as to many biological systems with extrapolation from physical and chemical studies. It was, therefore, a challenge to survey the information applicable to the liver cell, with special emphasis on its plasma membrane. However, this had to include also some necessarily limited information on the membranes of organelles in the cell about which much more data have been produced. A symposium dedicated to the role of membrane alterations in pathological processes of hepatocytes was convened where experts on various aspects of membranes and on their specific components presented data which, albeit derived from other systems, might be relevant to the hepatocellular membranes and particularly to their alterations in disease. While the scope and amount of information offered were surprisingly large and also complex, evaluation of the relevance to liver disease remains preliminary, yet it encourages further studies on specific models, not only of experimental hepatic injury but also of human disease.

Since the content of the presentations could not be predicted in the planning of the conference, the coverage does not claim to be comprehensive, but, hopefully, the reader will see a panorama of the state of the art, appreciate the potential of the field, and be stimulated in both theoretical and clinical considerations.

As in any symposium report, it is difficult to reconcile the different forms of presentations. Some of them are comprehensive reviews, others are papers usually published by specialty journals, some are indeed prolonged abstracts, and a few are best designated as progress reports. The editors were confronted with the dilemma of enforcing some uniformity without altering more than necessary the original manuscripts of the authors, to whom we are grateful for rapidly providing us with their papers. Where we did more violence to their papers than they would have wanted, we offer our sincere apologies for our attempt to obtain uniformity and cohesion in order to facilitate reading. We are also aware of frequent duplications of statements in the different papers. They were retained not only to permit the authors the original structures of their papers, but also because it was felt that the reader may consider the individual papers as self-contained units, particularly if they wish to read only one of them. The conference was enlivened by active discussion, but only few discussion remarks were incorporated, to provide actual data not covered in the papers.

It is the editors' opinion that symposium reports like this serve a specific purpose beyond adding to the publication lists of the authors or to 'paper pollution'. They could be a successful tool of communication if they represent the state of the art in reviewing and evaluating data either of the author or of others and if they present new concepts based on the data as well as describe targets of new research. They will have fulfilled their purpose if they review the field for the worker and the one with sometimes tangential interest in it, but they will have served their purpose even more if they attract scientists or physicians to the field. It can be argued whether research reports structured in standard fashion belong in such volumes since they are apt to appear usually in peer-reviewed journals, often faster than in the symposium reports, and thus are duplications, disturbing at a time when the bulk of the literature is becoming overwhelming. They are not indexed if they appear in such volumes and thus their usefulness for both the author and the scientific community is questionable since many journals, especially in the USA, accelerate publication particularly of preliminary reports. All this created a quandary reflected in the editing of this volume.

The conference would not have been possible without the legendary generosity and organizational talent of Dr Herbert Falk of Freiburg, whose help and innovating spirit in this as in many other endeavours is appreciated by the international community of hepatologists.

Finally, we wish once more to thank the contributors, the audience, and the most understanding publisher, and we hope that the reader will appreciate the flavour of this new, fascinating and indeed barely charted field of the role of membrane injury in hepatocellular disease.

The Editors

Introduction

HANS POPPER

The history of this symposium explains its purpose and philosophy. When the opportunity arose five years ago to select a topic for a symposium to precede the major International Congress on Drugs and the Liver[1] in 1973, Collagen Metabolism in the Liver was selected. The intention was to bring experts in the collagen field together with hepatologists, a few of whom had worked on fibrosis, while most were to be shown the potential of research on collagen in the pathogenesis, diagnosis and management of liver diseases. The Collagen meeting resulted in the presentation of interesting data[2], but the main consequence, and the desired one, was that in the years following the symposium, experts in connective tissue metabolism became interested in the liver as a model for their work and hepatologists started more and more to interest themselves in, and to work on, alterations of hepatic connective tissue.

When again a topic was to be chosen for this year's symposium, hopefully, to repeat the record of stimulation of research work rather than of presentation of data, the cell membranes of the hepatocyte were chosen. This was done to bring approach, information and technology in the rapidly expanding field of cell membranes together here and to apply it not only to the biology of the liver but, more importantly, to the understanding, diagnosis and eventual therapy of liver diseases. No attempt was made to duplicate the many recent reviews on the cell membranes[3–6].

Alterations of the cell membrane have several effects on the communication of the hepatocyte with the organism. They influence the transport of nutrients and waste substances in and out of the liver, they regulate the acceptance of hormonal and other messengers by the hepatocytes, and also permit hepatocytes to communicate with each other. Cell membranes play a role in phagocytosis and endocytosis and also in locomotion, though the latter is of little significance in the liver. Cell membranes and their lesions thus determine the social behaviour and restraints of the cell. This is particularly important in the regulation of hepatocellular function which depends little if at all on nervous innervation. Moreover, cell membrane alterations are also crucial in the two key phenomena of liver disease. In the mechanism of cholestasis the lesion of the canalicular membrane of the hepatocyte has moved into the centre of the pathogenesis and the last few years have yielded exciting observations. Far more problematic, however, is another key problem in liver diseases, namely the mysterious mechanism of death of the hepatocytes, which is the basic phenomenon in all liver diseases. Degeneration and necrosis of hepatocytes are the cause of death from liver injury in acute and often in chronic hepatic failure. Rational therapy for this failure will only be developed if this elusive mechanism is fully understood. While many initiating processes in this necrosis are being discovered, the terminal event common to all these processes is the functional failure and the subsequent destruction of the plasma membranes. This is the reason it is contemplated in the summary of this symposium, also to evaluate whether progress has been made or leads have been developed to understand the mechanism of liver cell death.

Again, experts in membrane research, of both the normal and the damaged membrane, participated. Some of them have used the hepatocytic membrane as a model, but most of the work on cell membranes was done on other models including artificial membranes, the membranes of bacteria, erythrocytes, cell organelles such as mitochondria and the chloroplast. In the case of hepatocytic membranes, much more is known about those of mitochondria and about the endoplasmic reticulum than about the plasma membrane and particularly its different portions, in view of the difficulties in making pure preparations of hepatocytic plasma membranes and fractions of the different parts[7]. Much has been learned from membrane investigations in clinical disciplines other than hepatology, such as the sister discipline, gastroenterology, but probably more from haematology, immunology, endocrinology and recently even neuroscience and anaesthesiology. The cell membrane alteration in neoplasma is of special interest[8] and much of our present knowledge of the normal and the abnormal function of the plasma membrane has been derived from the study of carcinogenesis and of the character of transformed cultured cells. Attempting to focus on the injury of the liver cells confronted us, however, with an abundance of often disjointed aspects, and the selection had to be difficult and arbitrary.

Human liver cell injury, aside from necrosis, in years past was exemplified by the cloudy swelling of the waterlogged organ which, as we know now, is associated with increased intracellular protein apparently from *de novo* synthesis, presumably as a result of excess penetration of amino acids through the altered membrane[9]. Then came Roessle's and Eppinger's concept of serous inflammation[10] as the basis of liver injury, when altered permeability of the capillary membrane was incriminated (Figure 1). The importance of this concept here is the emphasis on the processes in the perisinusoidal space, where fluid exchange and transport occur. Recent investigations suggest the presence of collagenous basement membrane components in this space also in the normal liver[11], and their dynamic formation and disappearance after acute injury[12]. These basement membranes were originally thought to play a role only in physical support, just as the cell membrane was originally considered to be only a device to support the cell and to create a compartment.

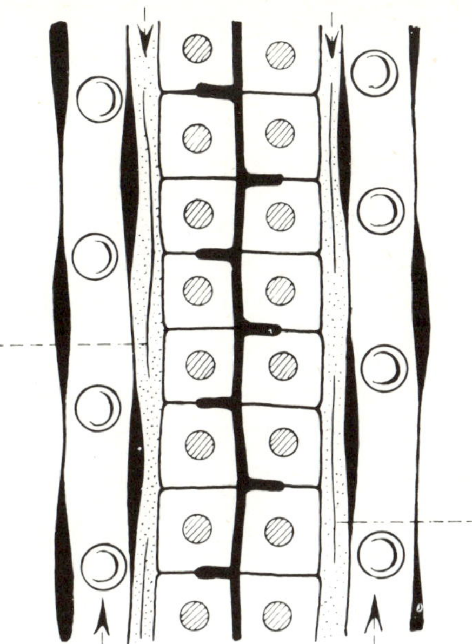

Figure 1 The historical concept of hepatocytes/blood relations as the basis of serous inflammation. Reproduced from Eppinger, Kaunitz and Popper[10]

The basement membrane of the nerve cell may be, however, associated with acetylcholine esterase[13], which means it participates in metabolic activity. The future may therefore detect transport mechanisms in the fluid milieu around the hepatocytes. Oligosaccharide extensions of receptors extend into this perisinusoidal space. At this meeting we plan to put less emphasis on the collagen-traversed interstitial space but more on the mechanisms involved in integrity and transport in the cell membrane itself in order to draw conclusions about the failure of these mechanisms.

In the hepatocytic plasma membrane, several domains are conventionally recognized (Figure 2). The first is the perisinusoidal portion, the greatest one in length, enlarged by the formation of irregular microvilli and endowed with receptors. It lines the perisinusoidal space where the exchange between hepatocytes and tissue fluid and, in turn, blood takes place. Its chemical constitution as well as its anatomy and function have been studied more thoroughly than the other domains. Its length varies in that it may extend between hepatocytes to the intercellular cleft as a pericellular space, also endowed with some microvilli, depending on function. In some conditions, such as in regeneration, in cirrhosis[14], or in cholestasis[15], only a short intercellular cleft remains. In these clefts the cell membranes are almost completely straight except for a few pegs and holes, serving probably less the structural integrity than communication between neighbouring cells, the nature of which is poorly understood. This communication is probably localized in

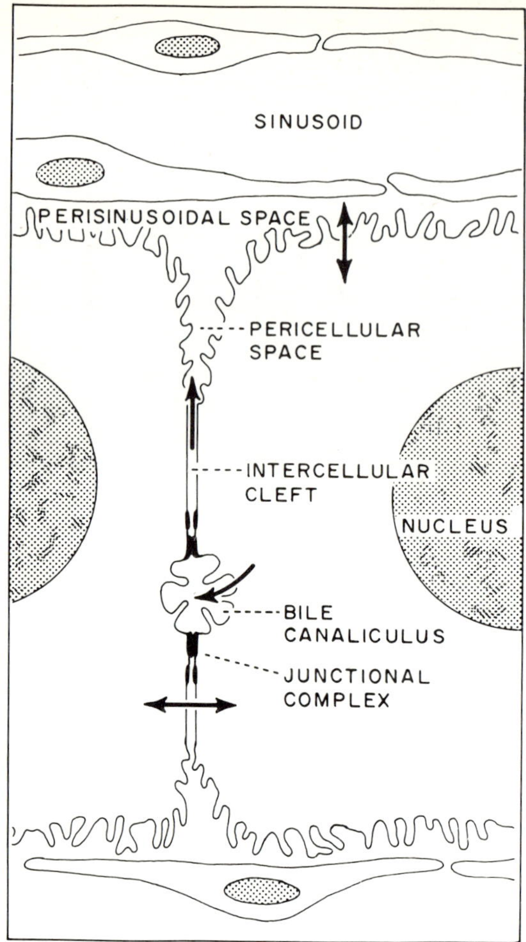

Figure 2 Schematic drawing of various domains of the plasma membranes of the hepatocytes

specialized areas of the cleft, the so-called gap junctions. The intercellular cleft is separated from the bile-canalicular lumen by the structurally elaborate junctional complex, which is a tight junction and might serve support by stabilizing the structure, but there is increasing evidence that these tight junctions are permeable in either direction for electrolyte-containing fluid but more important probably in transfer from the perisinusoidal space to the bile[16]. The regular presence of a mitochondrion attached to the desmosome suggests that energy may be required for this transfer[17]. The fourth distinct domain is the bile canaliculus, which normally is almost completely filled by regular finger-like microvilli. The microvilli around the tight junction, however, seem to differ from those on the roof of the canaliculi since they are less deformed in cholestasis[18]. At this conference the emphasis will be on the functional alterations of the perisinusoidal membrane and the one lining the bile canaliculi, on both of which the best information is available, and which are known to differ in function and transport mechanism despite many common structural and functional features.

In introducing the problem of the membrane, emphasis has to be placed on its dynamic behaviour, ever-changing in response to the needs of the organism or the cell, resulting from a balance of contending forces. While it separates external from the intracellular milieu, it allows for transmembranous control. Two key processes in the membrane are movement as well as stabilization of its components. Determination of the chemical constitution thus reflects little of the crucial topographic arrangement of the membrane components, which either have a random or an ordered distribution. We are concerned here mainly with the function and submicroscopic structure of three different layers (Figure 3). The middle one, the membrane itself, consists of two leaflets of phospholipids into which macromolecules of different shape and assembly are incorporated. These integral membrane proteins, including lipoproteins and

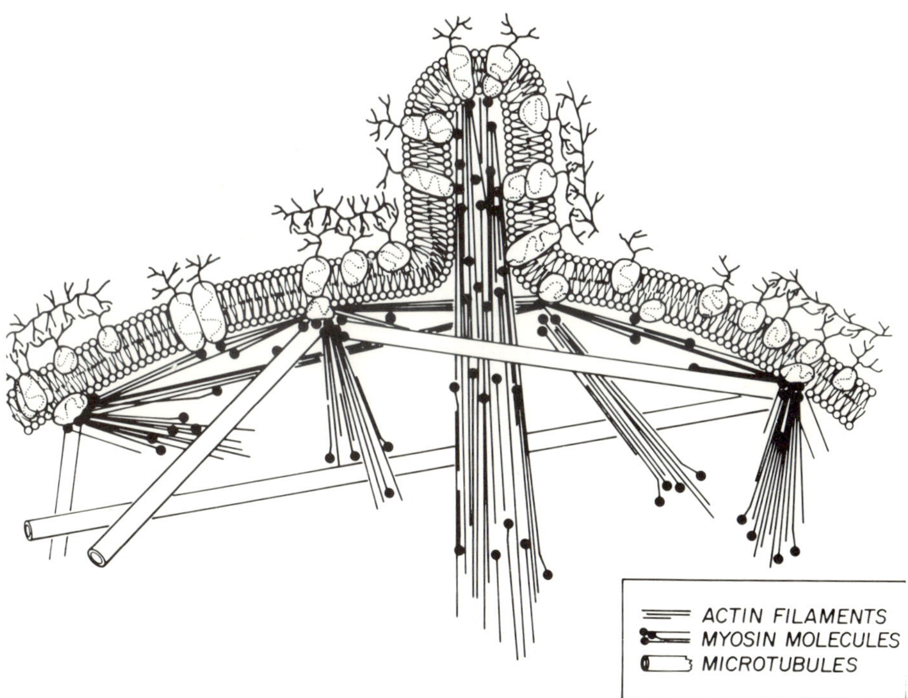

Figure 3 Schematic relation of the receptors, their oligosaccharide extensions to the plasma membrane and their hypothetical interaction with the cytoskeleton. Redrawn, by permission, from Nicolson, G. L. (1976) *Biochim Biophys. Acta,* **457**, 57, and from Nicolson and Poster[8]

glycoproteins, have specific functions both for the structural integrity of the dynamic membrane and for transport. The latter, usually energy-dependent, concerns both ions, mainly with the help of pumps, and various organic substances including drugs, bile acids and metabolites; the latter is usually facilitated by carrier proteins and its action requires particular consideration since simple flip-flop rotation is now doubted. The arrangement of central hydrophobic and peripheral hydrophilic areas of both integral membrane proteins and phospholipids assists in transport. In the hydrophobic area, glycoproteins and phospholipids are linked to produce a semiliquid consistency allowing for lateral movement. The biogenesis of the various components which have a different lifespan and their assembly as well as degradation are important. The coordination of all these factors provides rapid and reversible modulations of the surface and allows for confirmational changes as well as for steady states required by orderly function.

The second, the external or peripheral layer, can be removed by chelating agents without destroying the membrane. This outer layer is the carbohydrate-rich glycocalix of the morphologist. Oligosaccharide extensions of the receptors, anchored in the membrane in an asymmetrical distribution, extend into this layer. The receptors bind peptide hormones, neurotransmitters, drugs, toxins, viruses and antibodies[19]. Their key functions are recognition, reversible binding which changes the receptors, and also movement. The mechanisms of their action, their influence on hepatocytic function and the transmission of the message they receive into the interior of the cell vary, and sometimes interaction with effectors is required. Alteration of the receptor function may be a key process in cell dysfunction and death but also in irregular cell proliferation and even in cancer.

The third layer, on the inner surface of the leaflet, is the cytoskeleton consisting of at least two elements. One is the contractile microfilaments, 6–8 nm in diameter, of double helical actin nature, and combined with myosin. The role of these contractile elements, originally studied in the muscle cell, is now widely recognized in other cells including hepatocytes. The other elements are rigid microtubules, 25 nm in diameter and consisting of a glycoprotein. This microskeleton varies in distribution below the cell membrane. It serves not only for support and maintenance of the membrane tone but also for secretion including vesicular transport. It fixes the receptors in a dynamic fashion with microfilaments and microtubules serving antagonistic functions. It participates in endocytosis, internalizing membrane components, as well as in the shedding of cell components which, through a process called apoptosis[20] may result in elimination of the cell. The effects of various injurious agents, including local anaesthetics[21], permit the analysis of function and failure of the cytoskeleton. We expect that in addition to colchicine, cytochalasin B, and concanavalin A special emphasis will be given to phalloidin, one of two groups of phallotoxic polypeptides isolated from the poisonous mushroom *Amanita phalloides*. The other group consists of amanitines which are mainly responsible for the toxic effects in man[22]. The cytoskeleton is, however, by no means restricted to the submembranous location and fulfils an important transport and channelling function within the cell.

Although concentrating on the three layers around the membrane, we will have to consider, albeit in passing, the associated role of intracellular systems, for instance the second messengers, the cyclic nucleotides, or the cytoplasmic calcium concentration, which both influence and are influenced by membrane function, as well as the pattern of intracellular synthesis and transport of membrane components. We will be less concerned with the intracellular carrier proteins because that would widen the scope of this symposium unduly. The emphasis should thus be on disturbance of the selective permeability of the membrane, on the polarity of the cell, and on the altered distribution of floating proteins, lipids and carbohydrates.

We therefore would like to obtain a better understanding of liver injury and at least some answer to four problems: (1) hepatic effects of altered composition of the dynamic membrane and of its integral macromolecules, (2) hepatic consequences of receptor lesions, (3) hepatic sequelae of cytoskeleton changes, and (4) functional differences between perisinusoidal and bile-canalicular membranes in liver injury.

References

1. Gerok, W., and Sickinger, K. (eds.) (1975). *Drugs and the Liver* (Stuttgart: F. K. Schattauer Verlag)
2. Popper, H. and Becker, K. (eds.) (1975). *Collagen Metabolism in the Liver* (New York: Stratton Intercontinental Medical Book Corporation)
3. Singer, S. J. (1974). The molecular organization of membranes. *Ann. Rev. Biochem.*, **43**, 805
4. Gross, W., Hülsen, W. and Ring, K. (eds.) (1975). *Biologische Membranen* (Köln: Kiepenheuer and Witsch)
5. Mora, P. T. (ed.) (1976). *Cell Surfaces and Malignancy* (Washington, D.C.: DHEW Publication (NIH) 75-796)
6. Hughes, R. C. (1976). *Membrane Glycoproteins: A Review of Structure and Function* (Sevenoaks, Kent: Butterworths)
7. Toda, G., Oka, H., Oda, T. and Ikeda, Y. (1975). Subfractionation of rat liver plasma membrane. Uneven distribution of plasma membrane-bound enzymes on the liver cell surface. *Biochim. Biophys. Acta,* **413**, 52
8. Nicolson, G. L. and Poste, G. (1976). The cancer cell: dynamic aspects and modifications in cell surface organization. *N. Engl. J. Med.,* **295**, 197
9. Fonnesu, A. (1960). Changes in energy transformation as an early response to cell injury. In: H. B. Stoner and C. I. Threlfall (eds.), *The Biochemical Response to Injury* (Springfield, Ill.: Charles C. Thomas)
10. Eppinger, H., Kaunitz, H. and Popper, H. (1935). *Die seröse Entzündung. Eine Permeabilitäts-Pathologie* (Wien: Julius Springer)
11. Chung, E., Rhodes, R. K. and Miller, E. J. (1976). Isolation of three collagenous components of probable basement membrane origin from several tissues. *Biochem. Biophys. Res. Commun.,* **71**, 1167

12. Cossel, L. (1966). Über akutes Auftreten von Basalmembranen an den Lebersinusoiden (Beitrag zur Kenntnis der kapillären Basalmembran). *Beitr. Pathol. Anat.*, **134**, 103
13. Lwbuga-Mukasa, J. S., Lappi, S. and Taylor, P. (1976). Molecular forms of acetylcholinesterase from *Torpedo californica:* Their relationship to synaptic membranes. *Biochemistry*, **15**, 1425
14. Phillips, M. J., and Steiner, J. W. (1965). Electron microscopy of liver cells in cirrhotic nodules. I. The lateral cell membrane. *Am. J. Pathol.*, **46**, 985
15. Compagne, J., and Grisham, J. W. (1974). Electron microscopy of extrahepatic biliary obstruction. *Arch. Pathol.*, **97**, 348
16. Diamond, J. M. (1974). Tight and leaky junctions of epithelia: a perspective on kisses in the dark. *Fed. Proc.*, **33**, 2220
17. Sternlieb, I. (1969). Mitochondrion–desmosome complexes in human hepatocytes. *Z. Zellforsch.*, **93**, 219
18. Vial, J. D., Simon, F. R. and Mackinnon, A. M. (1976). Effect of bile duct ligation on the ultrastructural morphology of hepatocytes. *Gastroenterology*, **70**, 85
19. Kahn, C. R. (1976). Membrane receptors for hormones and neurotransmitters. *J. Cell Biol.*, **70**, 261
20. Kerr, J. F. R., Wyllie, A. H. and Currie, A. R. (1972). Apoptosis: A basic biological phenomenon with wide-ranging implications in tissue kinetics. *Br. J. Cancer*, **26**, 23
21. Poste, G., Papahadjopoulos and Nicolson, G. L. (1975). Local anesthetics affect transmembrane cytoskeletal control of mobility and distribution of cell surface receptors. *Proc. Natl. Acad. Sci. USA*, **72**, 4430.
22. Wieland, Th. (1968). Poisonous principles of mushrooms of the genus *Amanita*. Four-carbon amines acting on the central nervous system and cell-destroying cyclic peptides are produced. *Science*, **159**, 946

1
Principles of Membrane Differentiation

W. W. FRANKE and J. KARTENBECK

The past decade of research on membranes has been dominated by three at the first sight seemingly opposing concepts. (1) Membranes have been classified according to their intracellular location and have been discussed as morphologically continuous, *ergo* compositionally homogeneous structures. (2) Another concept has emphasized the fluidity of the membrane interior and the lateral mobility of membrane constituents, in particular membrane proteins[1,2]. (3) On the other hand, numerous studies have demonstrated the existence and formation of a variety of intramembranous domains characterized by a specific selection of one or several components which frequently occur in precisely localized positions and are obviously excluded from lateral movements by intramembranous or exogenous (cytoplasmic and extracellular and intravesicular, respectively) constraints. Examples of this latter type of organization are presented by the clusters of viral (glyco) proteins that are produced during bud formation at intracellular or plasma membranes (Figure 1; for references see below), by special intercellular junctional complexes (e.g. Figure 3; for review see[3]), by various forms of paracrystalline membrane plaques[4], etc. Taking together these concepts and the present experimental data, a membrane would thus have to be regarded as a specific mosaic of both fixed and mobile domains embedded in a morphological continuum of hydrophobic nature. Many membrane phenomena may primarily reflect alterations in the pattern of the arrangement of membrane components rather than chemical changes of the individual molecular constituents of the membrane as such. In discussions of membrane changes resulting in the formation of a different and/or differentiated membrane the following situations of 'membrane differentiation' should be distinguished.

CELL DIFFERENTIATIONS THAT ARE CHARACTERIZED BY CHANGES OF THE QUANTITIES OF MEMBRANES

In the course of many cell differentiation processes large increases of total surfaces of one or several membranes are observed, either in terms of absolute mass or relative to other cell membranes. Examples include (1) the development of rough surfaced endoplasmic reticulum (RER) in cells particularly active in protein secretion; (2) the formation of an enlarged smooth ER (SER) in cells producing steroid hormones or, in hepatocytes, after treatment with xenobiotic compounds; (3) the development of the Golgi apparatus (GA) in cells secreting carbohydrate-rich product; (4) the transitory accumulation or storage of specific vesicles within the cell, including increases in peroxisomes, lysosomes, secretory vesicles, etc.; (5) an enlargement of the cell surface; (6) increases in the numbers or sizes of mitochondria and plastids as well as changes in the ratio of the areas of inner and outer membranes of these organelles; and (7) increases of the nuclear surface, i.e. of the nuclear envelope (NE). Correspondingly, other cell differentiations are characterized by the reduction and, in some cases, complete disappearance of specific membranes[5]. Critically seen, however, the use of the term 'membrane differentiation' to designate such situations is not quite appropriate since these processes represent rather a change of the cell than a change of the specific membrane itself, which may remain unaltered.

DE NOVO APPEARANCE OR COMPLETE LOSS OF SPECIFIC MEMBRANE STRUCTURE IN THE COURSE OF CELL DIFFERENTIATION PROCESSES

In some cell differentiations novel membrane structures appear such as vacuole formation in plants, formation of paired cisternae and annulate lamellae (AL) in various animal and plant cell systems, and the formation of storage or secretory vesicles characteristic for a specific differentiated cell state. Correspondingly, losses of membrane structures such as plastids, ER, GA, and even NE, are observed in other cell differentiation processes[5].

QUANTITATIVE CHANGES OF SPECIFIC PARAMETERS AND PROPERTIES IN MEMBRANES

Examples of quantitative changes of certain substructures in a specific membrane system include the polyribosome density of RER, the proportions of tubular and vesicular components of SER or GA, the density of pore complexes in NE and AL, the density of ATPase-containing particles on mitochondrial cristae, and the frequency of cell surface specializations. Variations in the density of the interaction of non-membranous chemical moieties with membranes such as the association of membrane surfaces with peripheral proteins and cytoplasmic proteins including tubulin, spectrin, actin, alpha-actinin, myosin, and clathrin[6] also fall into this category. Again, such quantitative changes do not necessarily 'differentiate' the membrane character but it is obvious that, from a certain extent onwards, they might result in the formation of special clusters and domains and thus change the membrane nature itself, i.e. result in true differentiation (see below).

INTRAMEMBRANOUS DIFFERENTIATIONS (LATERAL MEMBRANE DIFFERENTIATIONS)

Some examples of lateral intramembrane heterogeneities are presented in Figures 1 to 3[5]. Regions of membranes which are of limited size and different from other regions of the specific membrane are represented not only by the relatively long-lived intercellular junction complexes but also by transitory differentiations. Especially clear examples are the formation of viral envelopes by budding which results in the specific enrichment and assembly of viral (glyco)proteins in the membrane bud[7,8] and the production of milk fat globules[9,10]. Important in the maintenance and establishment of such differentiated plaques in membranes seems to be the constraint of lateral mobility of the specific membrane components[5]. Such a constraint seems often to involve the mutual selective interaction of various membrane constituents and/or the interaction of specific membrane constituents with membrane-associated structures and molecules of the nucleo- or cytoplasmic compartments and of intracisternal or extracellular spaces; in some situations membrane-to-membrane interactions might also be involved[5,11]. Such special forms of membrane morphology indicate that this lateral heterogeneity is functionally important. For example, in an epithelial cell the basal plasma membrane might be structurally and functionally different from the apical surface, and both might be different from the lateral plasma membrane sides which in addition contain various differentiated regions for intercellular connection (*cf.* also the article by Wisher and Evans (Chapter 8) in this book). A particularly clear example of a structurally and chemically differentiated domain within one region of a cell membrane is the thickened plaques of hexagonally packed subunits that occur in the luminal plasma membrane of the mammalian urinary tract epithelium[4]. It is also obvious that the turnover of different loci and components in specific membranes can greatly differ[5]; (*cf.* also the article by Coleman *et al.* (Chapter 8) in this book). Nuclear pore complexes disappear in some regions of the nuclear surface of spermiogenic cells such as in the acrosome- or centriole-associated regions but are maintained in others[5]. There exist regions in the surface membrane in which endo- or exocytotic events occur relatively frequently, and membrane heterogeneity resulting from such events is often

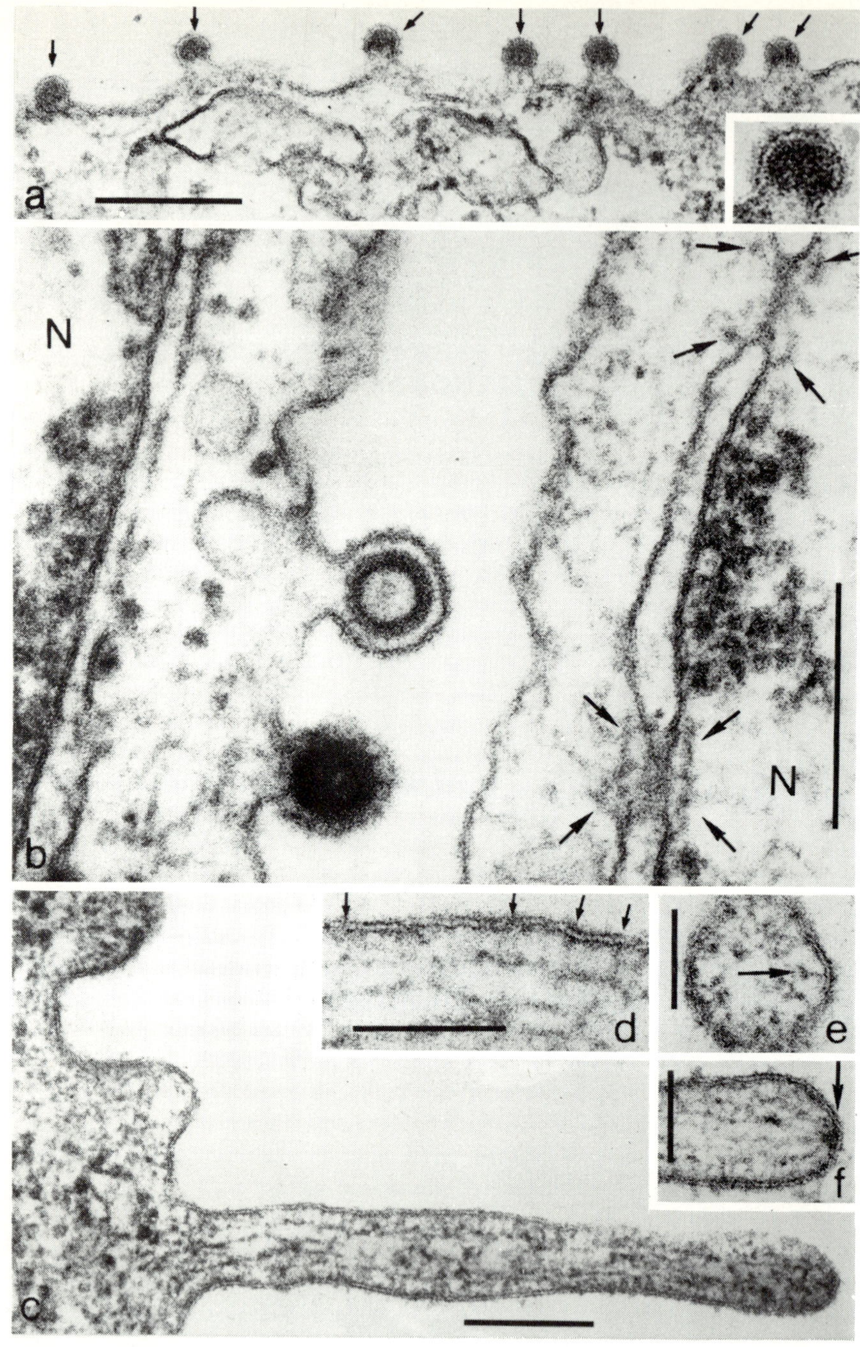
a
N
N
b
d
e
f
c

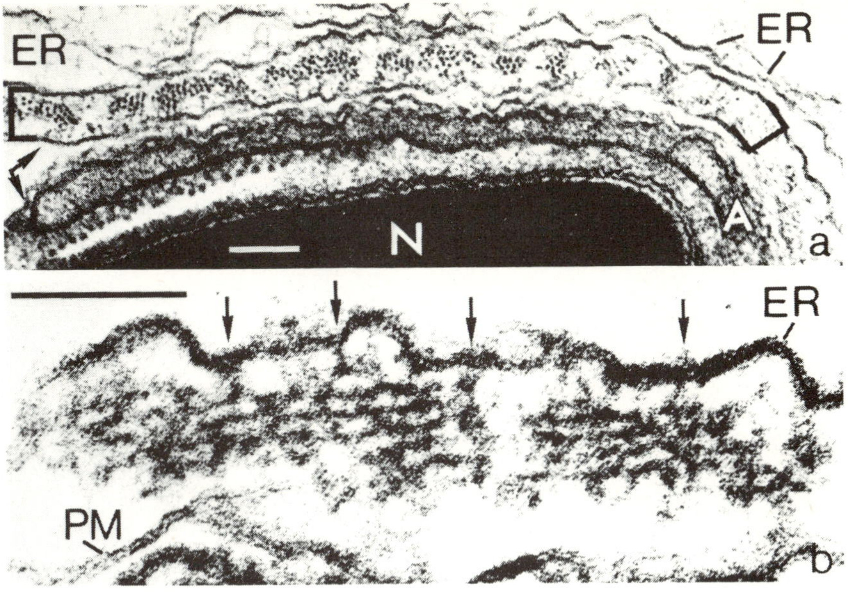

Figure 1 (left) Electron micrographs of ultrathin sections showing examples of local membrane differentiation of plasma membranes such as in 'buds' during the formation of the envelopes of togaviruses [(*a*); Semliki Forest virus produced in a BHK cell; note small arrows and insert] and C-type viruses [(*b*); endogenous C-type virus in a culture murine sarcoma 180 cell] and in microvilli [(*c*) to (*f*)]. During bud formation viral envelope proteins are selectively enriched in the evaginated portions of the membrane (see text). Note also, as another example of a localized membrane specialization, the association of the pores of the nuclear (N) envelope with granular and fibrillar components which thus constitute the pore complex (short arrows in (*b*)). The fingerlike membrane evaginations of the microvilli [(*c*) to (*f*) — present examples in the apical surface membrane of mammary epithelial cells of lactating rats (*cf.* reference [12])] are stabilized by and structurally and functionally interconnected with microfilaments via lateral cross-bridges (denoted by the short arrows in longitudinal and cross-section in the partial magnifications of (*d*) and (*e*)); these microfilaments seem to insert terminally at an electron-dense plaque which apparently contains alpha-actinin (denoted by the arrows in (*f*); for details see references [18, 19]). Bars denote 0.2 μm

Figure 2 (above) Association of bundles of microfilaments with specific regions of the plasma membrane (PM) and the endoplasmic reticulum (ER) system of Sertoli cells of rat testis as revealed in cross-sections (*a*) and longitudinal sections (*b*). These membrane-associated bundles of microfilaments occur specifically in surface regions which are attached to late spermatids (N denotes the nuclear interior, A denotes the acrosome of the spermatid, the two arrows in the left of (*a*) denote the small intercellular cleft between both cells) and at the specialized intercellular junctions between adjacent Sertoli cells[20, 21]. The brackets in (*a*) indicate the cortical region of the Sertoli cell which is limited by the subsurface cisterna of the ER and is characterized by the occurrence of paracrystalline bundles of microfilaments. The short arrows in (*b*) point to lateral cross-bridges between microfilaments and the ER membrane. Bars denote 0.1 μm

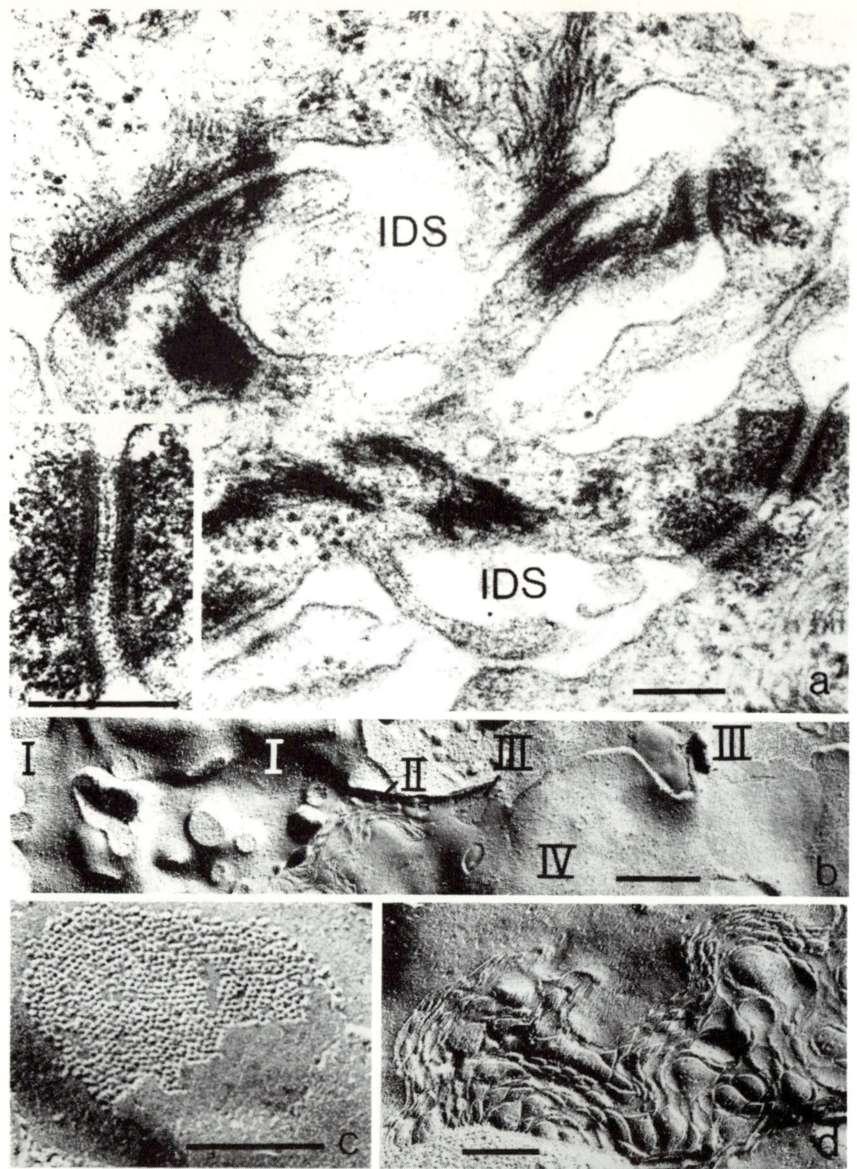

Figure 3 Examples of stable structural differentiations (intercellular junctions) in the plasma membrane of rat epidermal cells ((*a*) presents desmosomes with associated tonofilaments in an ultrathin section through teat nipple epidermis; IDS, interdesmosomal space; the insert shows a desmosome at high magnification) and lactating rat mammary epithelium ((*b*) to (*d*) present freeze-fracture images of I apical plasma membrane, II and (*d*) tight junctions, III and (*c*) gap junctions in two different fracture planes, and IV lateral cell surface). Scales indicate 0.2 μm (*a, c, d*) and 0.5 5 μm (*b*)

recognized in the form of pieces of membrane that have just been incorporated or are in the process of fusion (i.e. as exocytotic 'caveolae', *cf.* Figure 5[12,13]); and *vice versa*, in surface invaginations that are in the process of endocytotic vesicle formation. In both processes specific membrane differentiations are frequently correlated with the appearance of typical 'bristle coat' structures[5,12,13]. Surface protrusions and bud detachment processes such as during virus-budding and milk fat globule formation (see references quoted) are also not distributed at random and represent intramembranous islets of different turnover kinetics and fat rate. Therefore, experiments which use total membranes, either native or isolated, or gross subfractions are not necessarily relevant to the functional situations present *in situ* which might be controlled by specific, perhaps small intramembranous domains (*cf.* the article by Wisher and Evans (Chapter 8) in this book).

As to the question of how such local and sometimes very selective and specific local heterogeneities within a membrane are produced, recent analyses of 'capping' and 'clustering' phenomena provide some information and suggest that such differentiated domains within a membrane might be induced and stabilized by the association with extramembranous components[5,14,15]. Examples for such interactions are probably the surface attachment of certain extracellular materials that induce endocytotic invaginations or local membrane fusion, the association of viral cores with the cytoplasmic aspect of plasma membranes and endomembranes which may thus induce the local clustering of virus-specific membrane components and the subsequent formation and detachment of the viral envelope (for references see above), and the associations of membranes with microtubules and/or microfilaments (Figures 1 to 3) that have been shown to be cross-linked to specific membrane regions and, for example, influence the positioning of integral components in a variety of membranes. The apparently widespread occurrence of such structural interactions suggests that certain membranes or membrane domains which are structurally and functionally not confined to the membrane proper but are integrated into a cytoskeletal network are thus linked and coupled to a variety of non-membranous structures.

MEMBRANE DIFFERENTIATION BY CHEMICAL ALTERATION OF COMPONENTS AND CHANGES IN THE CHEMICAL COMPOSITION

A different category of membrane differentiation involves changes in the overall chemical composition or in the formation of specific novel molecules, i.e. true chemical changes that occur in one and the same membrane system in different stages of cellular development. Such changes pertain to lipid moieties as well as to proteins and glycoproteins and have been particularly intensely studied with respect to surface antigens and the immune response, cell transformation and developmental processes[5]. Especially intensely studied phenomena are the reduction of oligosaccharide chains in glycolipids[16] and the oversialylation of fucose-containing glycopeptides, and perhaps also fucoglycolipids, that occur in a variety of transformed cells (for reference see the contribution of Emmelot *et al.* (Chapter 12) in this book). It would be important to know whether such qualitative changes can occur by a modification of existing membrane components *in situ*, for example by surface-bound sugar transferases, or whether they require the coordinate synthesis of whole novel complex membrane units (*cf.* the article by Coleman *et al.* (Chapter 9) in this book). In view of the intimate interplay of membrane constituents, it is obvious that changes in the chemical composition of a specific membrane can then result in considerable effects on other membrane activities as well.

MEMBRANE DIFFERENTIATION DURING MEMBRANE TRANSLOCATION (MEMBRANE FLOW)

As has been shown in various cell systems the different membrane systems of the cell display differences in membrane structure and composition (*cf.*[5] and the articles by Fleischer and Fleischer (Chapter 3) and by Morré (Chapter 2) in this book). This is true not only for relatively unrelated membranes such as RER and inner mitochondrial membrane, but also differences are noted among membranes which can exist in direct morphological continuity and/or appear to be related with each other via vesicle flow processes such as in the secretory pathway (Figures 4 to 6). Two sorts of differences must be distinguished, quantitative and qualitative ones. Gradual changes in the centrifugal direction of secretory membrane flow (as schematically arranged in Figure 6) have been noted with respect to structural organization, lipid composition, patterns of polypeptides, carbohydrate composition and enzyme activities (see references quoted above). In this connection the term 'membrane differentiation' has been used for the 'change in composition or organization which accompanies the conversion from one type of membrane to another'[17], in order to explain the difference in the various membrane systems that participate in

---

Figure 4 Schematic presentation of diverse membrane types involved in the secretory pathway and their possible interrelationship via membrane continuities and membrane flow processes. (*a*) RER; (*a'*) nuclear envelope; (*b*) SER Ag (*b'*) transitional elements in RER and regions equivalent to transitional elements in the outer nuclear membrane; (*c*) ER-derived vesicles such as peroxisomes; (*d*) vesicles detached from ER; (*d'*) vesicles detached from nuclear envelope; (*e*) dictyosomes of the Golgi apparatus; (*f*) smooth-surfaced secretory vesicles derived from Golgi apparatus; (*f'*) relatively large Golgi apparatus-derived vesicle partially covered by bristle coat ('condensing vacuoles'); (*g*) and (*g'*) free secretory vesicles of the smooth (*g*) or coated (*g'*) type which are frequently smaller than (*f*) and (*f'*), respectively; (*h*) and (*h'*) represent exocytotic stages of (*g*) and (*g'*), respectively; (*i*) lysosomal vesicles derived from Golgi apparatus or SER; (*j*) secondary lysosome or phagocytotic vacuole; (*k*) large smooth-surfaced endocytotic vesicle during formation; (*l*) free endocytotic vesicle; (*m*) fusion of endocytotic vesicles with lysosome;

(*n*) small smooth-surfaced endocytotic vesicle; (*n'*) coated endocytotic vesicle; (*o*) stable evaginations of the plasma membrane such as microvilli; (*p*) plastid; (*q*) mitochondrion. Direct membrane continuities are observed between NE and ER, ER and Golgi apparatus (rarely observed), ER and outer mitochondrial membrane (rare in most cell systems), ER or outer nuclear membrane and outer plastidal membrane (rare)

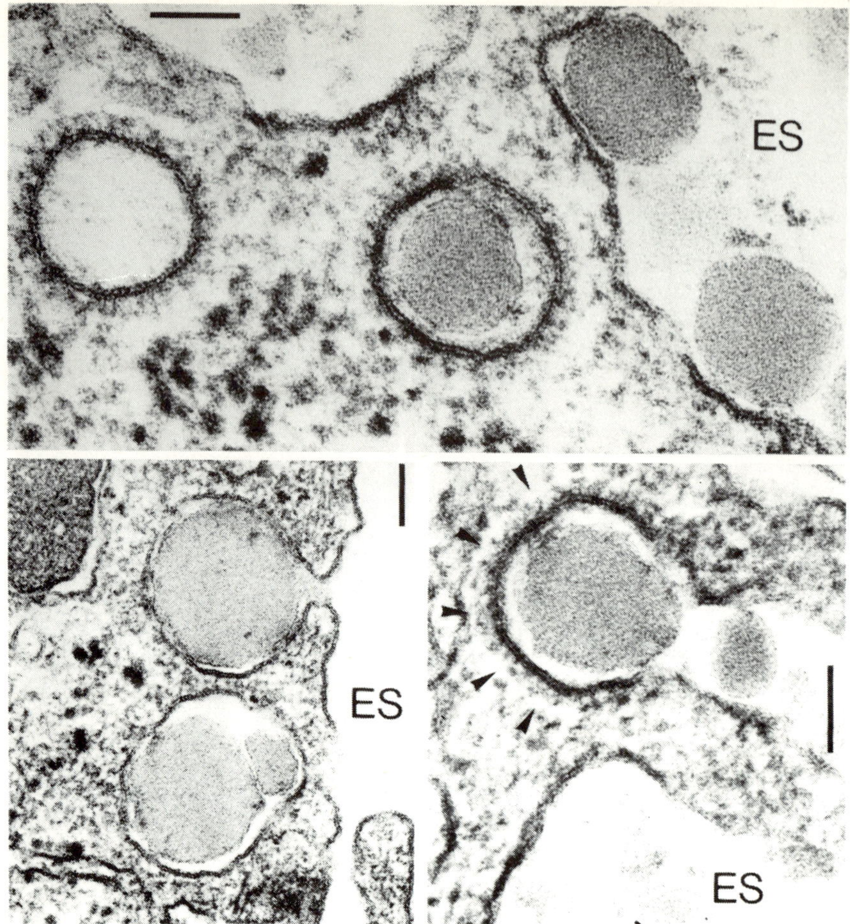

Figure 5(a)–(c) Micrographs showing the periphery of isolated rat hepatocytes (for details see reference[13]). Many of the secretory vesicles that contain lipoprotein aggregates show a polygonal coat (*a*). Some vesicles are in positions suggestive of fusion with the plasma membrane (*b*, *c*) and release of material into the extracellular space (ES). Figure 5(*c*) presents an apparently completely coated secretory vesicle (arrowheads denote coat structures) in the process of exocytotic release. Scales indicate 0.1 μm

the 'membrane flow' of the morphologically defined membrane units surrounding the secretory products. Morphological observations suggest the existence of specific sites of vesicle formation in endomembranes (e.g. Figure 6) as well as of specific sites of fusion with other membranes. This is indicative of a functional heterogeneity: some regions exist within membranes which more frequently, if not exclusively, participate in membrane flow processes than other regions. We propose that membrane flow is not at random but is selective for specific membrane components and excludes others (Figure 6; *cf.* [5]). This selection of domains involved in flow processes might again be induced by extramembranous components, including the secretory products themselves. Selective membrane flow would then necessarily result in gradual or abrupt changes in membrane composition and thus in 'membrane differentiation'.

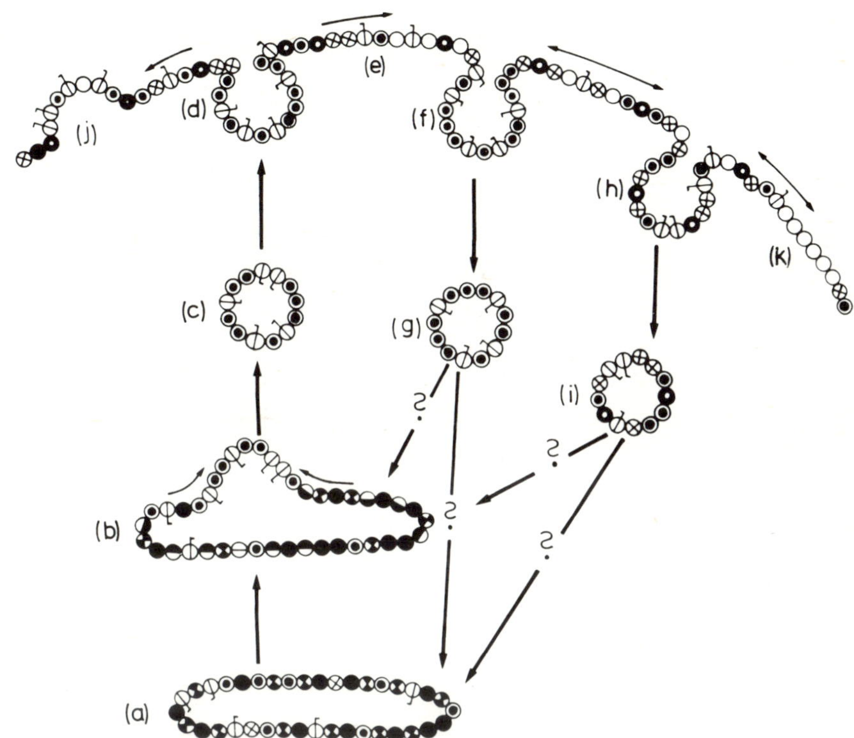

Figure 6 Schematic illustration of the concept of selective flow of membrane components. The concept assumes that the participation in vesicle formation and flow phenomena is different for each type of membrane component, particularly proteins and glycoproteins (components are represented by different symbols). In this concept a specific membrane continuum is heterogeneous with respect to the turnover and the translocation frequency of its constituents. For example, specific components of an endomembrane cisterna (*a*) aggregate into a semistable (transitory) cluster and locally alter the character of the membrane and might increase the plasticity and/or the fluidity. (*b*) Components enriched in such clustered components then are contained in a vesicle bud which finally detaches and becomes a vesicle. (*c*) After fusion of such a vesicle with another membrane (*d*), the components might become randomly distributed (*e*), again participate in aggregation and budding processes (e.g. as denoted by *j*), or remain together and recycle (*f* and *g*). Other possibilities would be that the components mix to a certain degree with components characteristic for their new 'host' membrane and recycle together in randomly or specifically aggregated moieties (*h* and *i*) or remain as stable clusters of specific components (*k*). Thin arrows indicate lateral mobility: thick arrows indicate vesicle translocation

Acknowledgement

We thank Dr H. Zerban for cooperation (Figures 3(*b*) to (*d*)) and the Deutsche Forschungsgemeinschaft for financial support.

References

1. Singer, S. J. and Nicolson, G. L. (1972). The fluid mosaic model of the structure of cell membranes. *Science*, **175**, 720
2. Nicolson, G. L. (1974). The interactions of lectins with animal cell surfaces. *Int. Rev. Cytol.*, **39**, 89
3. Staehelin, L. A. (1974). Structure and function of intercellular junctions. *Int. Rev. Cytol.*, **39**, 191
4. Hicks, R. M., Ketterer, B. and Warren, R. C. (1974). The ultrastructure and chemistry of the luminal plasma membrane of the mammalian urinary bladder: A structure with low permeability to water and ions. *Phil. Trans. Roy. Soc. Lond.*, **268**, 23
5. Franke, W. W. and Kartenbeck, J. (1976). Some principles of membrane differentiation. In: N. Müller-Bérat *et al.* (eds.), *Progress in Differentiation Research*, pp. 213–243. (Amsterdam: North-Holland)
6. Pearse, B. M. F. (1976). Clathrin: A unique protein associated with intracellular transfer of membrane by coated vesicles. *Proc. Natl. Acad. Sci. USA*, **73**, 1255
7. Simons, K., Garoff, H. and Helenius, A. (1977). The glycoproteins of the Semliki Forest virus membrane. In: R. A. Capaldi (ed.), *Membrane Techniques, Vol. 1* (New York: Marcel Dekker)
8. Lenard, J. and Compans, R. W. (1974). The membrane structure of lipid-containing viruses. *Biochim. Biophys. Acta*, **344**, 51
9. Patton, S. and Keenan, T. W. (1975). The milk fat globule membrane. *Biochim. Biophys. Acta*, **415**, 273
10. Jarasch, E.-D., Bruder, G., Keenan, T. W. and Franke, W. W. (1977). Redox constituents in milk fat globule membranes and rough endoplasmic reticulum from lactating mammary gland. *J. Cell Biol.*, **73**, 223
11. Franke, W. W., Kartenbeck, J., Zentgraf, H., Scheer, U. and Falk, H. (1971). Membrane-to-membrane cross-bridges. *J. Cell Biol*, **51**, 881
12. Franke, W. W., Lüder, M. R., Kartenbeck, J., Zerban, H. and Keenan, T. W. (1976). Involvement of vesicle coat material in casein secretion and surface regeneration. *J. Cell. Biol.*, **69**, 173
13. Kartenbeck, J., Franke, W. W., and Morré, D. J. (1977). Polygonal coat structures on secretory vesicles of rat hepatocytes. *Cytobiologie.*, **14**, 284
14. Yahara, I. and Edelman, G. M. (1975). Electron microscopic analysis of the modulation of lymphocyte receptor mobility. *Exp. Cell Res.*, **91**, 125
15. Yahara, I. and Edelman, G. M. (1975). Modulation of lymphocyte receptor mobility by locally bound concanavalin A. *Proc. Natl. Acad. Sci. USA*, **72**, 1579
16. Brady, R. O., and Fishman, P. H. (1974). Biosynthesis of glycolipids in virus-transformed cells. *Biochim. Biophys. Acta*, **355**, 121
17. Morré, D. J., Keenan, T. W. and Huang, C. M. (1974). Membrane flow and differentiation: Origin of Golgi apparatus membranes from endoplasmic reticulum. In: B. Ceccarelli, F. Clementi and J. Meldolesi (eds.), *Advances in Cytopharmacology*, pp. 107–125. (New York: Raven Press)
18. Mooseker, M. S. and Tilney, L. G. (1975). Organization of an actin filament–membrane complex. Filament polarity and membrane attachment in the microvilli of intestinal epithelial cells. *J. Cell Biol.*, **67**, 725

19. Rodewald, R., Newman, S. B. and Karnovsky, M. J. (1976). Contraction of isolated brush borders from the intestinal epithelium. *J. Cell Biol.*, **70,** 541
20. Toyama, Y. (1976). Actin-like filaments in the Sertoli cell junctional specializations in the swine and mouse testis. *Anat. Rec.*, **186,** 477
21. Franke, W. W., Grund, C., Fink, A., Weber, K., Jockusch, B. M., Zentgraf, H. and Osborn, M. (1977). Location of actin and α-actinin in the microfilament bundles associated with the Sertoli cell junctional specializations. *J. Microsc. Biol. Cell.* (In press)

2
Membrane Flow and its Contribution to Surface Formation

D. J. MORRÉ

Many of the events of liver pathology which are the main subject of this symposium depend on membrane change, with successful therapy frequently depending on membrane renewal. It is to these aspects of membrane dynamics in hepatocytes that this report is addressed. Three points will be stressed:

(1) That membrane flow does occur and that it may be important to liver pathology.
(2) That membrane flow is rapid and that the amounts of membrane delivered by this mechanism are quantitatively significant.
(3) That membrane flow is selective so that some membrane constituents are transferred while others are not.

MEMBRANE FLOW ACCOMPANIES HEPATIC SECRETION OF SERUM LIPOPROTEINS

It is well-established that in all eukaryotic cells at some stage of their development, secretory vesicles produced at the Golgi apparatus are released into the cytoplasm. They then migrate vectorially to the cell surface where the vesicle membranes fuse with the plasma membrane to discharge secretory product and add new plasma membrane for growth or membrane renewal during homeostasis or differentiation.

In rat liver (Figure 1), the secretory vesicles contain particles of very low density lipoproteins (VLDL) and low density lipoproteins (LDL). These lipoproteins are presently considered the principal vehicles for triglyceride and cholesterol transport from liver to extrahepatic tissue. In recent years, they have been consistently implicated in atherogenesis[1].

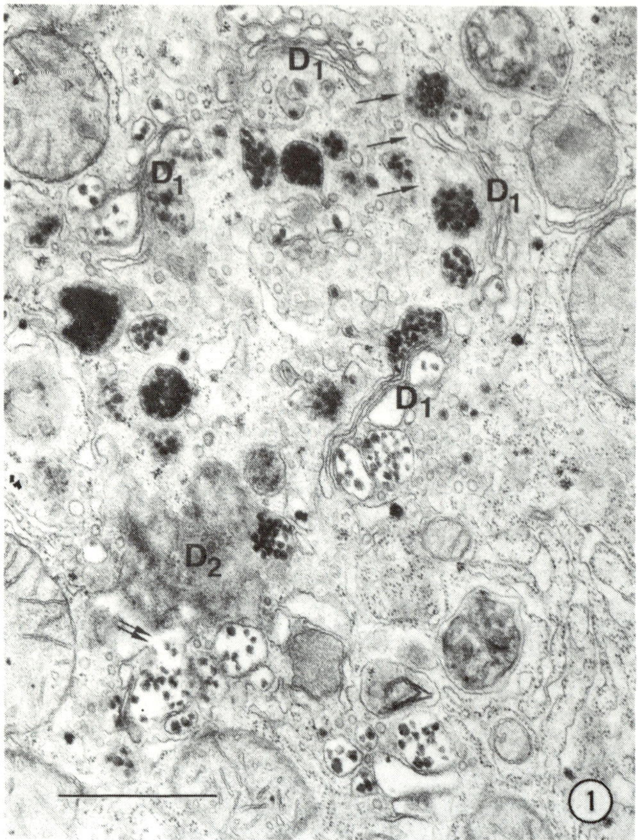

Figure 1 Portion of a Golgi apparatus of rat liver consisting of several stacks of cisternae (dictyosomes). Dictyosomes are sectioned tangentially (D_1) and in face view (D_2). The latter (D_2) show the flattened, central plate-like portions of the cisternae (saccules). In this particular fixation, the 30–60 nm diameter lipoprotein particles were strongly osmiophilic. During their formation, secretory vesicles are attached to the saccules via direct tubular connections (double arrows). Microtubules occur in the Golgi apparatus zone (arrows), usually aligned at the face occupied by mature secretory vesicles. Scale marker = 1 μm. From a study with W. D. Merritt, Purdue University

The apoprotein moieties of the lipoprotein are synthesized on polyribosomes and associated messenger RNA at the rough endoplasmic reticulum[2]. Electron microscope studies show that in isolated livers, strongly osmiophilic 30–80 nm particles which approximate the size range of plasma VLDL appear within minutes after perfusion with free fatty acids[3]. These appear first within the smooth endoplasmic reticulum[3] and Golgi apparatus[4] but within 15 min are found in the space of Disse[3,4]. From there, they proceed directly to the circulation[2] (Figure 2). Correlative biochemical, immunochemical and morphological studies, made possible by the development of procedures to iso-

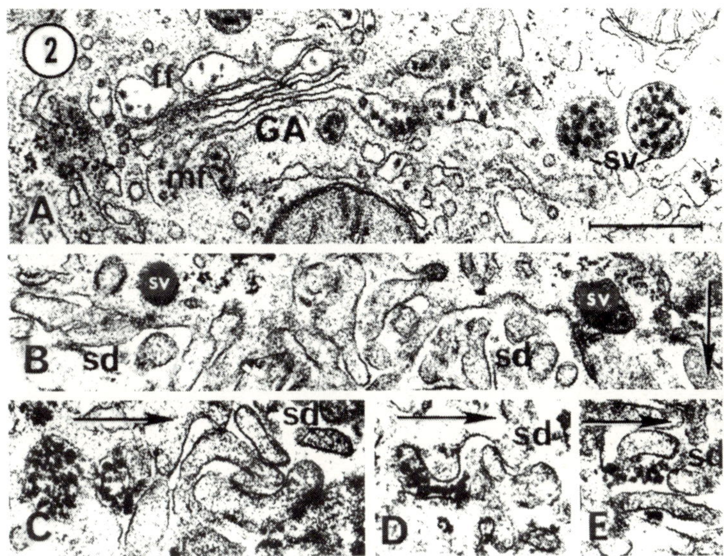

Figure 2 Golgi apparatus–secretory vesicle–plasma membrane associations in rat liver. A. Portion of a Golgi apparatus (GA) showing the dominant types of secretory vesicles. Those of the forming face (ff) appear swollen with an electron-transparent matrix. Secretory vesicles of the mature face (mf) appear more condensed with an electron-dense matrix. B. Secretory vesicles (sv) near the space of Disse at the cell surface. C–E. Clusters of particles, derived from secretory vesicles, are discharged into the space of Disse (sd) by fusion of the vesicle membranes with the plasma membrane. Arrows indicate the location of the free surface of the cell. Scale marker = 0.5 μm. Adapted from Morré and Van Der Woude[16]

late intact and functional Golgi apparatus from liver[5,6], verify that the lipoprotein particles of the Golgi apparatus are precursors of plasma VLDL[7–9] and contain the same major lipoprotein apoprotein as plasma VLDL[7,10].

More recently, lipoprotein-containing secretory vesicles associated with the hepatic Golgi apparatus were dissociated and isolated in pure form free of other Golgi apparatus elements[11]. These studies[11], coupled with extensive morphological investigations of lipoprotein secretion in rat and rabbit hepatocytes[12,13], showed two classes of lipoprotein-containing vesicles: immature vesicles with particles in the size range of VLDL and mature vesicles with particles of the VLDL range as well as particles in the LDL range (Figure 2). These observations are consistent with derivation of LDLs from VLDLs and suggest that secretory vesicles of the Golgi apparatus are one of the subcellular sites of this transformation. It is now generally accepted that LDL is derived from larger, less-dense VLDL through hydrolytic cleavage of fatty acids[1].

In parallel studies with rat intestine, Golgi apparatus was studied both *in situ* and in isolated fractions[14]. Here, two classes of mature secretory vesicles were observed. One contained VLDL-sized particles and another contained chylomicron-sized particles. In this tissue, the Golgi apparatus apparently functions as a sorting centre so that only vesicles containing one type of particle are transported.

Direct transfer of lipoprotein particles from smooth endoplasmic reticulum to the cell surface also remains a definite possibility[3,12,15–17]. By either route, the VLDL polypeptides of the rat and other species are synthesized primarily on rough endoplasmic reticulum, then sequentially modified in rough and smooth endoplasmic reticulum (addition of lipids; some non-terminal glycosylation) or in secretory vesicles (some modification of lipids and terminal glycosylation, see Figure 3). Depending on the route of secretion, particles differing in lipid composition or degree of glycosylation might be produced.

The secretory vesicle, derived either from the Golgi apparatus or an equivalent transition element, is a well-documented example of a cell component derived from one membrane structure which migrates to and coalesces with

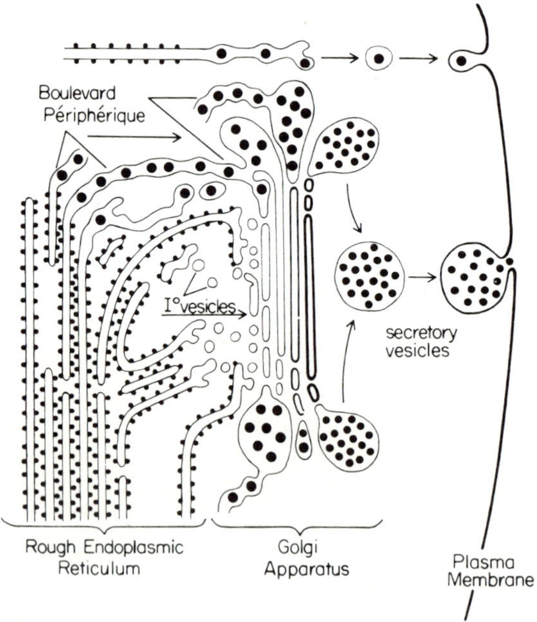

Figure 3 Diagram summarizing routes of synthesis and secretion of serum lipoproteins in rat liver. For details see text, Ovtracht *et al.*[17] and Morré *et al.*[12]

another membrane structure (the plasma membrane) to effect the physical transfer of membrane, i.e. membrane flow. In this manner, secretory vesicles provide structural and functional continuity between the internal cytoplasmic membranes and the cell surface[16,18] (Figures 2 and 3).

WHAT IS THE SOURCE OF NEW PLASMA MEMBRANE PROVIDED BY SECRETORY VESICLES?

The source of new membrane required for production of secretory vesicles is not known with certainty. One hypothesis is that endomembrane cisternae, or at least constituents of these cisternal membranes, are the immediate source. Secretory vesicles are attached to cisternae of Golgi apparatus or other transition elements by means of direct tubular connections during their formation[19] (Figure 1). A dynamic concept of Golgi apparatus functioning has evolved where utilization of cisternal membranes occurs on one face of the Golgi apparatus and is balanced by input of new membrane materials on the opposite or forming face[19] (Figure 3).

A morphological basis for membrane input into the Golgi apparatus is provided by small 50 nm transition vesicles which in liver and all eukaryotes appear to bleb off smooth portions of endoplasmic reticulum[19] or nuclear envelope and refuse to form new Golgi apparatus cisternae (Figure 4A). These observations, taken together with evidence for membrane differentiation summarized by Franke[20], lead to the following scheme of hepatic membrane flow through the Golgi apparatus:

(1) Membrane materials are received from endoplasmic reticulum in a selective pinching-off process (Figures 3 and 4A).

(2) Certain constituents, such as sugar prosthetic groups, and specific proteins, are added at the Golgi apparatus while other constituents, such as marker enzymes for endoplasmic reticulum, are either removed or not transferred to account for membrane differentiation.

(3) Finally, the finished product moves to the plasma membrane in the form of a secretory vesicle to contribute a new membrane to the cell surface (Figures 2, 4B). It is important to emphasize that the two processes, membrane flow and membrane differentiation, must occur in parallel if a true conversion of endoplasmic reticulum to plasma membrane is to be achieved.

The fate of the newly delivered plasma membrane is problematic. Mature, secretory hepatocytes are not growing and the membranes delivered by the secretory vesicles can contribute only to membrane renewal. A compensating amount of membrane must be removed and, presumably, delivered back to the cytoplasm for degradation and reuse in redirected synthesis (*cf.* [21]). A shuttle mechanism, as described for pancreas[22], or cyclic membrane flow, has been proposed but has not been investigated in liver. During removal of excess

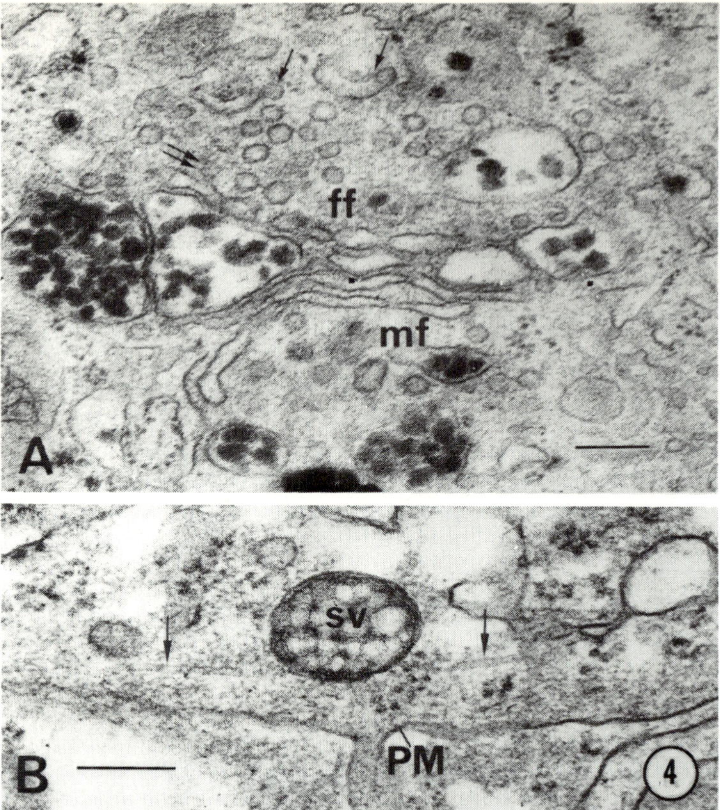

Figure 4 Morphological basis for flow dynamics of Golgi apparatus membranes. A. Numerous transition vesicles at the forming face (ff). The origin of these vesicles is presumed to be from endoplasmic reticulum (arrows). The vesicles are then thought to migrate to the Golgi apparatus where they fuse with the forming dictyosome cisternae (double arrows). B. Migrating secretory vesicles (sv) in the cytoplasm or near the cell surface (PM = plasma membrane) are frequently, if not always, associated with microtubules (arrows). This particular micrograph is from a hepatoma. Microtubules may serve as guide elements to account for vectorial migration of secretory vesicles to sites of fusion with the plasma membrane at the cell surface. Scale marker = 0.2 μm. From a study with W. D. Merritt, Purdue University

smooth endoplasmic reticulum membranes produced in response to phenobarbital, the number of autophagic vacuoles increases markedly[23] to account for membrane breakdown during turnover by a completely separate pathway.

In any event, Golgi apparatus possess only a limited capacity for biogenesis of membrane constituents[24]. The complete biosynthetic pathway for an important glycosphingolipid antigen of the cell surface, the so-called theta antigen or ganglioside G_{M1} important to tumorigenesis and toxin interactions[25,26], is present in endoplasmic reticulum and Golgi apparatus but lacking from the plasma membrane in rat liver[27]. It is one example of a plasma membrane constituent synthesized exclusively by internal membranes. Other constituents of plasma membranes including proteins and glycerophosphatides also appear to be synthesized by internal membranes, chiefly endoplasmic reticulum, so that some flow mechanism to deliver these constituents to the cell surface must exist.

HOW ARE MEMBRANE CONSTITUENTS TRANSPORTED FROM SITES OF SYNTHESIS TO SITES OF UTILIZATION?

Among the possibilities to account for movement, transport or migration of membrane constituents are:

(1) Lateral migration within the plane of the membrane

(2) Bulk migration in the form of vesicles, and

(3) Movement through the cytosol either free or complexed with cytoplasmic carriers or as micellar aggregates.

It is probable that all three mechanisms operate in rat liver (Figure 3). Evidence for operation of cytoplasmic carriers has recently been summarized[24]; lateral migration of membrane constituents via direct tubular connections has also been discussed[12,17,28]. Normally, it is not possible to distinguish among these various possibilities in kinetic experiments where radioactive membrane precursors are administered and followed through the endomembrane system.

Kinetics of hepatic membrane flow

On the premise that intrinsic membrane proteins are less subject to post-synthesis modification than are lipids or heteroglycans, this discussion will be limited to information on the flow kinetics of these membrane proteins. In the preparation of intrinsic proteins, the membranes are extracted with high salt and detergent to remove secretory and extrinsic proteins[29,30] (Table 1).

Table 1 Preparation of intrinsic membrane proteins of rat liver: evidence for specificity of the extraction procedure (from Elder and Morré[34])

1. Proteins extracted by treatment with KCL + DOC have mobilities on SDS polyacrylamide gel electrophoresis distinct from proteins not extracted.
2. No correlation between apparent molecular weight and extractibility by KCl + DOC.
3. Soluble supernatant or serum proteins yield no insoluble material after KCl + DOC treatment (not protein denaturation).
4. Immunoprecipitation with antisera against intrinsic protein fraction: not cross-reactive with either soluble or serum proteins.
5. Secretory proteins such as albumin and apo-VLDL appear in the soluble fraction after KCl + DOC treatment.
6. More than 95% of ribosomal proteins removed by KCl + DOC.

If [^{14}C]labelled amino acids (arginine or leucine) are given to rats as a pulse, the livers removed and fractionated, the kinetics are as follows (Figure 5). Intrinsic membrane proteins of the endoplasmic reticulum are labelled maximally in about 10 min after which a rapid turnover component disappears. Intrinsic membrane proteins of the Golgi apparatus are labelled in parallel with maximum incorporation after about 20 min. Eventually label appears in proteins of the plasma membrane.

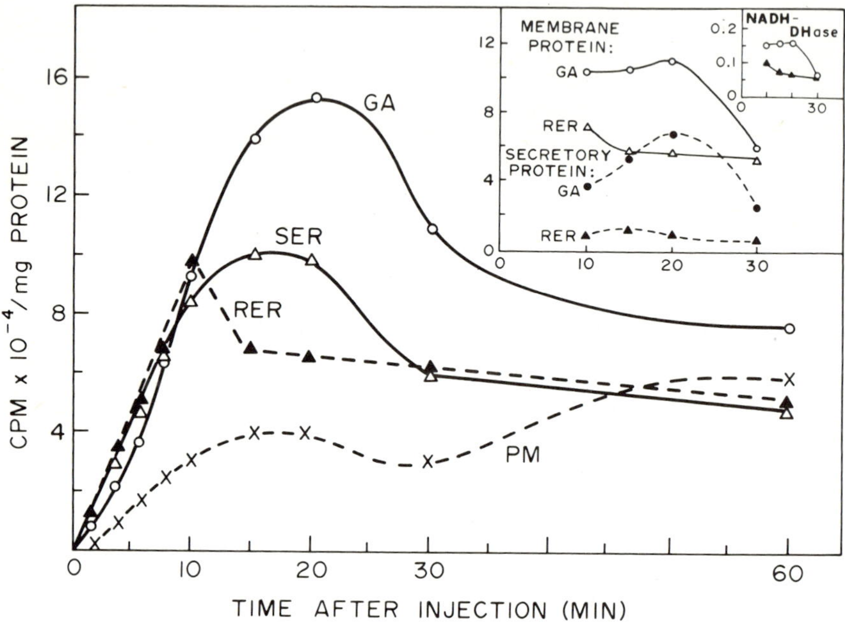

Figure 5 Kinetics of incorporation of radioactivity from [^{14}C]arginine into a fraction of intrinsic membrane proteins from rat liver[12,30]. Insets compare kinetics for secretory protein with those of total intrinsic membrane proteins and an electrophoretic band corresponding to NADH-dehydrogenase[12]

If one separates unextracted endoplasmic reticulum and Golgi apparatus by polyacrylamide gel electrophoresis, the secretory proteins, VLDL and albumin[12] (Figure 5 inset) become labelled in the Golgi apparatus more slowly than either the total intrinsic membrane proteins or the band corresponding to NADH dehydrogenase and parallel the kinetics of labelling of the triglycerides of VLDL with [^{14}C]glycerol[13]. These results show that movement of secretory proteins and of membrane proteins to the Golgi apparatus are not closely coupled but that they do exit together in the form of mature secretory vesicles.

One puzzle, recently explained, is how the specific activity of membrane proteins of the Golgi apparatus can exceed that of the endoplasmic reticulum if endoplasmic reticulum is, in fact, the precursor of the membranes of the Golgi apparatus. There are polyribosomes associated with the Golgi apparatus. First reported in plants by Franke, Scheer and co-workers[31,32] and in *Euglena*[33], they also occur in rat liver[34]. These polyribosomes are not membrane-bound in the usual sense but are a special class of free polyribosome embedded in a matrix material or zone of exclusion which surrounds the Golgi apparatus. In carefully isolated Golgi apparatus, these polyribosomes remain associated within this matrix and can be recovered from the Golgi apparatus pellet. They are active in protein synthesis (Table 2). On an RNA basis, Golgi apparatus poly-

Table 2 Incorporation *in vitro* of [^{14}C]leucine into protein insoluble in 1·5 M KCl and 0·1% deoxycholate by cell fractions from rat liver (calculated from Elder and Morré[34])

Fraction		*Radioactivity*	
	Protein/RNA	*c.p.m./mg protein*	*c.p.m./mg RNA*
Rough endoplasmic reticulum	2.5	810	2025
Golgi apparatus	8.2	340	2780
Plasma membrane	*	0	—

* No RNA was detected

ribosomes are even more active in the synthesis of high salt + detergent-insoluble proteins, the presumptive intrinsic fraction of membrane proteins, than are polyribosomes of rough endoplasmic reticulum. The products of synthesis by Golgi apparatus-associated polyribosomes are distinct and comigrate on polyacrylamide gel electrophoresis with certain proteins of the plasma membrane not present in endoplasmic reticulum[34]. With protein synthesis occurring both at the Golgi apparatus and at the endoplasmic reticulum, it is possible to understand how incorporation of radioactive amino acids into Golgi apparatus begins without an apparent lag (the Golgi apparatus contribution) and after a few minutes doubles in rate (the arrival of the endoplasmic reticulum contribution). The combined contributions lead to the situation where the peak specific activity of intrinsic membrane proteins of the Golgi apparatus exceeds that of those from the endoplasmic reticulum (Figure 5).

To evaluate the quantitative aspects of membrane flow, the data of Figure 5 have been combined with morphometric data from the laboratory of E. J. Weibel, University of Berne, and recovery data from cell fractionation experiments in our own laboratory (Table 3). There is good agreement between the

Table 3 Radioactive labelling of membrane proteins: a balance sheet for rat liver

Component	*mg/10 g Liver*	*Specific radioactivity*	*Total radioactivity*	*Rapid turnover* %	*c.p.m.*
Plasma membrane	50	1000	50 000	100	50 000
Rough ER	230	2000	460 000	20–25	90 000
Golgi apparatus	15	3000	45 000	25–50	20 000
				100	45 000

portion of the endoplasmic reticulum that turns over rapidly and the total incorporation into plasma membrane. At least 20–25% can be accounted for by passage through the Golgi apparatus — more if one assumes that the Golgi apparatus is turning over as predicted by the morphological model.

Membrane flow is selective

Membrane flow must be selective to account adequately for membrane differentiation[35,36]. This notion has been implicit in the membrane flow concept since its inception. Evidence for selective membrane flow is provided by the data of Kartenbeck[36] who found that different proteins are synthesized and inserted into the membranes at different rates, and from the work of Elder[34], who demonstrated that membrane proteins may be synthesized and inserted at different times. The same is true for lipids and sugars so that membrane biogenesis is sequential.

Some constituents that leave the endoplasmic reticulum never reach the plasma membrane. For example, the phenobarbital-induced O-demethylase appears first in rough endoplasmic reticulum, then in smooth endoplasmic reticulum, and finally at the Golgi apparatus but is never detected enzymatically at the plasma membrane[12]. Other constituents may never leave the endoplasmic reticulum but are simply excluded during the flow process, e.g. the bulk of the glucose-6-phosphatase.

If similar comparisons are made for a variety of markers, some are excluded early whereas other are excluded late or not at all (Tables 4 and 5). Some appear to be transferred while others are retained. Some are apparently acquired en route. These various possibilities are summarized in Figure 6 for enzymatic activities.

Perhaps some of the confusion over membrane flow has a semantic basis. We speak of membrane differentiation. We speak of membrane flow. It is as if

Table 4 Electron transport activities of rat liver endomembranes (from Morré *et al.*[35] and unpublished data of C. Frantz and F. L. Crane, Purdue University, and E. D. Jarasch, Heidelberg)

		Specific activity or amount/mg protein		
Constituent	*Units*	*ER*	*GA*	*PM*
NADH-cyt. *c* reductase	μmol/h	75	18	8
NADH-ferricyanide reductase	μmol/h	208	86	62
NADH-juglone reductase	μmol/h	42	23	17
NADH-oxidase (KCN)	μmol/h	1.3	4.6	3.8
Xanthine oxidase	nmol/min	0	2	17
Cytochrome b_5	mμmol	0.6	0.3	0.25
Cytochrome P_{450}	mμmol	0.6	0.2	0
NADPH-cyt. *c* reductase	μmol/h	6.0	0.8	n.d.

n.d. = not detected

Table 5 Summary of evidence for selectivity of membrane flow from studies of membrane differentiation in rat liver

Constituent	*NE*	*RER*	*SER*	*GA*	*SV*	*PM*
Nuclear pore complexes	×	±				
Ribosome binding sites	×	×	±			
Glucose-6-phosphatase	×	×	×	±		
NADPH-cyt. *c* reductase	×	×	×	×	?	
Cytochrome P_{450}	×	×	×	×	?	
CMP-NAN: G_{M1} sialyltransferase		×	×	×	×	
NADH oxidase	×	×	×	×	×	×
Cytochrome b_5	×	×	×	×	?	×
Xanthine oxidase				×	?	×
Na^+-K^+-Mg^{2+}-ATPase					?	×

× = major subcellular site; ± = minor subcellular site; ? = not determined

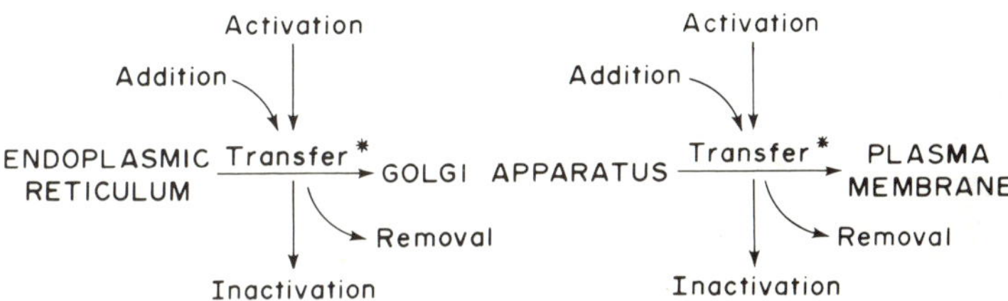

Figure 6 Summary of some of the types of transformations which might account for changes in enzyme and protein content of endomembranes during flow-differentiation[37]

the two processes were mutually exclusive when, in fact, they are not. Perhaps we should speak of *flow-differentiation* of membranes, *flow* to imply movement and the dynamic aspect, *differentiation* to account for change and selectivity. Without the process of membrane flow, the process of membrane differentiation is relegated to the role of a passive manifestation of membrane specialization.

SUMMARY

Processes of membrane flow and membrane differentiation, when combined, account for a contribution of endomembrane-derived secretory vesicles to the surface membranes of hepatocytes during the secretion of serum lipoproteins. Transfer is selective so that some constituents are transferred while others are not. Evidence for selective transfer comes from biochemical and morphological studies of membrane differentiation which show a gradual conversion of membranes from endoplasmic reticulum-like to plasma membrane-like across the stacked cisternae of the Golgi apparatus.

Membrane flow, the physical transfer of membranes from one cell component to another, is evidenced from: (1) short-term labelling and turnover of membrane proteins, (2) early drug-induced changes in enzymatic activities, and (3) kinetic analyses of electron transport chain constituents. One hypothesis is that membrane materials from endoplasmic reticulum are used to form the Golgi apparatus or other transition elements, and subsequently to form or replace constituents of the cell surface. Thus, those surface characteristics which determine self and non-self, immunological specificity, and certain forms of membrane dysfunction, as in the oncogenic transformation, may be partly determined by activities of internal endomembranes. An important question to pose for future meetings of this type is how surface characteristics, such as those which may be imparted to the plasma membrane by the Golgi apparatus and other internal membranes, are translated into regulatory signals important to normal or abnormal growth, development, or other aspects of cell function.

Acknowledgements

This work was supported in part by grants from the Indiana Heart Association and The American Heart Association.

References

1. Eisenberg, S. and Levy, R. I. (1975). Lipoprotein metabolism. In: R. Paoletti and D. Kritchevesky (eds.). *Advances in Lipid Research, Vol. 13*, pp. 1–89. (New York: Academic Press)
2. Schumaker, V. N. and Adams, G. H. (1969). Circulating lipoproteins. *Ann. Rev. Biochem.*, **38,** 113
3. Jones, A. L., Ruderman, N. B. and Herrera, M. G. (1967). Electron microscopic and biochemical study of lipoprotein synthesis in the isolated perfused rat liver. *J. Lipid Res.*, **8,** 429
4. Hamilton, R. L., Regan, D. M. and LeQuire, V. S. (1967). Lipid transport in liver. Electron microscopic identification of very low density lipoproteins in perfused rat liver. *Lab. Invest.*, **16,** 305
5. Morré, D. J., Merlin, L. M. and Keenan, T. W. (1969). Localization of glycosyltransferase activities in a Golgi apparatus-rich fraction isolated from rat liver. *Biochem. Biophys. Res. Commun.*, **37,** 813
6. Morré, D. J., Hamilton, R. L., Mollenhauer, H. H., Mahley, R. W., Cunningham, W. P., Cheetham, R. D. and LeQuire, V. S. (1970). Isolation of a Golgi apparatus-rich fraction from rat liver. I. Method and morphology. *J. Cell Biol.*, **44,** 484
7. Mahley, R. W., Hamilton, R. L. and LeQuire, V. S. (1969). Characterization of lipoprotein particles isolated from the Golgi apparatus of rat liver. *J. Lipid Res.*, **10,** 433
8. Mahley, R. W., Bersot, T. P., LeQuire, V. S., Levy, R. I., Windmueller, H. G. and Brown, W. V. (1970). Identity of very low density lipoprotein apoproteins of plasma and liver Golgi apparatus. *Science*, **168**, 380
9. Morré, D. J., Keenan, T. W. and Mollenhauer, H. H. (1971). Golgi apparatus function in membrane transformation and product compartmentalization: studies with cell fractions isolated from rat liver. *Adv. Cytopharmacol.*, **1,** 159
10. Alexander, C. A., Hamilton, R. L. and Havel, R. J. (1976). Subcellular localization of β-apoprotein of plasma lipoproteins in rat liver. *J. Cell Biol.*, **69,** 241

11. Merritt, W. D. and Morré, D. J. (1973). A glycosyltransferase of high specific activity in secretory vesicles derived from Golgi apparatus rat liver. *Biochim. Biophys. Acta*, **304,** 397
12. Morré, D. J., Keenan, T. W. and Huang, C. M. (1974). Membrane flow and differentiation: origin of Golgi apparatus membranes from endoplasmic reticulum. *Adv. Cytopharmacol.*, **2,** 107
13. Glaumann, H., Berstsrand, A. and Ericsson, J. L. E. (1975). Studies on the synthesis and intracellular transport of lipoprotein particles in rat liver. *J. Cell Biol.*, **64,** 356
14. Mahley, R. W., Bennett, B. D., Morré, D. J., Gray, M. E., Thistlethwaite, W. and LeQuire, V. S. (1971). Lipoproteins associated with the Golgi apparatus isolated from epithelial cells of rat small intestine. *Lab. Invest.*, **25,** 435
15. Jones, A. L. (1968). Some electron microscopic observations on lipoprotein and chylomicron metabolism in perfused liver. In: G. Cowgill, D. L. Estrich and P. D. Wood (eds.). *Proceedings of the 1968 Deuel Conference on Lipids on the Turnover of Lipids and Lipoproteins*, pp. 33–40. (Washington, D.C.: U.S. Department of Health, Education and Welfare)
16. Morré, D. J. and Van Der Woude, W. J. (1974). Origin and growth of cell surface components. In: E. D. Hay, T. J. King and J. Papaconstantionou (eds.). *Macromolecules Regulating Growth and Development*, pp. 81–111. (New York: Academic Press)
17. Ovtracht, L., Morré, D. J., Cheetham, R. D. and Mollenhauer, H. H. (1973). Subfractionation of Golgi apparatus from rat liver: method and morphology. *J. Microsc.*, **18,** 87
18. Hamilton, R. L. (1968). Ultrastructural aspects of hepatic lipoprotein synthesis and secretion. In: G. Cowgill, D. L. Estrich and P. D. Wood (eds.). *Proceedings of the 1968 Deuel Conference on Lipids on the Turnover of Lipids and Lipoproteins*, pp. 3–31. (Washington, D.C: U.S. Department of Health, Education and Welfare)
19. Morré, D. J., Mollenhauer, H. H. and Bracker, C. E. (1971). The origin and continuity of Golgi apparatus. In: T. Reinert and H. Ursprung (eds.). *Results and Problems in Cell Differentiation.* II. *Origin and Continuity of Cell Organelles*, pp. 82–126. (Berlin: Springer Verlag)
20. Franke, W. W. (1977). Chapter 1 of this volume
21. deDuve, C. and Wattiaux, R. (1966). Functions of lysosomes. *Ann. Rev. Physiol.*, **28,** 435
22. Palade, G. E. (1975). Intracellular aspects of the process of protein secretion. *Science*, **189,** 347
23. Bolender, R. P. and Weibel, E. R. (1973). A morphometric study of the removal of phenobarbital-induced membranes from hepatocytes after cessation of treatment. *J. Cell Biol.*, **56,** 746
24. Morré, D. J. (1977). The Golgi apparatus and membrane biogenesis. In: G. Poste and G. L. Nicholson (eds.). *Cell Surface Reviews, Vol. 4* (Amsterdam: Elsevier–North Holland) (In press)
25. Richardson, C. L., Baker, S. R., Morré, D. J. and Keenan, T. W. (1975). Glycosphingolipid synthesis and tumorigenesis: a role for the Golgi apparatus in the origin of specific receptor molecules of the mammalian cell surface. *Biochim. Biophys. Acta, Cancer Rev.*, **417,** 175
26. Cuatrecasas, P. (1974). Membrane receptors. *Ann. Rev. Biochem.*, **43,** 169
27. Keenan, T. W., Morré, D. J. and Basu, S. (1974). Ganglioside biosynthesis: Concentration of glycosphingolipid glycosyltransferases in Golgi apparatus from rat liver. *J. Biol. Chem.*, **269,** 310
28. Claude, A. (1970). Growth and differentiation of cytoplasmic membranes in the course of lipoprotein granule synthesis in the hepatic cell. I. Elaboration of elements of the Golgi complex. *J. Cell Biol.*, **47,** 745
29. Sabatini, D. D. (1974). Procedures for the selective release of content from microsomal vesicles without membrane disassembly. *Meth. Enzymol.*, **31,** 215
30. Franke, W. W., Morré, D. J., Deumling, B., Cheetham, R. D., Kartenbeck, J., Jarasch, E. D. and Zentgraf, H.-W. (1971). Synthesis and turnover of membrane-proteins in rat liver: an examination of the membrane flow hypothesis. *Z. Naturforsch.*, **26b,** 1031

31. Franke, W. W. and Scheer, U. (1972). Structural details of dictyosomal pores. *J. Ultrastruct. Res.*, **40,** 132
32. Franke, W. W., Kartenbeck, J., Krien, S., Van Der Woude, W. J., Scheer, U. and Morré, D. J. (1972). Inter- and intracisternal elements of the Golgi apparatus: a system of membrane-to-membrane cross-links. *Z. Zellforsch.*, **132,** 365
33. Mollenhauer, H. H. and Morré, D. J. (1974). Polyribosomes associated with the Golgi apparatus. *Protoplasma*, **79,** 333
34. Elder, J. H. and Morré, D. J. (1976). Synthesis *in vitro* of intrinsic membrane proteins by free, membrane-bound, and Golgi apparatus-associated polyribosomes from rat liver. *J. Biol. Chem.*, **251,** 5054
35. Morré, D. J., Franke, W. W., Deumling, B., Nyquist, S. E. and Ovtracht, L. (1971). Golgi apparatus function in membrane flow and differentiation: origin of plasma membrane from endoplasmic reticulum. *Biomembranes*, **2,** 95
36. Franke, W. W. and Kartenbeck, J. (1976). Some principles of membrane differentiation. In: N. Müller-Bérat (ed.). *Progress in Differentiation Research*, pp. 213–243. (New York: American Elsevier)
37. Morré, D. J. and Mollenhauer, H. H. (1974). The endomembrane concept: a functional integration of endoplasmic reticulum and Golgi apparatus. In: A. W. Robards (ed.). *Dynamic Aspects of Plant Ultrastructure*, pp. 84–137. (New York: McGraw-Hill)

Dr K. Gibinski, Katowice-Ligota (Poland): Fluid may be transported across the cell from one pole to another. Then vesicles appear within the cell which seem to indicate the way of transport. What is the origin of their membranes? They appear as single layer and thus seem to be different from the double layer of the cell membrane.

Dr Morré, Indiana (USA): Fluid transport, either into or out of cells, via plasma membrane — or water expulsion vacuole-derived vesicles — is assumed as within the realm of normal cell function. These processes, however, seem to involve normal double-layered membranes. Single-layered membranes, while less common, seem to occur in prokaryotes and have been suggested to surround certain types of lipid bodies in eukaryotes

The nucleotide pyrophosphatases of endoplasmic reticulum and plasma membrane; a test of the membrane flow hypothesis

K. DECKER

Membrane flow is an attractive hypothesis to explain the continued renewal of plasma membrane proteins which are supposed to be synthesized on the rough endoplasmic reticulum. An obvious difficulty for its general acceptance, however, is the fact that most enzyme activities found to reside in the plasma membrane (PM) cannot be detected in the endoplasmic reticulum (ER) and in the Golgi field. In order to reconcile this lack of enzymatic continuity, specific activation and deactivation mechanisms must be invoked. On the other hand, autoradiographic studies after pulse-labelling with amino acids suggest a flow of polypeptides from the ER to the PM. What is required to settle this ambiguity is a defined protein which can be unequivocally identified by enzymological or immunological means in both compartments.

During our studies on the pyrophosphatatic cleavage of nucleotides we isolated in homogeneous form a glycoprotein, nucleotide pyrophosphatase (E.C. 3.1.4.1), from plasma membranes as well as from endoplasmic reticulum of rat liver[1]; both enzymes are covalently bound to larger membrane fragments from which they could be detached specifically by trypsin; they seem to be identical in molecular weight, carbohydrate content and enzymological performance; they cross-react with each other's antibodies. Their biomodal

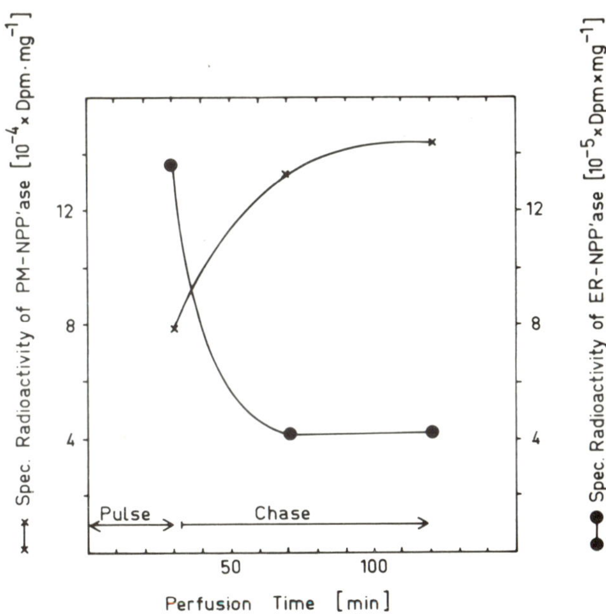

Figure 1 Change of the specific radioactivities of the nucleotide pyrophosphatases (NPP'ase) of ER and PM during a pulse-chase experiment. Livers were pulsed for 30 min with 400 μCi [^{3}H]leucine each. After a 2 min open perfusion with an isotope-free medium, they were perfused with a fresh medium containing 2 mM unlabelled leucine for periods between 40 and 90 min. Every point was obtained from six identically treated livers. Of the liver homogenate, 25% was used for the isolation of ER and 75% for PM. The values are *not* corrected for some cross-contamination of ER and PM which is, judging from marker enzyme activities, about 5% ER in the PM fraction

distribution was confirmed after enzymic iodination of intact liver cells and isotope analysis of the purified enzymes[2].

A possible precursor–product relationship of the ER and PM enzyme was investigated by T.-A. Tran-Thi in my laboratory. Using the isolated perfused rat liver, a 30 min pulse of [^{3}H]leucine was applied followed by a change of the medium and chases of different duration with unlabelled leucine. After separation of ER and PM, the nucleotide pyrophosphatase of either compartment was purified to polyacrylamide gel electrophoresis homogeneity and the specific radioactivity was determined. The enzyme from ER became rapidly labelled while the specific radioactivity of the PM enzyme rose during the period of sharply decreasing specific radioactivity of the ER enzyme (Figure 1). The increase of the specific radioactivity of the PM enzyme could be almost totally prevented if 100 μM vinblastine was added to the perfusion medium together with the labelled leucine. This labelling behaviour of the two proteins is suggestive of a precursor–product relationship and what one might expect if nucleotide pyrophosphatase is subject to membrane flow.

References

1. Bischoff, E., Tran-Thi, T.-A. and Decker, K. (1975). Nucleotide pyrophosphatase of rat liver. A comparative study of the enzymes solubilized and purified from plasma membrane and endoplasmic reticulum. *Eur. J. Biochem.*, **51,** 353
2. Bischoff, E., Wilkening, J., Tran-Thi, T.-A. and Decker, K. (1976). Differentiation of the nucleotide pyrophosphatases of rat liver plasma membrane and endoplasmic reticulum by enzymic iodination. *Eur. J. Biochem.*, **62,** 279

3
Membrane Diversity in the Rat Hepatocyte

S. FLEISCHER and BECCA FLEISCHER

The liver is the major metabolic centre for manufacture, interconversion and export of metabolites and complex macromolecules and as such serves a vital role in animal physiology. It consists predominantly of one cell type, the hepatocyte, which constitutes 78% of the volume. Non-hepatocytes, i.e., endothelial cells, Kupffer cells and fat-storing cells account for less than 7% of the liver volume; the extracellular compartments account for the remaining 15% of the volume[1].

The hepatocyte, like other mammalian cells, is subdivided into different membranous substructures, or organelles (Figure 1). In the past two or more decades, considerable effort has gone into the isolation of purified fractions referable to these organelles to characterize them as to composition and function. Such studies, together with ultrastructural approaches in conjunction with radioautography and histochemistry, form the basis of our knowledge of how cells function.

This review is concerned with (a) the isolation and characterization of purified subcellular fractions from liver. The organelles and their membranes differ in function and composition with considerable membrane diversity even between contiguous organelles such as endoplasmic reticulum, Golgi complex and plasma membrane[2]; (b) the state of the art of subcellular fractionation of liver; (c) its potential use in hepatic pathology.

PRINCIPLES OF ISOLATION

The isolation of purified subcellular fractions first requires the disruption of the cells in the tissue to form a 'homogenate'. The tissue is placed into several volumes of homogenization medium, cut into smaller pieces and then subjected to controlled shear in a tissue grinder. The Potter–Elvehjem glass–

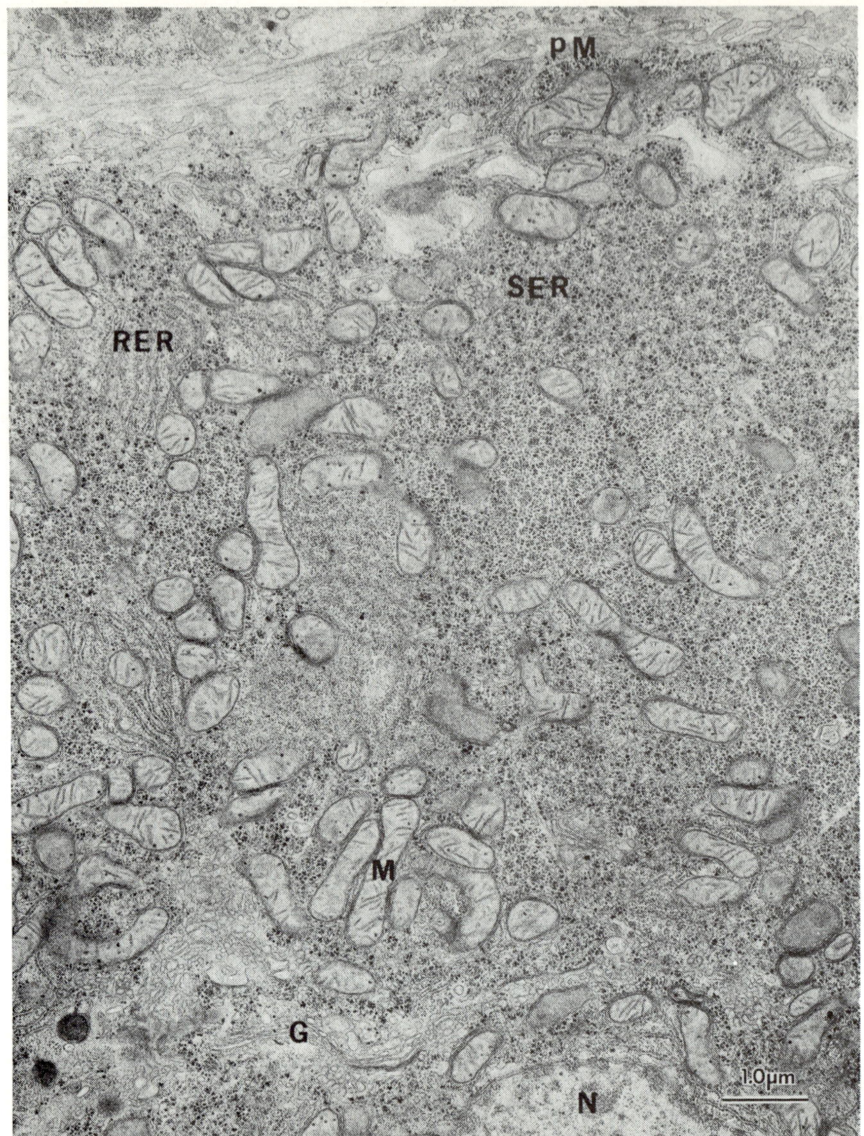

Figure 1 Electron micrograph of a section of a rat liver cell. The organelles designated are: N = nucleus; G = Golgi apparatus; M = mitochondria; RER = rough endoplasmic reticulum; SER = smooth endoplasmic reticulum; PM = plasma membrane

Teflon type is, in our experience, preferable for liver. The shear can be controlled by varying the rate of rotation of the pestle and the clearance between the Teflon pestle and the cylindrical walls of the glass homogenizer. The amount of shear applied is critical. Insufficient shear results in little or no cell disruption; excessive shear disrupts organelle structure with changes in sedimentation characteristics.

The isolation of purified subcellular fractions from the homogenate is usually achieved by centrifugation, under conditions which optimize the differential sedimentation velocity or differential density characteristics of the components. Each of the membranous organelles has been isolated with reasonable integrity with one exception, the endoplasmic reticulum. This organelle becomes disrupted forming membranous vesicles or 'microsomes'.

The development of methods for the isolation and characterization of highly purified subcellular fractions was first carried out one at a time. In this way, conditions most favourable for each organelle could best be applied[3–7]. The Golgi apparatus, also referred to as the Golgi complex, was the last membranous organelle to be isolated and characterized[8].

The Golgi apparatus mediates the intracellular transport of many types of secretory products from the rough endoplasmic reticulum (RER) to the plasma membrane, for release into the extracellular space[9]. RER, Golgi and plasma membranes are thus at least intermittently contiguous so that our first goal was to isolate and compare these membranes from a single source such as liver to better understand the function and interaction of these membranes.

Liver is particularly useful tissue for isolation and characterization of subcellular organelles since the purified fractions prepared from liver are derived mainly from a single cell type, the hepatocyte. Firstly, hepatocytes constitute most of the volume of liver[1] and are the major cell type broken by homogenization of the tissue[10]. Secondly, much information was already available on the metabolism, physiology and cell fractionation of liver[2–14]. Our early fractionation studies used beef liver, from which large quantities can be prepared[8]. For metabolic studies, however, rat liver offers an obvious advantage and our more recent studies used this source[15].

The comparison of the properties and composition of isolated organelles is meaningful only if the degree of purity can be established. If unique morphological features exist, examination of the fractions by electron microscopy is useful. Indeed, the earliest studies for the isolation of the Golgi complex depended primarily on morphological criteria[16,17]. However, membranous organelles frequently vesiculate during disruption of the cell forming small vesicles or 'microsomes'. 'Rough microsomes' are microsomes which are studded with ribosomes and can thus be identified as being derived from the RER. However, the origin of 'smooth microsomes' cannot readily be defined by morphological features. Thus, to evaluate the level of cross-contamination of a preparation of an organelle with other membranes, the activity of 'marker' enzymes is monitored. The term 'marker enzyme' refers to an enzyme that is localized, ideally, in one subcellular organelle. The detection of this activity in another subcellular fraction indicates contamination and the amount of contamination can be estimated. The percentage contamination on a protein basis may be estimated by dividing the specific activity of the marker enzyme in the fraction where it is a contaminant by the specific activity of the marker enzyme in the purified fraction where it is localized, multiplied by 100. Marker enzymes are particularly valuable in subcellular fractionation to guide purification and to estimate contamination of purified subcellular fractions. With the available technology, the use of a reliable marker enzyme enables one to monitor many fractions, usually with great sensitivity. Such a measurement is especially valuable for characterizing smooth microsomes where morphological identification is practically useless.

ISOLATION OF GOLGI APPARATUS

Although the Golgi apparatus of liver has unique morphological features (Figure 2), they are lost during the usual procedures used to homogenize tissues. Glutaraldehyde fixation was used prior to tissue disruption in an attempt to stabilize structure[17]. However, this approach also inactivates functional characteristics. We therefore attempted to develop biochemical assays for this organelle in fractions isolated from liver, by bringing together observations from several laboratories. Neutra and LeBlond showed, using radio-autography, that galactose injected into animals was incorporated first into the Golgi region of most cells including hepatocytes[18]. The enzymic glycosylation of proteins was being studied elsewhere[19–21]. The enzyme responsible for the addition of a galactose moiety, as the penultimate sugar, to serum glyco-proteins secreted by liver, is galactosyltransferase, i.e., UDP-galactose (UDP-Gal):*N*-acetylglucosamine (GlcNAc) galactosyltransferase (Equation 1)[19–21].

$$\text{UDP-Gal} + \text{protein-GlcNAc} \xrightarrow[\text{Mn}^{2+}]{\text{galactosyl-transferase}} \text{UDP} + \text{protein-GlcNAc-Gal} \qquad (1)$$

This enzyme can also utilize free *N*-acetylglucosamine as a substrate[20,21]. The reaction in Equation 2 formed the basis of our assay procedure for the enzyme in our subcellular fractions.

$$\text{UDP-[}^{14}\text{C]-Gal} + \text{GlcNAc} \xrightarrow[\text{Mn}^{2+}]{\text{galactosyl-transferase}} \text{UDP} + \text{GlcNAc-[}^{14}\text{C]-Gal} \qquad (2)$$

Thus, we suspected that galactosyltransferase might be localized in the Golgi apparatus. We purified a subfraction of bovine liver microsomes which was highly enriched in this activity and poor in enzymes present in purified preparations of RER, mitochondria, and plasma membranes[8]. The fraction showed unique morphological features, namely large sacs with attached tubules and it reacted heavily with osmium tetroxide on prolonged treatment, a property characteristic of Golgi apparatus *in situ*[22]. Similar results were obtained when the method was applied to rat liver[15]. We have since modified our method to both improve our yields and to obtain a more intact preparation

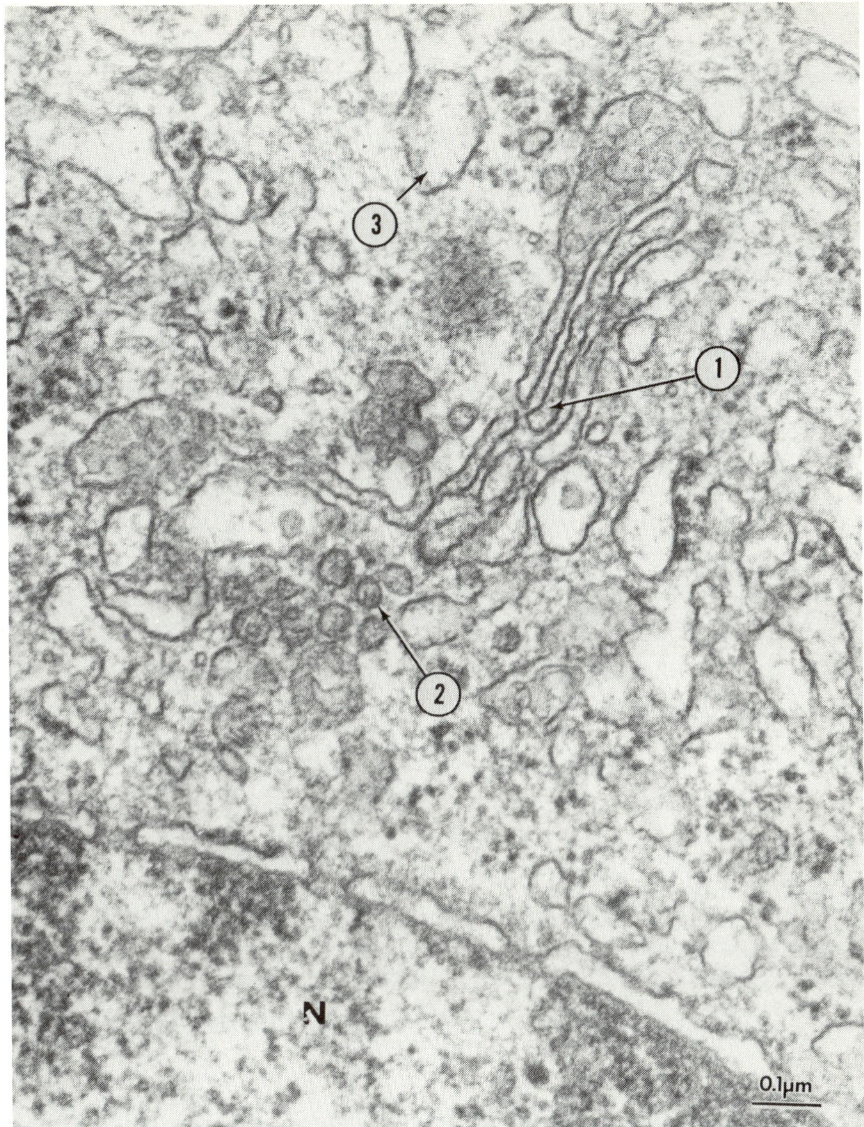

Figure 2 Morphology of the Golgi apparatus in a section of rat liver. The features indicated are: (1) Characteristic stacks of flattened cisternae; (2) peripheral vesicles or anastomizing tubules at the periphery of the cisternae; and (3) large secretory vesicles

of the Golgi apparatus. With the use of gentle homogenization coupled with a single flotation in a sucrose step-gradient we have isolated largely intact, recognizable, purified Golgi apparatus from rat liver[23]. About 20–35% of the galactosyltransferase activity of the homogenate is recovered in the fraction with about a 100-fold enrichment over the activity of the homogenate (Table 1). The galactosyltransferase appears to be localized exclusively in the Golgi complex. Smooth microsomes contain smooth membranes mainly referable to endoplasmic reticulum, but they also contain plasma and Golgi membranes. The detectable activity of galactosyltransferase in smooth microsomes is undoubtedly referable to Golgi membranes.

Table 1 Specific activity of some marker enzymes in subcellular fractions of rat liver

Fraction	*Bound P* (μg/mg *prot.*)	*Glu-6-phos-phatase**	*RI NADH-cyt.* c *red.*	*Succinate cyt.* c *red.*	*ATPase*	*5′-AMPase*	*Gal. transf.**
Homogenate	18.2	0.096	0.317	0.132	0.177	0.036	6
Mitochondria	10.4	0.003	0.185	0.457	0.420	0.000	0
Golgi complex	23.1	0.035	0.198	0.005	0.132	0.089	764
Rough microsomes	40.1	0.244	2.27	0.012	0.065	0.046	12
Smooth microsomes	33.6	0.197	2.53	0.005	0.088	0.192	41
Plasma membranes	23.1	0.040	0.139	0.020	1.24	0.990	5
Nuclei	40.8	0.036	0.075	0.002	—	—	2
Supernatant	14.9	0.003	0.004	0.002	—	—	0

Specific activity expressed as μmoles per min per mg protein except galactosyltransferase which is in nmoles per hour per mg protein.
* At 37 °C, all others at 30 °C.
The estimate of contamination in the Golgi complex preparation described in the table is 20% using appropriate marker enzymes. The contamination by endoplasmic reticulum (ER) is 9.6% obtained using glucose-6-phosphatase activity. In the calculation 'rough microsomes' are used as an index for pure endoplasmic reticulum membranes. The specific activity value of glucose-6-phosphatase for the 'rough microsomes' must be multipled by 3/2 to obtain the specific activity of ER membranes devoid of ribosomes since one-third of the protein in 'rough microsomes' is referable to ribosomes, i.e. 0.244 × 3/2 or 0.366. Thus, the contamination of the Golgi complex with ER is 9.6%, i.e. [0.035/0.366 × 100]. The contamination by mitochondria is about 1% using succinate-cytochrome *c* reductase]0.005/0.457 × 100]. The contamination by plasma membrane is 9% [0.089/0.990 × 100]. There was negligible contamination with lysosomes, peroxisomes and nuclei (not shown).

In addition to its high content of UDPGal:GlcNAc galactosyltransferase activity, the Golgi fraction differs from plasma membranes in its low ATPase, 5′-nucleotidase and UDPGalactose hydrolase activities (not shown), and from endoplasmic reticulum in its low glucose-6-phosphate and rotenone-insensitive NADH-cytochrome *c* reductase activities (Table 1). The purity of the Golgi fraction varies from 80 to 90%, as judged by the activities of various marker enzymes for other organelles.

The characteristic morphological features of the Golgi complex have been retained in the isolated preparation (Figure 3A). Three-dimensional distinguishing features have been preserved, such as flattened cisternae aligned in characteristic stacks, numbers of small vesicles or tubules attached at the periphery of the cisternae and large secretory vesicles full of particles (Figure 3B). The particles in the secretory vesicles are mainly precursors of serum low density lipoproteins secreted by the liver via the Golgi complex[24].

The highly purified subcellular fractions have been characterized by composition as well as functional parameters. It is thus possible to associate specific cell functions with individual subcellular organelles, as well as to assess the similarity or difference between membranes within a single cell type. The strength of this approach depends both on the availability of purified fractions as well as on the determination of contamination by the other subcellular organelles. The extent of contamination limits the strength of the conclusions.

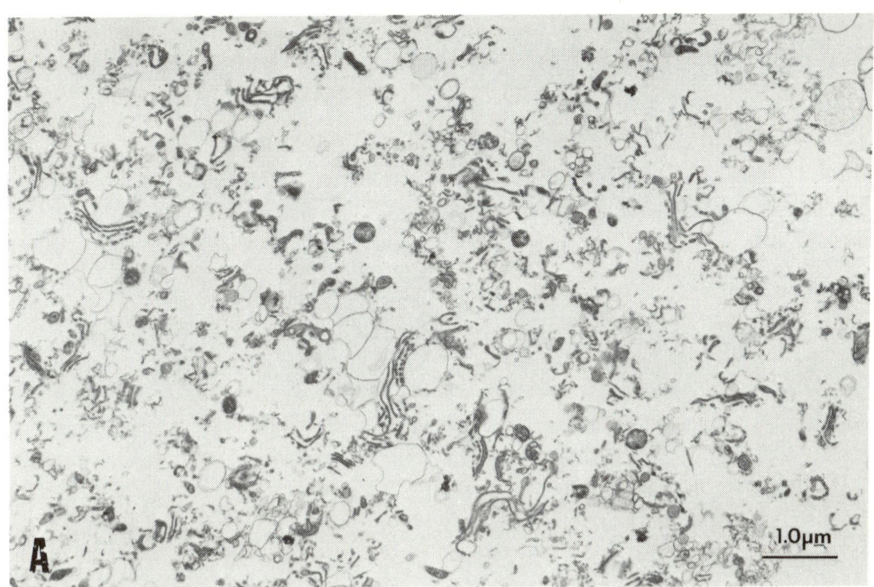

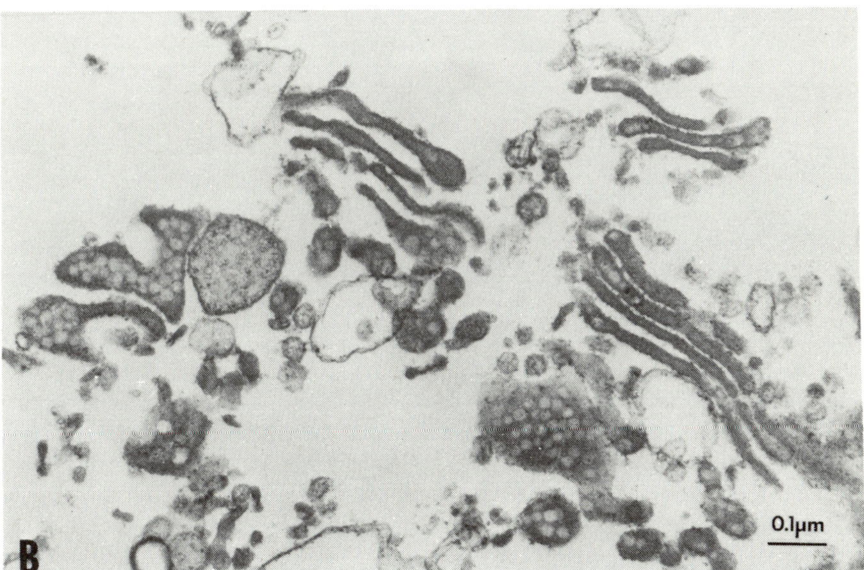

Figure 3 Electron micrographs of Golgi-rich fraction isolated from rat liver homogenate by centrifugation in a sucrose step-gradient. (A). Large field view; (B). Higher magnification of isolated Golgi complex from rat liver. The characteristic three-dimensional structure has been retained

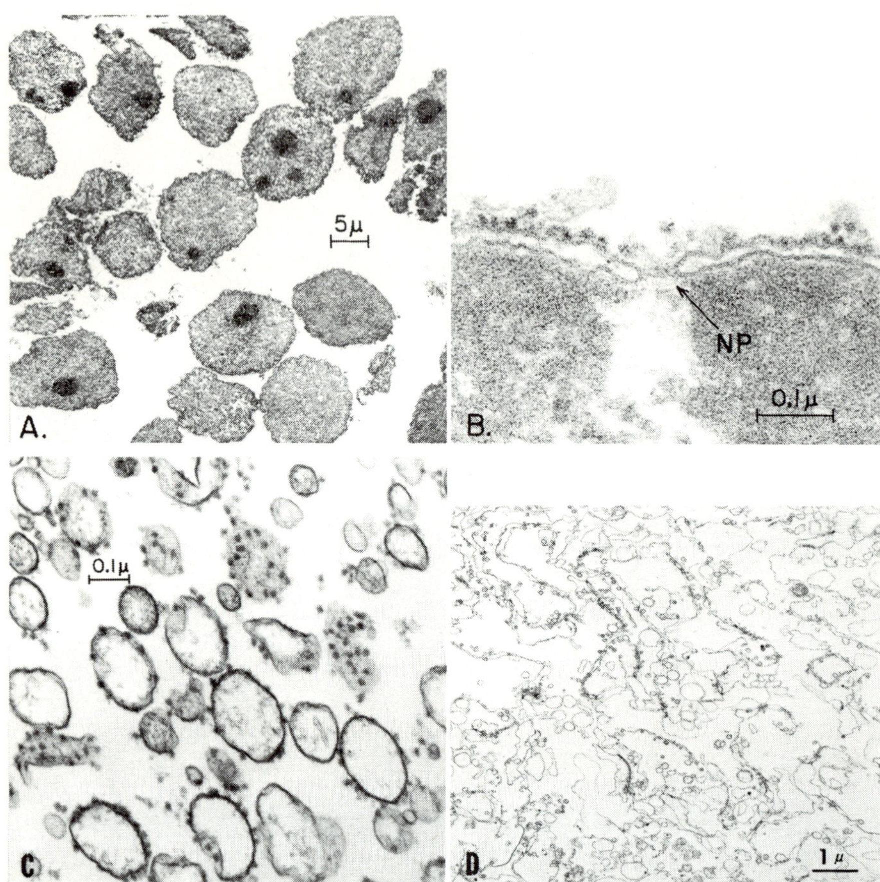

Figure 4 Electron micrographs of other purified subcellular fractions from rat liver. (A). Nuclei, low magnification. (B). Higher magnification of A to show details of nuclear membrane. Note outer leaflet of nuclear membrane with attached ribosomes and a nuclear pore (NP). (C). Rough endoplasmic reticulum. Note ribosomes attached to the outer face (cytoplasmic side) of most of the vesicles. (D). Plasma membranes. Numerous desmosomes, tight junctions and bile canaliculi can be observed

CHARACTERIZATION OF PURIFIED SUBCELLULAR FRACTIONS

Chemical and enzymatic characteristics of the Golgi apparatus of liver[2,25] have been compared with the other organelles including endoplasmic reticulum and plasma membranes, with which Golgi interacts to mediate the transport of secretory products out of the cell. The differences reflect the unique functions carried out by each organelle within the cell (*cf.* Tables 1 to 5).

Proteins

The protein profiles of the purified subcellular fractions were compared by polyacrylamide gel electrophoresis using dissociating conditions. Each subcellular fraction has a characteristic pattern[15,26] (Figure 5). The Golgi fraction differs from both endoplasmic reticulum and plasma membranes. Key differences persist even after the contents of the compartments have been removed or when the ribosomes have been separated from rough microsomes (not shown in the figure).

The cytochrome content of purified subcellular fractions from bovine and rat liver has been studied together with Drs Angelo Azzi and Britton Chance[27]. Cytochrome b_5 is present in the highest amount in the endoplasmic reticulum and in significant amounts in the Golgi complex. A key difference is the absence of cytochrome P_{450} from the Golgi complex. The purified plasma membrane fraction is devoid of cytochromes[27].

Lipids

The lipid content and composition varies in the purified subcellular fractions (Table 2)[28]. The Golgi complex and the plasma membrane have the highest content of neutral lipid. The cholesterol content is greatest in the plasma membrane fraction. The Golgi complex contains about 60% of the cholesterol content of plasma membrane although it has the greatest amount of cholesterol ester. The cholesterol content in the other intracellular membranes is low. The Golgi complex can be disrupted by nitrogen decompression in a Parr bomb releasing its intracisternal contents[23]. The resulting Golgi membranes have decreased neutral lipid content, increased percentage of cholesterol, but decreased cholesterol ester content, so that the cholesterol composition of Golgi membranes, *per se*, becomes more similar to that of the plasma membranes. Golgi membranes contain more phospholipid than plasma membranes. Rat liver Golgi preparations have the highest concentration of ubiquinone per mg of protein while no ubiquinone was detected in rough endoplasmic reticulum fractions, confirming the observation by Nyquist *et al.*[29]

Table 2 Lipid content of rat liver subcellular fractions

Cell fraction	*Mitochondria*	*Rough microsomes*	*Golgi complex*	*Golgi membranes*	*Plasma membranes*
Neutral lipid	0.027	0.051	0.335	0.341	0.322
Phospholipid	0.175	0.374	0.670	0.825	0.672
Total lipid	0.202	0.425	1.005	1.166	0.994
% NL	13.4	13.6	33.3	29.3	32.4
Cholesterol	0.003	0.014	0.071	0.078	0.128
free (%)	—	—	64	86	92
esterified (%)	—	—	36	14	8
Ubiquinone	1.3	ND	4.7	6.2	0.18
Total phosphorus	9.6	38.7	23.7	34.0	23.6
Lipid phosphorus	6.9	13.8	20.8	30.1	23.4

All values expressed as mg/mg protein except for phosphorus and ubiquinone (Coenzyme Q) which are in μg/mg protein. Golgi membranes were prepared by disruption of the Golgi apparatus using a Parr bomb[23]. Total lipid is the sum of the neutral lipid and phospholipid values. % NL is (neutral lipid/total lipid) × 100. ND, not detected. Dashes indicate values not determined. The ubiquinone value in plasma membrane can be accounted for by contamination of the fraction with mitochondria.

Table 3 Phospholipid composition of rat subcellular fractions

Phospholipid	*Liver*					*Kidney*		
	Mitrochondria	*Rough microsomes*	*Golgi complex*	*Golgi membrane*	*Plasma membranes*	*Plasma membranes*	*Golgi complex*	*Rough microsomes*
Origin	0.9	0.9	1.0	0.8	0.9	1.5	1.1	1.1
PE	34.6	21.8	23.5	19.6	23.3	31.4	23.1	22.4
PC	40.3	58.4	54.0	49.6	39.3	33.2	52.1	37.9
Sph	0.5	2.5	7.8	7.6	16.0	16.6	6.7	19.5
DPG	17.8	1.1	1.0	1.2	1.0	0.0	0.0	0.0
PI	4.6	10.1	8.6	12.2	7.7	1.7	7.1	1.7
PS	0.7	2.9	3.0	5.6	9.0	10.0	5.4	11.0
LPE	0.6	1.1	0.3	1.6	1.3	0.5	0.3	1.6
LPC	ND	0.5	0.4	1.4	1.0	4.5	3.6	4.0
PA	ND	0.8	0.4	0.4	0.5	0.6	0.7	0.7

All values expressed as % of total phospholipid phosphorus. The abbreviation of the phospholipids are as follows: PE = phosphatidylethanolamine; PC = phosphatidylcholine; Sph = sphingomyelin; DPG = diphosphatidylglycerol; PI = phosphatidylinositol; PS = phosphatidylserine; LE = lysophosphatidylethanolamine; LPC = lysophosphatidylcholine; PA = phosphatidic acid. The values for the kidney fractions were corrected for a small amount of mitochondrial contamination[28].

who prepared Golgi apparatus from rat liver by a different procedure. Mitochondria contain about the same amount of ubiquinone as the Golgi apparatus when normalized per mg phospholipid. When the intracisternal contents of the Golgi complex are removed (about 35% of the protein) the ubiquinone (Coenzyme Q) content of the Golgi membranes increases about 30%, indicating that the ubiquinone is primarily membrane-bound. There is practically no Coenzyme Q in the plasma membrane.

The phospholipid composition of purified subcellular fractions from rat liver and kidney are compared in Table 3[28]. In liver, the Golgi complex resembles RER in its phospholipid composition except for sphingomyelin which is intermediate in content between that of endoplasmic reticulum and plasma

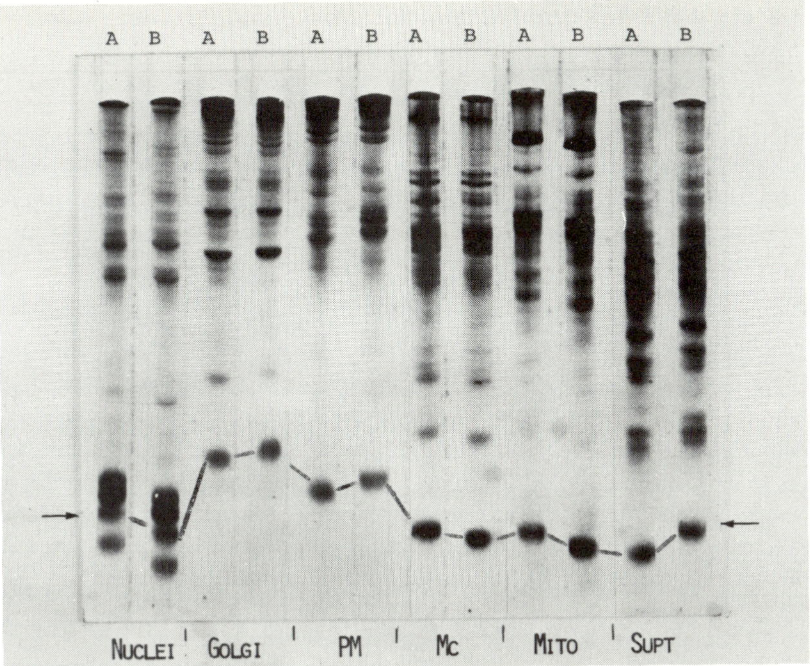

Figure 5 Protein patterns of purified subcellular fractions from fresh and frozen liver. Polyacrylamide gel electrophoresis was carried out using dissociating conditions in phenol-urea-acetic acid[26]. Each subcellular fraction has a distinctive pattern. Ribonuclease was added as an internal standard (arrow). From left to right are nuclei, Golgi complex, plasma membrane, microsomes (mainly ER), mitochondria and supernatant. A and B refer to fractions derived from fresh and frozen samples respectively. The frozen sample was stored for 9 months prior to subcellular fractionation. There are no observable differences between subcellular fractions from fresh or frozen samples

membranes. Golgi apparatus from rat kidney, however, has a phospholipid composition which resembles liver Golgi complex much more closely than it does any other renal cell fraction[28].

Studies with Dr George Rouser's laboratory, on the phospholipid composition of purified subcellular fractions from several tissues including liver and from different animal sources[28,30] revealed that mitochondria, nuclei, and Golgi complex each have characteristic phospholipid composition while plasma membrane and endoplasmic reticulum vary greatly with the tissue and source. The generalization for Golgi complex is thus far based on analysis of only two different tissues, liver and kidney, and should therefore be regarded as preliminary.

Thus the composition and function of the membranes of endoplasmic reticulum, Golgi complex and plasma membrane differ although they appear to be contiguous.

ROLE OF LIVER GOLGI COMPLEX IN SECRETION

The liver secretes many serum proteins, the most abundant of which is serum albumin. In collaboration with Dr Theodore Peters, Jr., we determined first the serum albumin content of the Golgi apparatus and compared it with rough and smooth microsomes of rat liver[31]. The vesicles were disrupted using deoxycholate and the amount of serum albumin was determined after precipitation with specific antibodies. Albumin was present in smooth and rough microsomes but was most abundant in the Golgi apparatus where it constituted 3–6% of the total protein.

Next, we followed the time-course for the appearance of newly synthesized albumin in endoplasmic reticulum and Golgi apparatus of rat liver after pulse labelling with [^{14}C]leucine[31]. The label first appears in the rough endoplasmic reticulum (maximum level at about 5 minutes). This is expected in view of its synthesis there. The level peaks in the smooth endoplasmic reticulum at 15 minutes, and in the Golgi at 20 minutes. Release of radioactive albumin into the blood began after 15 minutes and continued steadily for 55 minutes concurrent with the decline in the Golgi fraction. These data support the view that albumin is transported by way of the Golgi apparatus during secretion.

Table 4 The ratio of proalbumin to albumin in subcellular fractions of rat liver

Subcellular fraction	*Proalbumin* (%)	*Albumin* (%)	*Proalbumin albumin*
Rough endoplasmic reticulum	95	5	19
Smooth endoplasmic reticulum	51	49	1.0
Golgi apparatus	33	67	0.5
Supernatant fraction	0	100	0.0

Studies of K. Edwards, B. Fleischer, H. Dryburgh, S. Fleischer and G. Schreiber[33].

Serum albumin is synthesized first as proalbumin, which contains an additional oligopeptide as the N terminus[32]. With Dr Gerhard Schreiber's laboratory, we measured proalbumin and albumin in purified subcellular fractions of rat liver[33]. (The results are summarized in Table 4.) In RER, proalbumin is the main form detected; significant amounts of free albumin were not found. The ratio of proalbumin to albumin decreased in smooth endoplasmic reticulum and still further in the Golgi apparatus. The supernatant fraction contained only albumin probably originating from the blood which was not removed from the liver prior to homogenization. These results indicate that proalbumin is converted to albumin within the compartments of the smooth endoplasmic reticulum and the Golgi apparatus.

Since the hepatic Golgi apparatus takes part in the secretion of serum lipoprotein we investigated, with Dr Bert van Golde, the possibility that the Golgi apparatus might be involved in lipid biosynthesis by studying the subcellular distribution of enzymes involved in the biosynthesis of these lipids[34,35]. We found that the endoplasmic reticulum is the main centre for lipid biosynthesis in liver. Lecithin, sphingomyelin, phosphatidylethanolamine (via the diglyceride pathway) triglycerides, phosphatidylinositol and phosphatidylserine are formed exclusively in the endoplasmic reticulum. Mitochondria synthesize cardiolipin, and phosphatidylethanolamine via decarboxylation of phosphatidylserine.

Thus, the Golgi complex of rat liver is not capable of synthesizing lecithin, sphingomyelin, or triglycerides, the major lipid components of serum lipoproteins. Similar results have also been obtained with purified bovine liver subcellular fractions. We conclude that in hepatic biosynthesis of lipoproteins, both the lipid and protein components are made in the endoplasmic reticulum and are transported to the Golgi complex. The transport may occur as lipid–protein complexes but the final packaging seems to occur in the Golgi complex since it is within its secretory vesicles that lipoprotein particles can first be visualized. These results confirm the *in vivo* studies of Stein and Stein, who showed by means of radioautography, that labelled glycerol and palmitic acid are incorporated first into rough and smooth endoplasmic reticulum of rat liver and later appear in the Golgi apparatus[36]. These observations indicate also that intracellular transport of phospholipids must occur to supply phospholipid to membranes of organelles such as the Golgi complex and the plasma membrane which do not synthesize their own phospholipids. Thus, organelles in liver depend on the endoplasmic reticulum for their continued supply of both phospholipid and protein constituents.

SUBCELLULAR FRACTIONATION OF LIVER – STATE OF THE ART

There is good basis for the suggestion that membrane heterogeneity exists within a single organelle. In a cell which has functional polarity or directionality, the plasma membranes at opposite sides of such a cell would be expected to be different[37,38]. For example, intestinal brush-border cells communicate between two compartments, the lumen of the intestine and the blood, transferring nutrients across themselves from apical (mucosal) to basolateral (serosal) face. Two fractions of plasma membranes have been purified from these cells which are believed to be derived from the apical and basolateral faces, respectively. The former is rich in the alkaline phosphatase and digestive enzymes while the latter has high (Na^+–K^+) ATPase activity[38].

In liver several studies point to suborganelle heterogeneity in membranes. The difference between mitochondrial inner and outer membrane has been documented. Both membranes have been separated from one another and are entirely different in function and composition[39,57]. Plasma membrane from the rat hepatocyte has been subfractionated and contains three distinct functional domains[40]. Some progress in the subfractionation of Golgi complex from rat liver has been reported[41]. Thus, the diversity in hepatocyte membranes seems to be more extensive than merely between contiguous membranes, extending at least in part to the suborganelle level.

We have developed a procedure for the isolation of each of the purified subcellular fractions from a single homogenate. This procedure incorporates the accumulated experience from the procedures developed individually for each organelle (Figure 6). The procedure has been described in great detail

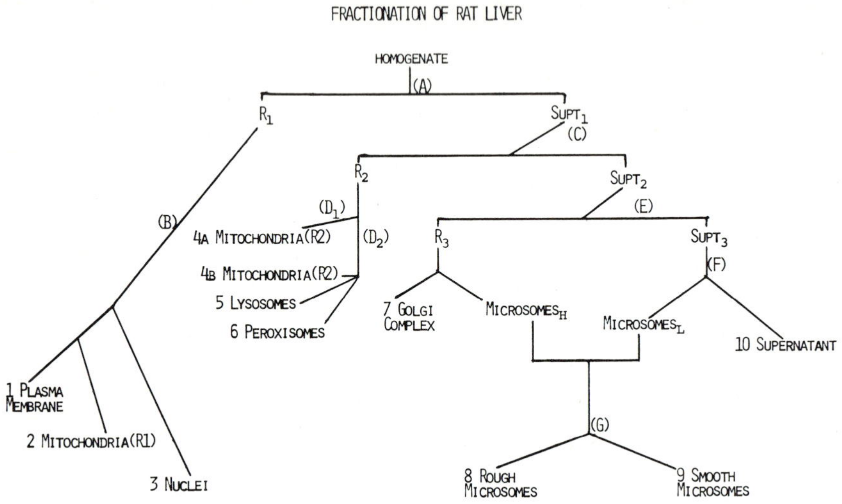

Figure 6 Flow diagram for the subcellular fractionation of rat liver. The letters in parentheses refer to preparatory separation steps and the numbers refer to the purification steps for the subcellular fractions[42]

elsewhere[42]. Electron micrographs of several of the purified subcellular fractions using this procedure are included in Figure 4. The Golgi fraction obtained by this procedure is similar to that shown in Figure 3.

We have also developed a procedure for the storage of rat liver homogenates for prolonged periods of time (many months)[43]. The procedure consists of: (1) preparation of a concentrated liver homogenate in buffered isotonic sucrose containing 10% dimethylsulphoxide (a cryoprotective agent). (2) quick freezing of the homogenate in liquid nitrogen. (3) storage in a liquid nitrogen refrigerator (~ -195 °C) until ready for use. For subsequent subcellular fractionation, the sample is rapidly thawed, diluted and the homogenate is then fractionated into purified subcellular fractions by the same procedure as for freshly prepared homogenates (*cf.* Figure 6).

The technology is now available for the complete fractionation of rat liver into its known component membranous organelles, and much is known about the characteristics of such purified fractions. It is now possible to look for aberration on the subcellular level in 'membrane alterations as the basis of liver injury'. The methodology needs to be simplified and more extensive characterization of 'normal' purified subcellular fractions is warranted. The time seems ripe for the formal emergence of a new field of biochemical pathology whose aim it is to assess pathology in terms of aberration at the subcellular level.

The development of technology to store samples for prolonged periods facilitates such study. Samples can be collected with the decision to work up the sample postponed until further information is available. Samples obtained in one laboratory can be shipped to another for subfractionation and characterization.

CONCLUDING REMARKS

There is considerable diversity in hepatocyte membranes. Such differences in membrane composition and function reflect the specific roles carried out by the subcellular organelles and their membranes. Some of the characteristics of the organelles of liver are useful in their isolation and characterization (Table 5).

The technology is now available for the isolation, from a single homogenate,

Table 5 Selected diagnotic properties of subcellular organelles of rat liver

Cell fraction	*Characteristic*	*Reference*
Plasma membranes	5'Nucleotidase*	44
	Alkaline phosphodiesterase I	45
	Prostaglandin E_2 receptor	46
Mitochondria†	Cytochrome oxidase	47
	Succinate-cytochrome *c* reductase	15
	Cytochrome a + a_3	27
Nuclei	DNA‡	48
	DNA-dependent RNA polymerase	48
	CMP-sialic acid synthetase	49
Lysosomes	Acid hydrolases (e.g. arylsulphatase)	47, 50
Peroxisomes	Catalase	51
	Urate oxidase	52
Golgi apparatus	UDP-Gal:*N*-acetylglucosamine galactosyltransferase	15, 23
Endoplasmic reticulum	Rotenone-insensitive DPNH-cytochrome *c* reductase§	15, 23
	Glucose-6-phosphatase ¶	15, 23
	Choline phosphotransferase	34, 35
	RNA ‖	53, 54
Supernatant (soluble phase of the cell)	Phosphoglucomutase	56
	Lactic dehydrogenase	55

* This enzyme is also present at very low specific activity in endoplasmic reticulum[56].
† Mitochondria are composed of two different membranes. The components shown are characteristic of inner membranes. A characteristic enzyme for the outer membrane is monoamine oxidase[57].
‡ Mitochondria also contain small amounts of DNA and DNA-dependent RNA polymerase[58].
§ Although localized mainly in endoplasmic reticulum, this activity is also present in the outer membranes of the mitochondrion[57].
¶ Also present in nuclear membranes[59].
‖ Two types of endoplasmic reticulum are present in liver. Rough endoplasmic reticulum has attached ribosomes and thus a very high content of RNA. Smooth endoplasmic reticulum appears to be very similar in composition and properties to rough endoplasmic reticulum but has no ribosomes and consequently only a small amount of RNA. Smooth endoplasmic reticulum is the major component of the smooth microsome fraction of liver[60].

of purified subcellular fractions referable to each organelle. Procedures for long-term storage of liver for subsequent fractionation into purified subcellular organelles further makes feasible the association of defects at the subcellular level with pathology. Samples can be collected and stored for later study or shipped to another laboratory which is more specialized in the technology of subcellular fractionation.

ACKNOWLEDGEMENTS

The authors thank Mr Akitsugu Saito for the electron micrographs used in this paper. This work was supported in part by Grants AM 14632 and AM 17223 of the US Public Health Service.

References

1. Blouin, A., Bolender, R. P. and Weibel, E. R. (1977). Distribution of organelles and membranes between hepatocytes and non-hepatocytes in the rat liver Parenchyma. *J. Cell Biol.*, **72,** 441
2. Fleischer, B. and Fleischer, S. (1971). Comparison of cellular membranes of liver with emphasis on the Golgi complex as a discrete organelle. In: L. Manson (ed.), *Biomembranes*, Vol. 2. p. 75. (New York: Plenum Press)
3. Neville, D. M. (1960). The isolation of a cell membrane fraction from rat liver. *J. Biophys. Biochem. Cytol.*, **8,** 413
4. Cheveau, J., Moulé, Y. and Rouiller, C. (1956). Isolation of pure and unaltered liver nuclei: morphology and biochemical composition. *Exp. Cell Res.*, **11,** 317
5. Dällner, G. (1963). Studies on the structural and enzymic organization of the membranous elements of liver microsomes. *Acta Pathol. Microbiol. Scand., Suppl.*, **166,** 1
6. Leighton, F., Poole, B., Beaufay, H., Baudhuin, P., Coffey, J. W., Fowler, S. and de Duve, C. (1968). The large-scale separation of peroxisomes, mitochondria and lysosomes from the livers of rats injected with Triton WR-1339. *J. Cell Biol.*, **37,** 482
7. Stein, Y., Widnell, C. and Stein, O. (1968). Acylation of lysophosphatides by plasma membrane fractions of rat liver. *J. Cell. Biol.*, **39,** 185
8. Fleischer, B., Fleischer, S. and Ozawa, H. (1969). Isolation and characterization of Golgi membranes from bovine liver. *J. Cell Biol.*, **43,** 59
9. Palade, G. (1975). Intracellular aspects of the process of protein synthesis. *Science*, **189,** 347
10. Dällner, G., Siekevitz, P. and Palade, G. E. (1966). Biogenesis of endoplasmic reticulum membranes. *J. Cell Biol.*, **30,** 97
11. Claude, A. (1947). Studies on cells: morphology, chemical constitution, and distribution of biochemical functions. *Harvey Lectures*, **43,** 121
12. de Duve, C. (1971). Tissue fractionation past and present. *J. Cell Biol.*, **50,** 20D
13. de Duve, C. and Baudhuin, P. (1966). Peroxisomes (microbodies and related particles). *Physiol. Rev.*, **46,** 323
14. de Duve, C. and Wattiaux, R. (1966). Functions of lysosomes. *Ann. Rev. Physiol.*, **28,** 435
15. Fleischer, B. and Fleischer, S. (1970). Preparation and characterization of Golgi membranes from rat liver. *Biochem. Biophys. Acta* **219,** 301
16. Kuff, E. L. and Dalton, A. J. (1959). Biological studies of isolated Golgi membranes. In: T. Hayashi (ed.), *Subcellular Particles*, p. 114. (New York: Ronald Press)

17. Mollenbauer, H. H., Morré, D. J. and Bergmann, L. (1967). Homology of form in plant and animal Golgi apparatus. *Anal. Rec.*, **158,** 313
18. Neutra, M. and LeBlond, C. P. (1966). Radioautographic comparison of the uptake of galactose-H^3 and glucose-H^3 in the Golgi region of various cells secreting glycoproteins or mucopolysaccharides. *J. Cell Biol.*, **30,** 137
19. Roseman, S. (1970). The synthesis of complex carbohydrates by multiglycosyltransferase systems and their potential function in intercellular adhesion. *Chem. Phys. Lipids*, **5,** 270
20. McGuire, E. J., Jourdian, G. W., Carlson, D. M. and Roseman, S. (1965). Incorporation of D-galactose into glycoproteins. *J. Biol. Chem.*, **240,** 4113
21. Schanbacher, F. L. and Ebner, K. E. (1970). Galactosyltransferase acceptor specificity of the lactose synthetase a protein. *J. Biol. Chem.*, **245,** 5057
22. Friend, D. S. and Murray, M. J. (1965). Osmium impregnation of the Golgi apparatus. *Am. J. Anat.*, **117,** 135
23. Fleischer, B. (1974). Isolation and characterization of Golgi apparatus and membranes from rat liver. *Meth. Enzymol.*, **31,** 180
24. Hamilton, R. L., Regen, D. M., Gray, M. E. and LeQuire, V. S. (1967). Lipid transport in liver. I. Electron microscopic identification of very low density lipoproteins in perfused rat liver. *Lab. Invest.*, **16,** 305
25. Fleischer, B., Zambrano, F. and Fleischer, S. (1974). Biochemical characterization of the Golgi complex of mammalian cells. *J. Supramolec. Struct.*, **2,** 737
26. Zahler, W. L., Fleischer, B. and Fleischer, S. (1970). Gel electrophoresis patterns of the proteins of organelles isolated from bovine liver. *Biochim. Biophys. Acta*, **203,** 283
27. Fleischer, S., Fleischer, B., Azzi, A. and Chance, B. (1971). Cytochrome b_5 and P_{450} in liver cell fractions. *Biochim. Biophys. Acta*, **225,** 194
28. Zambrano, F., Fleischer, S. and Fleischer, B. (1975). Lipid composition of the Golgi apparatus of rat kidney and liver in comparison with other subcellular organelles. *Biochim. Biophys. Acta*, **380,** 357
29. Nyquist, S. E., Crane, F. L. and Morré, D. J. (1971). Vitamin A: concentration in the rat liver Golgi apparatus. *Science*, **173,** 939
30. Rouser, G., Nelson, G. J., Fleischer, S. and Simon, F. (1968). Lipid composition of animal cell membranes, organelles and organs. In: D. Chapman (ed.), *Biological Membranes*, p. 5. (New York: Academic Press)
31. Peters, T., Jr., Fleischer, B. and Fleischer, S. (1971). The biosynthesis of rat serum albumin. IV. Apparent passage of albumin through the Golgi apparatus during secretion. *J. Biol. Chem.*, **246,** 240
32. Urban, J., Inglis, A. S., Edwards, K. and Schreiber, G. (1974). Chemical evidence for the difference between albumins from microsomes and serum and a possible precursor–product relationship. *Biochem. Biophys., Res. Commun.*, **61,** 494
33. Edwards, K., Fleischer, B., Dryburgh, H., Fleischer, S. and Schreiber, G. (1976). The distribution of albumin precursor protein and albumin in liver. *Biochem. Biophys. Res. Commun.* **72,** 310
34. van Golde, L. M. G., Fleischer, B. and Fleischer, S. (1971). Some studies on the metabolism of phospholipids in Golgi-complex from bovine and rat liver in comparison to other subcellular fractions. *Biochim. Biophys. Acta*, **249,** 318
35. van Golde, L. M. G., Raben, J., Batenburg, J. J., Fleischer, B., Zambrano, F. and Fleischer, S. (1974). Biosynthesis of lipids in Golgi complex and other subcellular fractions from rat liver. *Biochem. Biophys. Acta*, **360,** 179
36. Stein, O. and Stein, Y. (1967). Lipid synthesis, intracellular transport, storage and secretion. *J. Cell. Biol.*, **33,** 319
37. Kyte, J. (1976). Immunoferritin determination of the distribution of (Na^+ + K^+) ATPase over the plasma membranes of renal convoluted tubules. *J. Cell Biol.*, **68,** 287
38. Fujita, M., Kawai, K., Asano, S. and Nakao, M. (1973). Protein components of two different regions of an intestinal epithelial cell membrane: regional singularities. *Biochim. Biophys. Acta*, **307,** 141

39. Greenawalt, J. W. (1974). The isolation of outer and inner mitochondrial membrane. *Meth. Enzymol.*, **31,** 301
40. Kremmer, T., Wisher, M. H. and Evans, W. H. (1976). The lipid composition of plasma membrane subfractions originating from the three major functional domains of the rat hepatocyte cell surface. *Biochim. Biophys. Acta*, **455,** 655
41. Ehrenreich, J. H., Bergeron, J. J. M., Siekevitz, P. and Palade, G. E. (1973). Golgi fractions prepared from rat liver homogenates. *J. Cell Biol.*, **59,** 45
42. Fleischer, S. and Kervina, M. (1974). Subcellular fractionation of rat liver. *Meth. Enzymol.*, **31,** 6
43. Fleischer, S. and Kervina, M. (1974). Long-term preservation of liver for subcellular fractionation. *Meth. Enzymol.*, **31,** 3
44. Widnell, C. C. and Unkeless, J. C. (1968). Partial purification of a lipoprotein with 5′-nucleotidase activity from membranes of rat liver cells. *Proc. Natl. Acad. Sci. USA*, **61,** 1050
45. Aronson, N. N. and Touster, O. (1974). Isolation of rat liver plasma membranes fragments in isotonic sucrose. *Meth. Enzymol.*, **31,** 90
46. Smigel, M. and Fleischer, S. (1974). Characterization and localization of prostaglandin E_1 receptors in rat liver plasma membranes. *Biochim. Biophys. Acta*, **332,** 358
47. Appelmans, F., Wattiaux, R. and de Duve, C. (1955). The association of acid phosphatase with a special class of cytoplasmic granules in rat liver. *Biochem. J.*, **59,** 438
48. Tata, J. R. (1974). Isolation of nuclei from liver and other tissues. *Meth. Enzymol.*, **31,** 253
49. Kean, E. L. (1972). CMP-Sialic acid synthetase of nuclei. *Meth. Enzymol.*, **28B,** 413, and unpublished observations of B. Fleischer and S. Fleischer
50. Stahn, R., Maier, K.-P. and Hannig, K. (1970). A new method for the preparation of rat liver lysosomes. *J. Cell Biol.*, **46,** 576
51. Baudhuin, P., Beaufay, H. B., Rahman-Li, Y., Sellinger, O. Z., Wattiaux, R., Jacques, P. and de Duve, C. (1964). Intracellular distribution of monoamine oxidase, aspartate aminotransferase, alanine aminotransferase, D-amino acid oxidase and catalase in rat liver tissue. *Biochem. J.*, **92,** 179
52. Leighton, F., Poole, B., Lazarow, P. B. and de Duve, C. (1969). The synthesis and turnover of rat liver peroxisomes. *J. Cell Biol.*, **41,** 521
53. Hers, H. G., Berthet, J., Berthet, L. and de Duve, C. (1951). Le systeme hexose—phosphatasique localisation intra-cellulaire des ferments par centrifugation fractionnée. *Bull Soc. Chim. Biol.*, **33,** 21
54. Anderson, N. G. and Green, J. G. (1967). The soluble phase of the cell. In: D. B. Roodyn (ed.), *Enzyme Cytology*, p. 475. (New York: Academic Press)
55. Lowry, O. H. (1957). Micromethods for the assay of enzymes. *Meth. Enzymol.*, **4,** 366
56. Widnell, C. C. (1972). Cytochemical localization of 5′-nucleotidase in subcellular fractions isolated from rat liver. *J. Cell Biol.*, **52,** 542
57. Sottocasa, G. L., Kuylenstierna, B., Ernster, L. and Bergstrand, A. (1967). Separation and some enzymatic properties of the inner and outer membranes of rat liver mitochondria. *Meth. Enzymol.*, **10,** 448
58. Nass, M. M. K. (1968). Properties of organelle-associated and isolated mitochondrial DNA. In: E. C. Slater, J. M. Tager, S. Papa and E. Quagliariello (eds.), *Biochemical Aspects of the Biogenesis of Mitochondria*, p. 27. (Bari, Italy: Adriatica Editrice)
59. Kashnig, D. M. and Kasper, C. B. (1969). Isolation, morphology, and composition of the nuclear membrane from rat liver. *J. Biol. Chem.*, **241,** 3786
60. Beaufay, H., Amer-Costesec, A., Thinès-Sempoux, D., Wibo, M., Robbi, M. and Berthet, J. (1974). Analytical study of microsomes and isolated subcellular membranes from rat liver. *J. Cell Biol.*, **61,** 213

4

Hepatocyte Endomembrane Proteins and their Assembly

G. DALLNER and L. C. ERIKSSON

Enzymes and enzyme systems localized on the endoplasmic reticulum (ER) of the liver cell mediate a large number of cellular functions. The complexity of certain of these functions would seem to require stability of the relative topology of the catalytic components; complete fluidity would disturb the numerous protein–protein and protein–lipid interactions which are involved in most complex membrane reactions. The stability of the topology of the ER membrane in the lateral plane is well documented by subfractionation studies. Analysis of subfractions of both rough and smooth microsomes and of submicrosomal particles have demonstrated that certain enzymes are segregated into smaller regions of the membrane[1]. Therefore, free movement of enzyme proteins in the lateral plane does not seem to occur; rather, the specialization of different membrane regions is maintained.

The specific and non-random arrangement of ER enzymes, on the other hand, does not exclude the rapid turnover of various macromolecules at different rates, as well as movement within the membrane of those proteins which are destined for another compartment[2]. Thus, the ER membrane may be used for investigating two different and important phenomena: membrane stability and membrane dynamics.

It is well-established that protein components of the ER can move within the lateral plane of the membrane from one place to another. This has been demonstrated by studies of newborn and adult animals with and without drug induction[3,4]. In the following discussion an example of a different pathway of protein biosynthesis will be given: a glycoprotein, completed and synthesized in a membrane compartment, is discharged to the cytoplasm in order to approach and to be incorporated into a new membrane compartment. Such biosynthetic processes related to the ER membranes may occur only if another type of heterogeneity exists: the asymmetric distribution of both protein and lipid components in the transverse plane of this organelle.

ASYMMETRY OF PROTEINS

The kinds and the amount of charged protein localized on the surface of the ER membrane are important, since they not only influence the properties of the membrane itself but also regulate the interaction of this structure with its surroundings. Smooth microsomes have a relatively high net negative surface charge which can be altered by a number of conditions (Table 1). Treatment

Table 1 Effect of enzyme treatment on the electrophoretic mobility of smooth microsomes

Treatment	*Mobility*	
	$\mu m\ s^{-1}\ V^{-1}\ cm$	*% of control*
None	0.55 ± 0.06	100
PLPase A	0.62 ± 0.04	112
PLPase C	0.54 ± 0.05	98
PLPase D	0.54 ± 0.03	98
Neuraminidase	0.56 ± 0.03	102
Trypsin	0.65 ± 0.08	119
Papain	0.68 ± 0.07	123
Trypsin + PLPase A	0.64 ± 0.03	117
Trypsin + PLPase C	0.56 ± 0.04	102
Trypsin + PLPase D	0.66 ± 0.04	120

Smooth microsomes were treated with the enzymes listed in the table and were subjected to electrophoresis using a free electrophoretic system and a stabilizing gradient[5]. The values are given as mean values ± SEM ($n = 8$)

with phospholipase removes a few proteins as well as phospholipids, while trypsin and papain remove other types of proteins. The negative surface charge increases to different extents after these two treatments. If trypsin-treated smooth microsomes are subjected to phospholipase C treatment, several new protein species are again liberated and the negative surface charge is decreased. If a pathological condition changes the kinds or the amount of proteins on the cytoplasmic surface of the ER, a similar increase or decrease in the net negative charge will occur.

The transverse distribution of various ER enzymes may be investigated using the low deoxycholate–KCl procedure, which allows penetration of macromolecules into the vesicle lumen without disassembling the membrane structure itself[6]. Some of the microsomal enzymes are obviously localized on the cytoplasmic surface, while others are solubilized or inactivated by proteolytic treatment of permeable vesicles and consequently localized at the luminal surface[7,8] (Table 2). Those enzymes which are not influenced by the trypsin digestion may be buried completely or partially by surrounding membrane components. The only ER enzyme which to date has not been found to display an asymmetric transverse distribution is cytochrome P_{450}.

Similar to the case for proteins with known enzymatic activities, glycoproteins with unknown functions are also asymmetrically distributed in the trans-

Table 2 Transverse localization of enzymes in the ER

Cytoplasmic surface		*Luminal surface*
Cytochrome b_5	ATPase	Glucose-6-phosphatase
NADH-cyt. b_5 reductase	AMPase	β-Glucuronidase
NADPH-cyt. *c* reductase	UDP-glucuronyltransferase	Nucleoside diphosphatase
GDP-mannosyltransferase	Epoxide hydrase	Acetanilide-hydr. esterase
Carboxylesterase		
	DT-diaphorase	DT-diaphorase
Cyt. P_{450}	Cyt. P_{450}	Cyt. P_{450}

The data are taken from earlier publications[7,8]

verse plane of the ER[9,10]. In rough microsomes and probably also in smooth microsomes and Golgi membranes the protein-bound sialic acid and a large part of the protein-bound mannose, galactose and glucosamine residues are found on the outer surface of the vesicles; none or only smaller amounts of these sugars can be found at the luminal surface (Figure 1). Clearly, protein asymmetry is a basic feature of ER membranes.

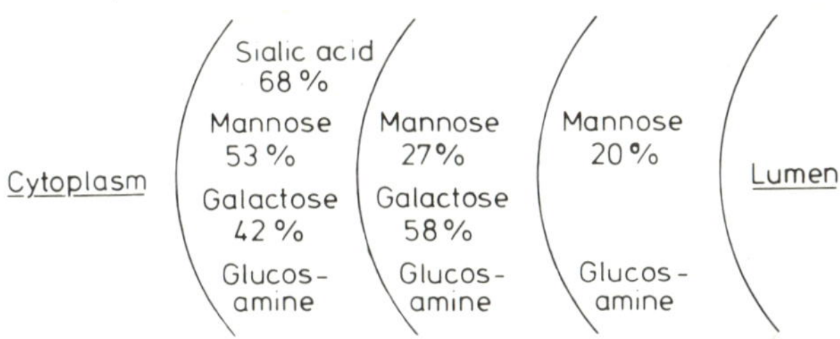

Figure 1 Protein-bound sugar residues in rough microsomes. The vesicles were treated with trypsin in the presence or absence of low concentrations of deoxycholate and the sugars were measured as described previously[9]. Data taken from references [9] and [10]

LIPID ASYMMETRY

Like proteins, phospholipids of intracellular hepatocyte membranes are asymmetrically distributed in the transverse plane[11]. In microsomes, as well as in Golgi membranes, the inner mitochondrial membrane and lysosomal membranes, only phosphatidylcholine can be found in both halves of the membrane bilayer (Figure 2). Phosphatidylethanolamine and phosphatidylserine when the latter is present, are found at the cytoplasmic surface. Phosphatidylinositol, sphingomyelin, and cardiolipin (in the case of mitochondria) may be assigned completely to the inner surface. The relative lipid homogeneity in each half of the lipid bilayer may affect membrane fluidity and thereby represent an important factor in translocation of proteins.

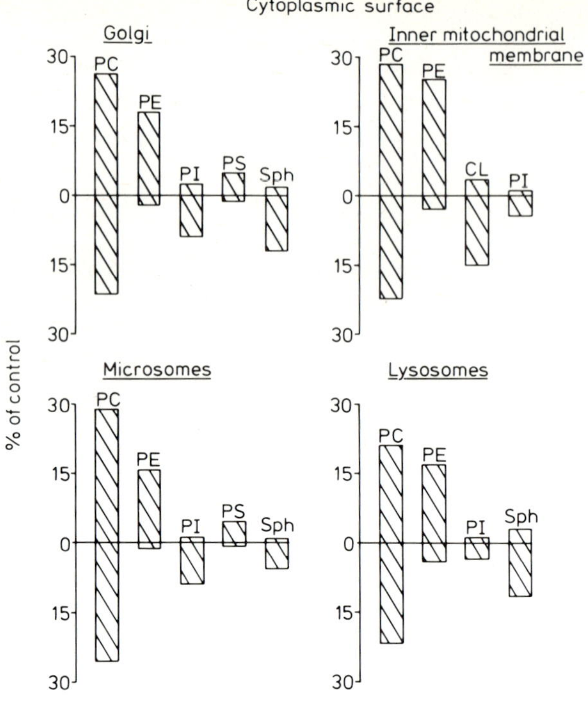

Figure 2 Transverse distribution of phospholipids in cytoplasmic membranes. Abbreviations: PC = phosphatidylcholine; PE = phosphatidylethanolamine; PI = phosphatidylinositol; PS = phosphatidylserine; Sph = sphingomyelin; CL = cardiolipin. Data taken from reference[11]

INTRAMEMBRANOUS LOCALIZATION OF GLYCOPROTEIN BIOSYNTHESIS

There are two types of glycoproteins which are synthesized in the ER. The secretory proteins which are transported to the blood enter the ER lumen after release from bound ribosomes and can later on be recovered as free proteins in the lumen. Obviously, in this case the sequential synthesis of the oligosaccharide chain takes place at the inner half of the ER membrane. The situation for membrane glycoproteins is different. When rough and smooth microsomes and Golgi membranes are pretreated with trypsin, *in vitro* incorporation of mannose from GDP-mannose into the lipid intermediates dolichol phosphate and dolichol pyrophosphate-oligosaccharide, as well as into the endogenous protein acceptor, is decreased (Figure 3). This indicates that some of the transferases are affected by proteolysis of the intact vesicles. When trypsin treatment was carried out either after *in vitro* incubation or after *in vivo* injection of mannose, lipid intermediates and mannosylated protein acceptors were

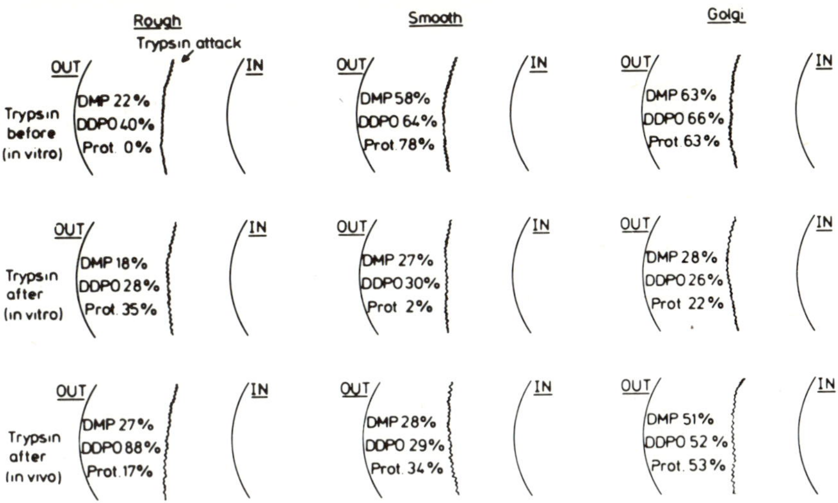

Figure 3 Distribution of mannose incorporated *in vitro* and *in vivo*. Trypsin treatment was performed using 50 μg trypsin/mg protein at 30 °C for 10 min. In experiments *in vivo* [^{3}H]mannose (100 μCi/100 g body weight) was injected into the portal vein 5 min before decapitation. DMP = dolichol monophosphate; DDPO = dolichol pyrophosphate-oligosaccharide. Data taken from reference[12]

removed. The latter proteins are immunologically distinct from serum proteins and represent newly synthesized membrane glycoproteins undergoing completion of their oligosaccharide chains.

INCORPORATION OF CYTOPLASMIC LIPOPROTEINS INTO MICROSOMES

Rough and smooth microsomes contain glycoproteins with terminal sialic acid; and since this sugar is added in the Golgi membranes, it is necessary to envisage a pathway for the transfer of certain glycoproteins from the Golgi to the ER[13–15]. When double-labelled cytoplasmic proteins from the void volume of a Sephadex G-25 or G-100 column were incubated with rough microsomes, both labels displayed a high degree of incorporation into the vesicles (Table 3). The incorporated radioactivity could not be removed by various washing procedures or with low concentration of detergents.

A lipoprotein could be purified from the supernatant by flotation of the G-25 pool in a KBr solution with a density of 1.21. This highly purified protein (LP fraction) proved to be most effectively incorporated *in vitro*. The isolated LP fraction contains two peptides as revealed by SDS gel electrophoresis, one with a molecular weight around 70 000 and the other around 11 000 (Figure 4). This gel electrophoretic pattern of the LP fraction is very different from

Table 3 Incorporation of glycoproteins from different fractions into rough microsomes

	G-25 pool	*G-100 pool*	*LP fraction*
Amount of NANA (μg NANA/mg protein)	0.30	0.96	38.1
Total radioactivity in the incubation mixture			
$[^3H]$GlN, c.p.m.	3610	7790	11 425
$[^{14}C]$Leucine, c.p.m.	4410		4277
Radioactivity transferred to microsomes			
$[^3H]$GlN, c.p.m.	843	2160	4650
$[^3H]$GlN, %	23.3	27.8	40.7
$[^{14}C]$Leucine, c.p.m.	554		840
$[^{14}C]$Leucine, %	12.5		19.6

Rough microsomes were incubated with G-25 pool, G-100 pool and LP fractions as described earlier[16]. Data taken from reference[16]

that of three blood lipoproteins, VLDL, LDL and HDL. This experiment, as well as immunological analyses, demonstrates that the presence of lipoprotein in the supernatant is not due to contamination by serum proteins.

The isolated and purified cytoplasmic lipoprotein complex was characterized in detail and the main results are given in Table 4. The intact complex has a molecular weight above 200 000 but is easily broken up into smaller components during purification and storage.

LIPOPROTEIN OF THE GOLGI COMPLEX

In order to investigate the origin of the cytoplasmic lipoprotein, Golgi fractions labelled *in vivo* were incubated in sucrose. When the released proteins were purified by flotation in a KBr solution with a density of 1.21, the fraction recovered contained only two peptides and only one of them was glycoprotein (Figure 5A). Upon incubation with microsomes this latter protein (with a molecular weight of 69 000) was incorporated to a high degree (Figure 5B). In these and other respects the protein released from the Golgi was similar to the supernatant lipoprotein.

IN VIVO EXPERIMENTS

In *in vivo* pulse labelling experiments performed by injection of $[^3H]$glucosamine, the protein-bound sialic acid of the Golgi membranes, and of the supernatant exhibited rapid labelling reaching a maximum after only 30 min (Figure 6). The pattern of incorporation into rough and smooth microsomes was very different; the radioactivity incorporated increased slowly

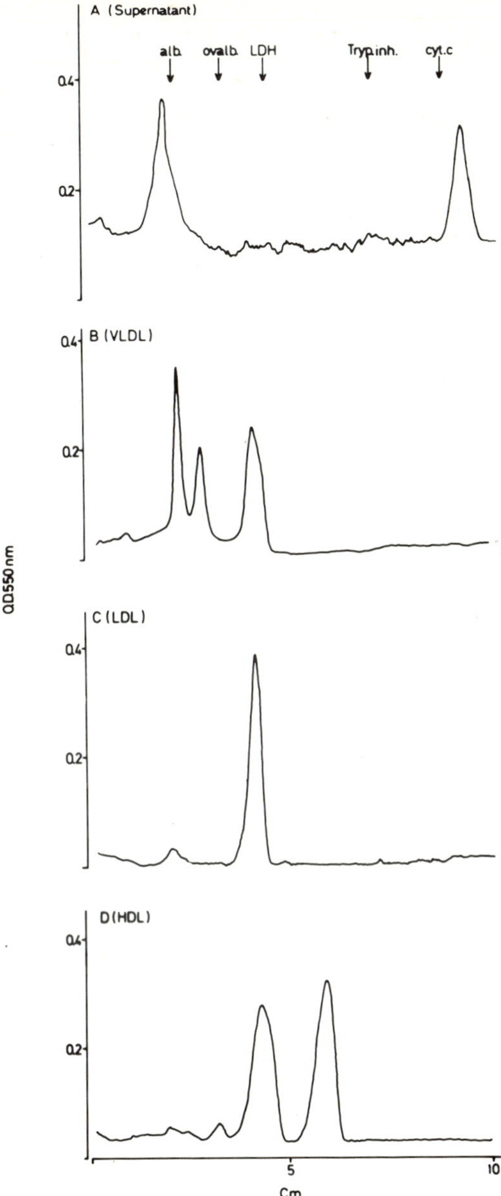

Figure 4 SDS-gel electrophoretic patterns of supernatant and serum lipoproteins. (A) Supernatant LP; (B) very low density lipoprotein (VLDL); (C) low density lipoprotein (LDL); (D) high density lipoprotein (HDL). Data taken from reference[14]

and reached a maximum only after 3 hours. These experiments suggest the possibility that even *in vivo* the glycoprotein can be transferred from the Golgi through the cytoplasm to the ER.

Table 4 Properties of the cytoplasmic lipoprotein complex

Equilibrium density	1.07–1.14 g/ml
Sedimentation coefficient	4.9 S
Calculated molecular weight	210 000 daltons
Peptides after SDS-gel electrophoresis	11–13 000 and 67–69 000 daltons
Amino acid composition	identical in the two peptides
Distribution in a $CHCl_3/CH_3OH/H_2O$ system	67% in the organic phase
Oligosaccharide composition	mannose, galactose, glucosamine, NANA
Lipid composition (44% by weight)	
neutral lipids (50%)	cholesterol, triglycerides
phospholipids (50%)	phosphatidylcholine lysophosphatidylcholine sphingomyelin phosphatidylethanolamine phosphatidylinositol

Data taken from reference[16]

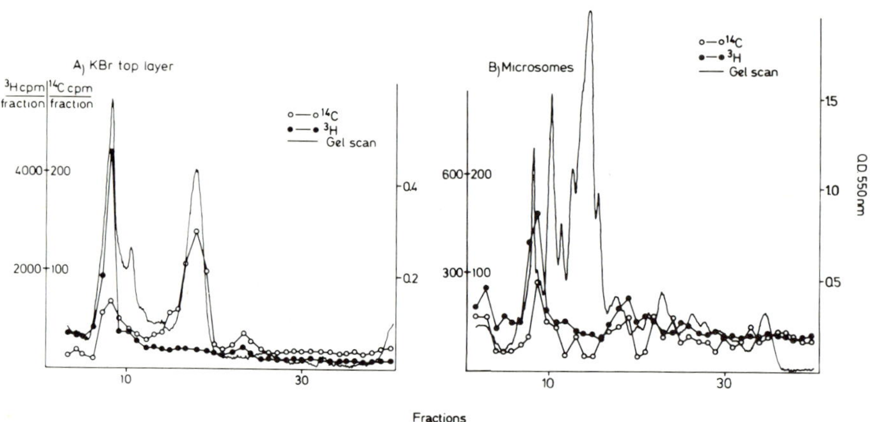

Figure 5 Incubation of microsomes with purified proteins released from Golgi vesicles. (A) The top layer after centrifugation of the protein released from Golgi vesicles in KBr solution, and (B) microsomes after incubation with the proteins shown in Figure 5A. ^3H = [^3H]Glucosamine; ^{14}C = [^{14}C]leucine. Data taken from reference[15]

The time-course of the *in vivo* incorporation of [^3H]glucosamine into glucosamine isolated from proteins of rough microsomal membranes was also determined (Figure 7). The initial rapid labelling of the total membrane glycoprotein agreed well with the core position of this sugar in the oligosaccharide chain. On the other hand, in the trypsin-sensitive portion of the membrane glycoprotein on the cytoplasmic surface of the rough ER, label

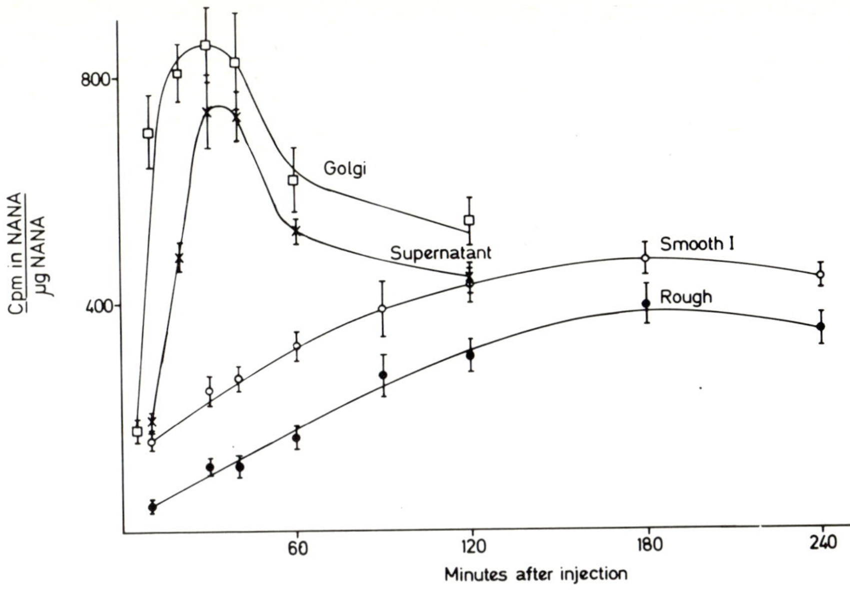

Figure 6 Incorporation of [^{3}H]glucosamine *in vivo* into protein-bound NANA of different subcellular fractions. Data taken from reference[13]

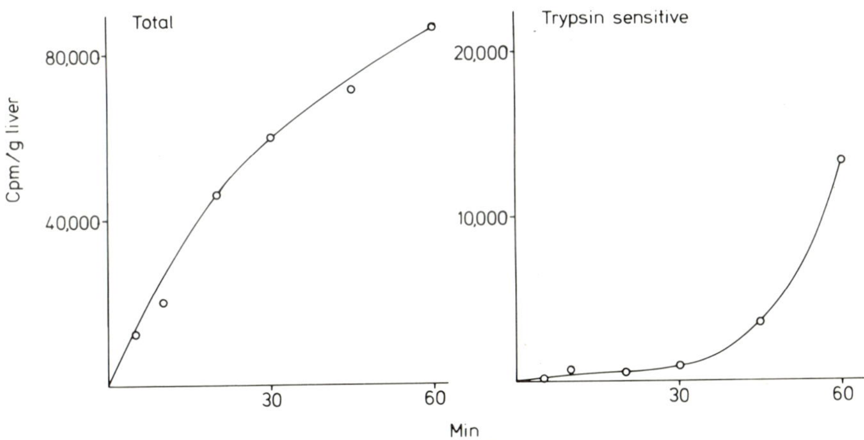

Figure 7 *In vivo* incorporation of [^{3}H]glucosamine into rough microsomes. [^{3}H]Glucosamine (150 μCi/100 g body weight) was injected into the portal vein and after various times the rats were sacrificed and rough microsomes prepared. Glucosamine was isolated both from trypsin-sensitive and trypsin-insensitive protein and the radioactivity in the isolated sugar was determined. Data taken from reference[17]

was not apparent in the first 30 min; but after this timepoint incorporation increased rapidly. Again, this experiment suggests that the trypsin-sensitive glycoprotein is incorporated from the cytoplasm.

CONCLUSION

The endoplasmic reticulum membrane of the hepatocyte displays a high degree of asymmetry of both proteins and lipids in the transverse plane. This may be an important factor for the transverse movement of some macromolecules. The protein moiety of ER membrane glycoproteins appears to be synthesized by bound ribosomes and the oligosaccharide chain is completed during the movement of the protein from the ER to the Golgi system. Glycosylation of these proteins seems to occur close to the cytoplasmic surface by a transferase system which is different from that involved in the glycosylation of the serum proteins. The completed protein is discharged from the Golgi membrane and appears in the cytoplasm as a lipoprotein complex. The basic unit of this complex is probably 10–12 peptides embedded in a lipid micelle. This glycolipoprotein complex is incorporated into microsomes upon *in vitro* incubation and the same process appears to occur *in vivo*.

Acknowledgements

The work described in this paper was carried out in collaboration with Drs F. Autuori, A. Bergman, Å. Elhammer, O. Nilsson, H. Svensson and L. Winqvist and was supported by grants from the Swedish Medical Research Council.

References

1. DePierre, J. W. and Dallner, G. (1975). Structural aspects of the membrane of the endoplasmic reticulum. *Biochim. Biophys. Acta*, **415,** 411
2. Omura, T. (1973). Biogenesis of endoplasmic reticulum membranes in liver cells. *Abstr. 9th Int. Congr. Biochem., Stockholm,* p. 249
3. Dallner, G., Siekevitz, P. and Palade, G. E. (1966). Biogenesis of endoplasmic reticulum membranes I and II. *J. Cell Biol.,* **30,** 73 and **30,** 97
4. Omura, T. and Kuriyama, Y. (1970). Influence of drugs on the turnover of microsomal proteins. *Proc. 4th Int. Congr. Pharmacol., Basel,* **4,** 94
5. Blad, D., Winqvist, L. and Dallner, G. (1977). Electrophoretic mobility of microsomes from rat liver. *J. Cell Sci.,* **22** (In press)
6. Kreibich, G., Debey, P. and Sabatini, D. D. (1973). Selective release of content from microsomal vesicles without membrane disassembly. *J. Cell Biol.,* **58,** 436
7. Omura, T., Kuriyama, Y., Takesue, S. and Oshino, N. (1970). Microsomes and drug-oxidising enzyme system. *Abstr. 8th Int. Congr. Biochem., Switzerland,* p. 260
8. Nilsson, O. and Dallner, G. (1975). Distribution of constitutive enzymes and phospholipids in microsomal membranes of rat liver. *FEBS Lett.,* **58,** 190
9. Bergman, A. and Dallner, G. (1976). Distribution of protein-bound sugar residues in microsomal subfractions and Golgi membranes. *Biochim. Biophys. Acta,* **433,** 496
10. Winqvist, L., Eriksson, L. C., Dallner, G. and Ersson, B. (1976). Binding of glycoproteins of microsomal and Golgi membranes to lectins. *Biochim. Biophys. Res. Commun.,* **68,** 1020

11. Nilsson, O. S. and Dallner, G. (1977). Transverse asymmetry of phospholipids in subcellular membranes of rat liver. *Biochim. Biophys. Acta*, **464,** 453
12. Dallner, G., Peterson, E., Nilsson, O. and Hemming, F. W. (1976). Intramembranous localization of mannosylated glycoproteins in microsomal membranes. *Abstr. 10th Int. Congr. Biochem., Hamburg,* p. 295
13. Autuori, F., Svensson, H. and Dallner, G. (1975). Biogenesis of microsomal membrane glycoproteins in rat liver. I. Presence of glycoproteins in microsomes and cytosol. *J. Cell Biol.,* **67,** 687
14. Autuori, F., Svensson, H. and Dallner, G. (1975). Biogenesis of microsomal membrane glycoproteins in rat liver. II. Purification of soluble glycoproteins and their incorporation into microsomal membranes. *J. Cell Biol.,* **67,** 700
15. Elhammer, Å., Svensson, H., Autuori, F. and Dallner, G. (1975). Biogenesis of microsomal membrane glycoproteins in rat liver. III. Release of glycoproteins from the Golgi fraction and their transfer to microsomal membranes. *J. Cell Biol.,* **67,** 715
16. Svensson, H., Elhammer, Å., Autuori, F. and Dallner, G. (1976). Biogenesis of microsomal membrane glycoproteins in rat liver. IV. Characteristics of a cytoplasmic lipoprotein having properties of a membrane precursor. *Biochim. Biophys. Acta*, **455,** 383
17. Bergman, A. (1976). Relation of glucosamine-containing glycoproteins to microsomal membranes. *Abstr. 10th Int. Congr. Biochem., Hamburg,* p. 295

5
Morphological Findings of Liver Protein Synthesis and Secretion

G. FELDMANN and M. MAURICE

It is well-established that most of the plasma proteins are produced by the liver[1]. The demonstration of this fact is based on biochemical results obtained with several techniques such as perfused isolated liver[2], total hepatectomy[3], incubation of tissue slices[4] or microsomal cell fractions[5] with radioactive amino acids. Although much information has been obtained on the synthesis and secretion of the proteins produced by the liver the techniques used do not contribute any knowledge of the nature, the number and the distribution of those cells which in the liver produce the plasma proteins.

Most of the morphological techniques available today are not capable of answering these questions. Apparently, on conventional optical or electron microscopy, all the hepatic cells are similar. Histochemical techniques which are said to locate proteins in the cells are for the most part unspecific. In fact, the only easy way to detect a protein is by immunofluorescence. This technique, proposed 25 years ago by Coons and Kaplan[6], is based on the use of specific antibodies against a given protein and on the labelling of these antibodies by fluorochrome in order to recognize in a tissue or a cell the complex composed of the protein and its antibody. With immunofluorescence, the first results for the human liver were obtained only in 1964 when Hamashima, Harter and Coons[7] managed to locate albumin and fibrinogen in the normal liver. They demonstrated that at a given time about 10 and 1% of the hepatocytes contained albumin or fibrinogen respectively. However, as immunofluorescence is not usable on electron microscopy, it is not possible to distinguish, among the albumin- or fibrinogen-containing cells, those cells which are synthesizing the protein from those accumulating it passively. Indeed, electron microscopy is the only morphological method which can demonstrate at a subcellular level where a protein is synthesized and secreted.

Immunofluorescent techniques were greatly improved in 1966 by Avrameas and Uriel[8] and by Nakane and Pierce[9] when these authors proposed the replacement of the fluorochrome used to label the antibodies by an enzyme. The principles of the immunoenzymatic techniques are similar to those of the immunofluorescent techniques: instead of demonstrating the protein by means of an ultraviolet examination of the tissue section, the protein is demonstrated by a specific enzyme histochemical reaction which is usable on light and electron microscopy. Of the numerous enzymes proposed, the most popular is horseradish peroxidase. The histochemical reaction[10] which demonstrates it gives dark-brown deposits on light microscopy, and on electron microscopy electron-dense precipitates which are easily recognizable in the cells.

Using immunoperoxidase techniques in our laboratory, we have investigated plasma proteins produced by the liver such as albumin, ceruloplasmin, α_1-antitrypsin (α_1-AT) and fibrinogen. Normal human or rat liver was studied to determine the nature, the number and the distribution of the cells which synthesize and secrete the plasma proteins. In addition the cellular location of these proteins was studied in pathological or experimental conditions where there is a defect in secretion or in synthesis.

The techniques used to demonstrate the proteins in the liver have been described in detail in other works[11,12] and reviewed during a recent symposium on immunoenzymatic techniques[13]. Briefly, specific antibodies against human albumin, α_1-AT or ceruloplasmin or rat fibrinogen were prepared by immunization of rabbits or sheep with purified proteins mixed with complete Freund's adjuvant. Purified antibodies were labelled with horseradish peroxidase[14]. The liver fragments were immediately fixed in paraformaldehyde, no matter which protein was being studied. Preliminary experiments have shown that this fixative gives the most satisfactory results for immunoperoxidase studies of the liver. After fixation, the liver fragments were cut with a cryostat and sections were incubated with labelled antibodies. Peroxidase was demonstrated by a specific histochemical reaction[10]. A number of the sections were examined under light microscopy; the remainder were post-fixed in osmium tetroxide, embedded in epoxy resin and ultrathin sections were prepared for electron microscopy. Normal γ-globulins labelled with peroxidase were used for control reactions. The sections were examined without further staining.

LOCALIZATION OF ALBUMIN, α_1-AT, CERULOPLASMIN AND FIBRINOGEN IN THE NORMAL LIVER

In the normal liver, no matter which protein was studied, similar results were obtained in both man and rat.

On light microscopy (Figure 1), the protein was always located in the hepatocytes and never in the Kupffer cells or in the biliary cells. In the

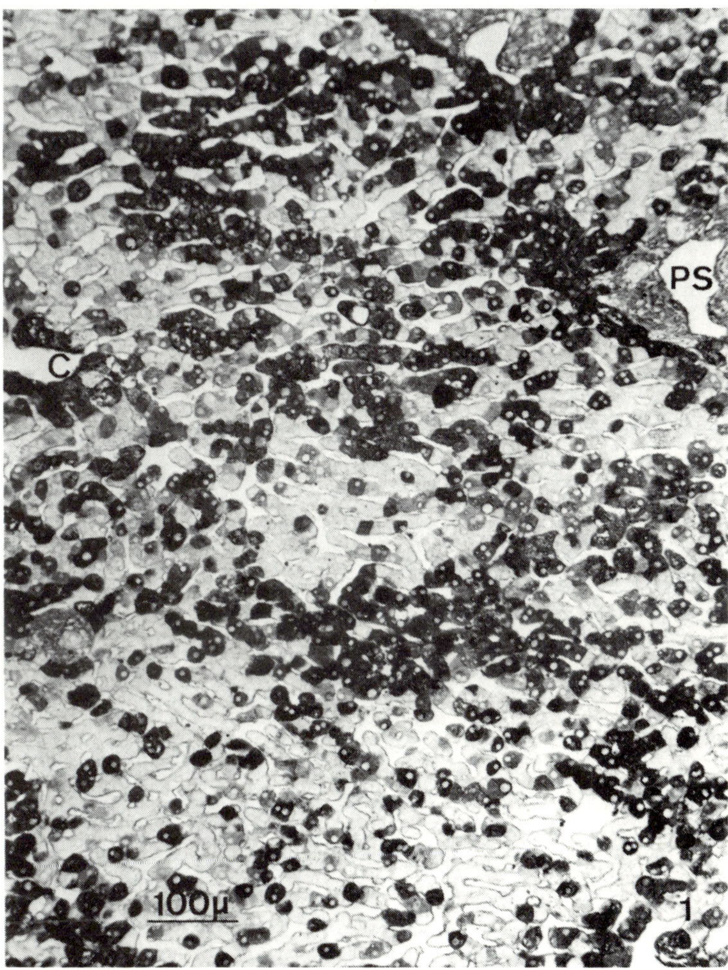

Figure 1 Light microscopic appearance of normal human liver. Localization of albumin. After incubation of the sections with antihuman albumin antibodies labelled with peroxidase and demonstration of the enzyme, dark brown deposits indicating the presence of albumin are visible in the cytoplasm of some hepatocytes randomly distributed within the hepatic lobule (PS: portal space; C: central vein) (× 120)

hepatocytes, the protein was visible only in the cytoplasm and never in the nucleus. The intensity of the reaction indicating the presence of the protein varied from one cell to another. In some cells only a part of the cytoplasm gave the reaction; in other cells, the reaction was diffuse throughout the cytoplasm. The protein-containing cells seemed to be randomly distributed within the hepatic lobule, around the portal space as well as the central vein or in the midlobular zone. These cells were sometimes more abundant just around the central vein; in some sections, they were located along the portal vessels in zone 1 of the liver acinus described by Rappaport[15]. However it was not usually possible to observe a preferential cellular localization in the hepatic lobule.

In addition to the random distribution, the percentage of protein-containing cells varied greatly from one hepatic lobule to another: some lobules did not contain any cell with detectable protein. Striking differences have been observed between two pieces of the liver taken from two different lobes. The main difference between the four proteins studied was the percentage of protein-containing cells in the normal liver. At a given time, only a part of the hepatocytes contained a protein detectable by immunoperoxidase. In the human, this part represented about 35% of the hepatocytes for albumin, 1 to 5% for α_1-AT and 10% for ceruloplasmin; in the rat, the fibrinogen-containing hepatocytes did not exceed 1%.

On electron microscopy, there was no difference between the four proteins. All of them were located on the ribosomes and the membranes of the rough endoplasmic reticulum (RER) of the hepatocytes (Figure 2). A small amount was located on the membranes of the smooth endoplasmic reticulum (SER) and the Golgi apparatus (GA) (Figure 3). Little or no protein was detectable in the lumina of these three organelles.

The other cytoplasmic organelles of the hepatocytes did not contain any detectable protein; in particular, no reaction was observed on the free cytoplasmic ribosomes.

Some comments can be made about these findings:

(1) The protein cytoplasmic localization is in agreement with some previous studies performed with immunofluorescence. In particular, albumin and fibrinogen were located, in humans[7] as well as in the dog[16] or in the rat[17], mainly in the hepatocytes and always in the cytoplasm of these cells, never in their nucleus. The localization of albumin or fibrinogen in the Kupffer cells claimed by some authors[7], is probably explained by technical errors, discussed in detail elsewhere[11].

(2) The location of the different proteins on the ribosomes and the membranes of the RER is consistent with their presence in the microsomal fraction obtained after ultracentrifugation of the liver[5,18]. This location in organelles which are known to be the intracellular sites of protein synthesis suggests that the hepatocytes which contain a given protein are in fact those which manufacture it. The variations in the intensity of the reaction noted between one cell

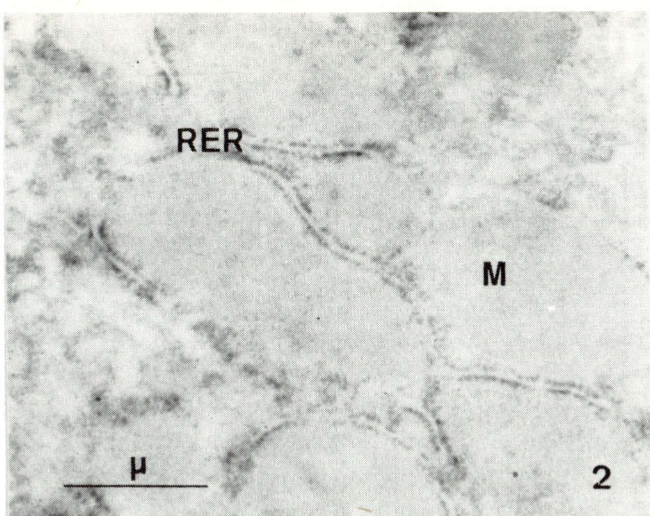

Figure 2 Normal human liver. Localization of α_1-antitrypsin. Electron microscopic appearance of a part of the cytoplasm of a hepatocyte after demonstration of the peroxidase; electron-dense precipitates, demonstrating α_1-antitrypsin on the ribosomes and the membranes of the rough endoplasmic reticulum (RER); the lumina of this organelle do not contain any detectable α_1-antitrypsin (M: mitochondria) (($\times$ 19 000)

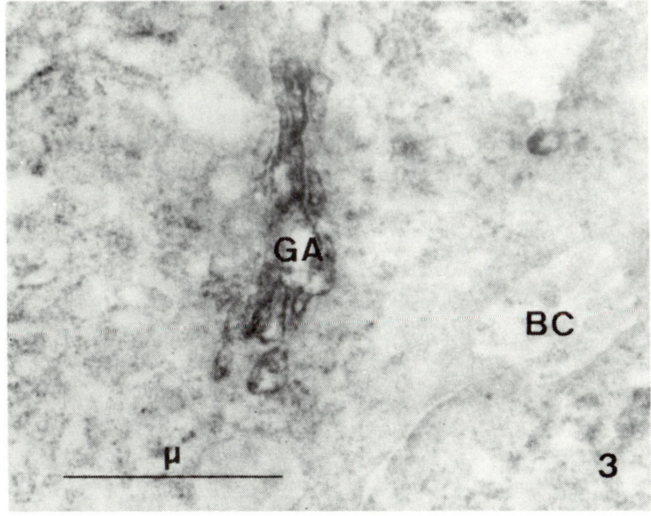

Figure 3 Normal human liver. Localization of α_1-antitrypsin. In some hepatocytes the protein is mainly located on the membranes of the Golgi apparatus (GA); a small amount of protein is present in some saccules (BC: bile canaliculus) (($\times$ 29 000)

and another are possibly due to inequality in the activity of protein-synthesizing cells. The failure to detect a protein on the free cytoplasmic ribosomes is compatible with the hypothesis[19,20] that the proteins secreted into the plasma are formed by the membrane-bound ribosomes of the RER, while the proteins retained in the cell are formed by the free ribosomes. Recently[21], in fast-growing rat LF hepatomas, immunoperoxidase has demonstrated an almost exclusive cytoplasmic localization of aldolase C, an enzyme which is not secreted into the plasma.

(3) The location of the proteins on the membranes of the SER and the GA is consistent with the data from ultracentrifugation and autoradiographic studies[22,23], and could be related to the passage of these proteins through these organelles. According to a current concept, the RER, the SER and the GA represent the successive portions of a canalicular network through which these secreted proteins are conveyed after their synthesis. The absence or the nearly complete absence of the proteins studied from the lumina of these three organelles is apparently not in agreement with this concept. Two explanations are possible: the absence could be the result of a poor penetration of the labelled antibodies into the tissue; more probably, as will be shown later, it is due to the very small amount of protein present.

(4) The fact that only a part of the hepatocytes contains a protein could be explained by two hypotheses: one, that all the hepatocytes are able to synthesize all the plasma proteins produced by the liver, but at a given time only a part of them, and in some cases a very small part, is at work, and two, that some hepatocytes are specialized in the production of some proteins.

With a static morphological method, it is not possible to choose between these two hypotheses. Some experimental results obtained with colchicine however (see later) are in agreement with the first hypothesis.

(5) The differences observed in the cellular percentage for each protein is not explained: no obvious relationship seems to exist between this percentage and the concentration of protein in the plasma or its half-life. The difference could be explained by the biochemical structure of the studied protein; α_1-AT, fibrinogen and ceruloplasmin are glycoproteins and the percentage of hepatocytes which at a given time contain one of these proteins is roughly similar (between 1 and 10%): by contrast, for albumin, which is not a glycoprotein, the percentage was about 35%. During the synthesis of a glycoprotein, the addition of the carbohydrate moieties takes place in the endoplasmic reticulum and the Golgi apparatus[24]. The final part of the glycosidation, in particular the attachment of sialic acid, would be an essential step before the secretion of a glycoprotein[25]. In the absence of sialic acid, as will be seen later in some pathological conditions, the protein accumulates. The two facts (the biochemical difference between albumin and the glycoproteins and the difference in the cellular percentage) suggest that the glycoproteins are quickly exported into the plasma, while albumin remains for a longer time in the hepatocyte. Consequently, it might be easier to detect it with the immunoperoxidase technique.

LOCALIZATION OF ALBUMIN, α_1-AT, FIBRINOGEN AND CERULOPLASMIN WHEN THE PLASMA CONCENTRATION IS DECREASED

In general, a decrease in the plasma concentration of a protein secreted by the liver is the consequence of a depressed synthesis of this protein. However, under certain circumstances, the decreased plasma protein concentration appears to be the consequence of a depressed secretion, hepatic synthesis remaining normal. Immunoperoxidase techniques seem to be a sensitive procedure for detecting such defects because on electron microscopy these defects are expressed by an accumulation of the protein in the lumina of the organelles engaged in their secretion (i.e. the endoplasmic reticulum and the Golgi apparatus). This accumulation is easily recognizable, since little if any protein is visible in the lumina of these organelles in the normal state.

One example of a depressed protein secretion was observed in man (α_1-AT in α_1-antitrypsin deficiency), the other in the rat (fibrinogen after colchicine administration). In both circumstances, biochemical studies have proven the defect. A pathological condition (alcoholic cirrhosis with disturbed albumin secretion) and some other examples where the defect could be possible but not yet proven, will also be presented. Finally, the results obtained with immunoperoxidase techniques, where there is a defect in secretion, will be compared with the results obtained in Wilson's disease, where a defect in ceruloplasmin synthesis has been suggested.

The deficiency of α_1-AT is a genetically transmitted disorder, characterized by a marked decrease in α_1-AT plasma concentration[26]. Light microscopy examination of the liver shows in some hepatocytes numerous round or oval globules which on conventional electron microscopy are associated with a dilatation of the endoplasmic reticulum; inside the dilated lumina, an amorphous material is present[27]. With anti-α_1-AT antibodies labelled with peroxidase, electron-dense precipitates were detected on this amorphous material (Figure 4), signifying that this material is composed of a protein antigenically similar to α_1-AT[28]. The accumulation of α_1-AT in the endoplasmic reticulum suggests that the decrease in plasma concentration of α_1-AT observed in α_1-AT deficiency is the consequence of a defect in secretion. In this disease the defect is probably due to changes in the structure of α_1-AT. Biochemical studies have shown that carbohydrates such as sialic acids are missing[29,30] and these carbohydrates, as noted earlier, are probably essential for the exportation of the glycoprotein. In their absence, the unsecreted α_1-AT would accumulate in the hepatocytes.

A defect in secretion can be induced experimentally with colchicine or vinblastine. When one of these drugs is administered in suitable dose to an animal, liver protein synthesis is unchanged, but the protein secretion is inhibited and consequently, the plasma protein concentration is decreased. This fact is now well-demonstrated for several proteins synthesized by the liver, such as lipoproteins[31], albumin[32] or coagulation factors[33]. Similar findings have been

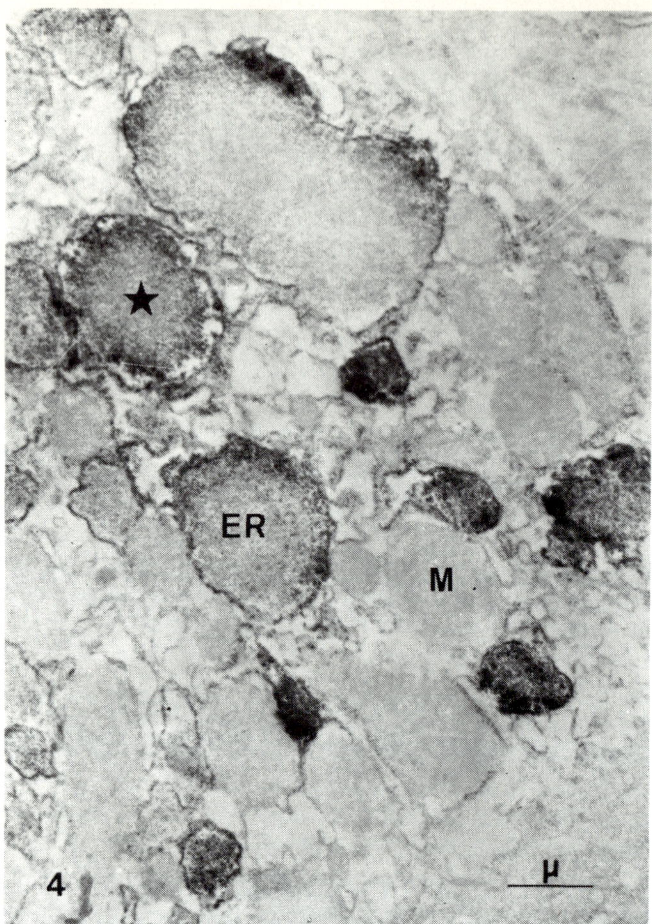

Figure 4 Electron microscopic appearance of a hepatocyte in a patient with α_1-antitrypsin deficiency. Electron-dense precipitates indicating the presence of α_1-antitrypsin (star) are seen on the material accumulated in the dilated lumina of the endoplasmic reticulum (ER) (M: mitochondria) (× 11 000)

observed in this laboratory for fibrinogen after intraperitoneal administration of colchicine[12]. However, the organelles in which fibrinogen accumulates were either the RER or the Golgi vesicles depending on the amount of protein produced by the liver and the time when the rats are sacrificed. In the normal state, the effect of colchicine was only observed 8 hours after drug administration and the protein accumulated mainly in the lumina of the RER (Figure 5). When the hepatic synthesis of fibrinogen was increased following ancrod administration (an agent which induces a rapid destruction of plasma fibrinogen and which is non-toxic[34]), the effect of colchicine was visible 4 hours after its administration and fibrinogen accumulated mainly in vesicles belonging to the GA (Figure 6).

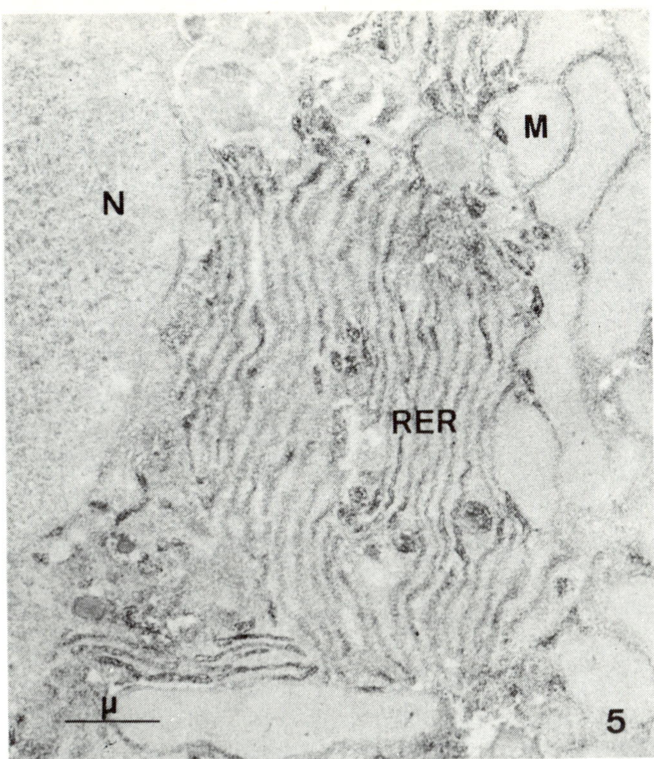

Figure 5 Colchicine administration in the rat. Localization of fibrinogen. When the hepatic fibrinogen synthesis is normal, the effects of colchicine on fibrinogen secretion are observed only 8 hours after the drug administration; the protein accumulates mainly in lumina of the rough endoplasmic reticulum (RER) (N: nucleus; M: mitochondria) (× 12 500)

These early and late effects of colchicine suggest that the drug inhibits the secretion of fibrinogen at different cellular levels. The first effect probably occurs at the end of the secretion between the GA and the plasma membrane and before the protein secretion. The late effect occurs at the beginning of the secretion in the RER immediately after the synthesis of fibrinogen by the membrane-bound ribosomes. Up to now, it has been difficult to explain the mechanism of these two effects. One of the cellular effects of colchicine is to inhibit the polymerization of tubulin[35], thus preventing the formation of microtubules. These organelles are probably important in protein secretion[36]. However, their part in the secretion of proteins by the hepatocytes is still debated[32]. For fibrinogen secretion, it seems difficult to explain the two effects in terms only of the microtubules. Biochemical studies have shown that the effect of colchicine on tubulin is rapid and reversible[35] and this probably explains the early effect. The late effect could be a binding of colchicine on the RER membranes, inhibiting the secretion of the new synthesized fibrinogen.

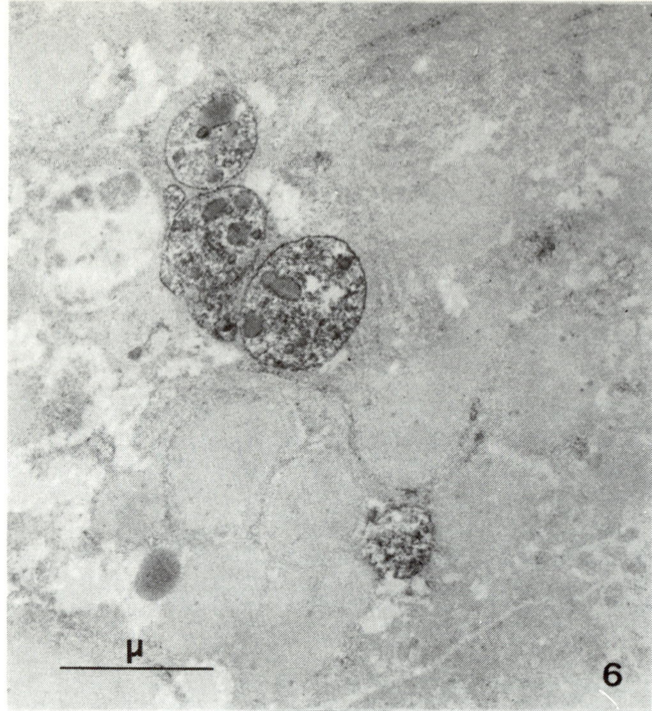

Figure 6 Colchicine administration in the rat. Localization of fibrinogen. When the hepatic fibrinogen synthesis is increased, the effects of colchicine on fibrinogen secretion are observed 4 hours after the drug administration and the protein accumulates mainly in cytoplasmic vesicles (× 20 000)

Biochemical studies have shown that colchicine binds easily to the liver microsomes[37].

Eight hours after colchicine administration, when the synthesis of fibrinogen is normal, most of the hepatocytes contained fibrinogen[12]; this fact suggests that the hepatocytes are probably not specialized for the synthesis of plasma proteins. However, it seems difficult to accept that the synthesis of one protein requires synthesis by all the hepatocytes. A cellular cyclic activity could be the best way to explain the results obtained in the normal liver.

Accumulation of albumin in the organelles engaged in its secretion has also been observed in another pathological circumstance, alcoholic cirrhosis in man. In this condition, the albumin plasma concentration is very often decreased. However, Rothschild *et al.*[38] have demonstrated that in several patients with cirrhosis albumin synthesis was normal or even elevated. In some liver biopsy specimens, albumin was demonstrated in large amounts either in the lumina of the RER (Figure 7) or in vesicles of the GA (Figure 8). These findings, too, could be interpreted as a depressed secretion. However, biochemical studies are necessary to prove that in alcoholic cirrhosis, a defect in

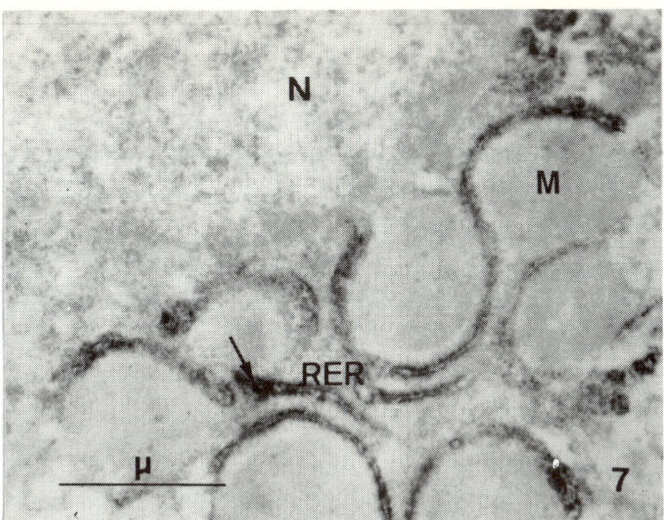

Figure 7 Alcoholic cirrhosis. Localization of albumin. Albumin accumulates in the lumina (arrow) of the rough endoplasmic reticulum (RER) (N: nucleus; M: mitochondria) (× 22 200)

secretion and not a defect in synthesis explains the decrease in albumin plasma concentration; the action of alcohol on albumin synthesis seems to be different depending on the time of administration: acute alcohol administration in the animal depresses albumin synthesis[39] while chronic administration has apparently no effect on its synthesis[40]. Recently it has been suggested that alcohol

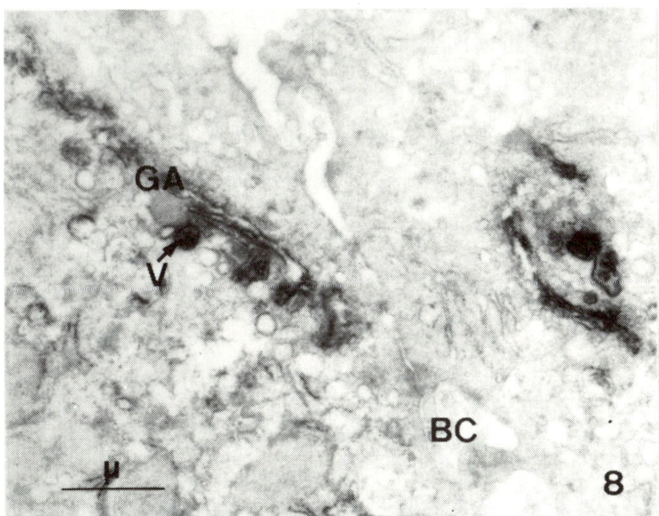

Figure 8 Alcoholic cirrhosis. Localization of albumin. In some hepatocytes the protein accumulates in the vesicles (V) of the Golgi apparatus (GA) (BC: bile canaliculus) (× 13 800)

could act on tubulin and microtubules, preventing the exportation of albumin into the plasma[41].

Protein accumulation in the hepatocyte organelles engaged in protein secretion does not seem to be rare. Two similar situations have recently been observed, one involving albumin in the rat during the physiological postnatal development[42] and the other involving albumin and fibrinogen in primary cultures of newborn rat liver tissue[43].

Finally, immunoperoxidase techniques have been used to demonstrate ceruloplasmin in the hepatocytes of patients with Wilson's disease. Three patients with little or no plasma ceruloplasmin were studied. No protein was observed on either the membranes or the ribosomes or in the lumina of the RER, and the other organelles did not contain any ceruloplasmin. This absence of ceruloplasmin is in agreement with greatly diminished ceruloplasmin synthesis in biochemical studies in Wilson's disease[44]. The negative results obtained with immunoperoxidase when there is a defect in protein synthesis strengthen the results obtained as to a defect in protein secretion: the accumulation of a protein in the organelles engaged in its secretion demonstrates such a defect.

SUMMARY

Immunoperoxidase techniques have been used to locate in hepatic cells several plasma proteins produced by the liver. In the normal liver, only a part of the hepatocytes contains the protein. On electron microscopy, the protein is shown to occur in the organelles responsible for its synthesis and also in the organelles responsible for its secretion. Some experimental results obtained after administration of colchicine in the rat suggest that the hepatocytes are not specialized in the production of a given plasma protein but are the seat of a cyclic activity. When the protein is demonstrated in large amounts in the organelles engaged in its secretion, this accumulation strongly suggests a defect in the secretion into the plasma. Several pathological or experimental conditions illustrating this situation are reported.

Acknowledgements

This work was supported by a grant from INSERM (contract no. 7650817).

References

1. Schultze, J. E. and Heremans, J. F. (1966). *Molecular Biology of Human Proteins with Special Reference to Plasma Proteins. Vol. I* (Amsterdam: Elsevier Publishing Company)

2. Miller, L. L. and Bale, W. F. (1954). Synthesis of all plasma protein fractions except gammaglobulins by the liver. The use of zone electrophoresis and lysine-ϵ-C^{14} to define the plasma proteins synthesized by the isolated perfused liver. *J. Exp. Med.*, **99,** 125
3. Kukral, J. C., Kerth, J. D., Pancner, R. J., Cromer, D. W. and Henagar, G. C. (1961). Plasma protein synthesis in the normal dog and after total hepatectomy. *Surg. Gynecol. Obstet.*, **113,** 360
4. Peters, T., Jr. and Anfinsen, C. B. (1950). Production of radioactive serum albumin by liver slices. *J. Biol. Chem.*, **182,** 171
5. Campbell, P. N., Greengard, O. and Kernot, B. A. (1960). Studies on the synthesis of serum albumin by the isolated microsome fraction from rat liver. *Biochem. J.*, **74,** 107
6. Coons, M. H. and Kaplan, A. H. (1950). Localization of antigen in tissue cells. II. Improvements in a method for the detection of antigen by means of fluorescent antibody. *J. Exp. Med.*, **91,** 1
7. Hamashima, Y., Harter, J. G. and Coons, A. H. (1964). The localization of albumin and fibrinogen in human liver cells. *J. Cell Biol.*, **20,** 271
8. Avrameas, S. and Uriel, J. (1966). Méthode de marquage d'antigènes et d'anticorps avec des enzymes et son application en immunodiffusion. *C.R. Acad. Sci. (Paris)*, **262,** 2543
9. Nakane, P. K. and Pierce, G. B. (1966). Enzyme labeled antibodies. Preparation and application for the localization of antigens. *J. Histochem. Cytochem.*, **14,** 929
10. Graham, R. C., Jr. and Karnovsky, M. C. (1966). The early stages of absorption of injected horseradish peroxidase in the proximal tubules of mouse kidney: ultrastructural cytochemistry by a new technique. *J. Histochem. Cytochem.*, **14,** 291
11. Feldmann, G., Penaud-Laurencin, J., Crassous, J. and Benhamou, J. P. (1972). Albumin synthesis by human liver cells: its morphological demonstration. *Gastroenterology*, **63,** 1036
12. Feldmann, G., Maurice, M., Sapin, C. and Benhamou, J. P. (1975). Inhibition by colchicine of fibrinogen translocation in hepatocytes. *J. Cell Biol.*, **67,** 237
13. Feldmann, G., Druet, P., Bignon, J. and Avrameas, S. (1976). *Immunoenzymatic Techniques* (Amsterdam: North-Holland Publishing Company)
14. Avrameas, S. (1969). Coupling of enzymes to proteins with glutaraldehyde. Use of the conjugates for the detection of antigens and antibodies. *Immunochemistry*, **6,** 43
15. Rappaport, A. M. (1963). Acinar units and the pathophysiology of the liver. In: Ch. Rouiller (ed.). *The Liver. Morphology, Biochemistry, Physiology. Vol. I*, pp. 265–328. (New York: Academic Press).
16. Barnhart, M. I. and Forman, W. B. (1963). The cellular localization of fibrinogen as revealed by the fluorescent antibody technique. *Vox Sang.*, **8,** 461
17. Lane, P. S. (1969). The cellular distribution of albumin in normal rat liver demonstrated by immunofluorescent staining. *Clin. Sci.*, **36,** 157
18. Williams, C. A. (1970). Analysis of 'acute phase' protein synthesis in the mouse by immunoelectrophoresis. In: M. A. Rothschild and T. Waldmann (eds.). *Plasma Protein Metabolism, Regulation of Synthesis, Distribution and Degradation*, pp. 383–392. (New York: Academic Press)
19. Campbell, P. N., Serk-Hanssen, G. and Lowe, E. (1965). Studies on the protein-synthesizing activity of the ribosomes of rat liver. The activity of free polysomes. *Biochem. J.*, **97,** 422
20. Redman, C. M. (1969). Biosynthesis of serum proteins and ferritin by free and attached ribosomes of rat liver. *J. Biol. Chem.*, **244,** 4308
21. Hatzfeld, A., Feldmann, G., Frayssinet, C. and Shapira, F. (1976). Cellular localization of a carcinofetal enzyme (aldolase C) and of the normal adult enzyme (aldolase B) in fast-growing rat LF hepatomas. In: W. H. Fisham and S. Sell (eds.). *Onco-developmental Gene Expression*, pp. 185–190. (New York: Academic Press)
22. Ashley, C. A. and Peters, T., Jr. (1969). Electron microscopic radioautographic detection of sites of protein synthesis and migration in liver. *J. Cell Biol.*, **43,** 237
23. Glaumann, H. and Ericsson, J. L. E. (1970). Evidence for the participation of the Golgi apparatus in the intracellular transport of nascent albumin in the liver cell. *J. Cell Biol.*, **47,** 555

24. Clauser, H., Herman, G., Rossignol, B. and Harbon, S. (1972). Biosynthesis of glycoproteins. Biosynthesis at the cellular and subcellular level. In: A. Gottschalk (ed.). *Glycoproteins, their Composition, Structure and Function*, Part B, pp. 1151–1169, 2nd ed. (Amsterdam: Elsevier Publishing Company)
25. Schachter, H., Jabbal, I., Hudgin, R. L., Pinteric, L., McGuire, E. J. and Roseman, S. (1970). Intracellular localization of liver sugar nucleotide glycoprotein glycosyltransferases in a Golgi-rich fraction. *J. Biol. Chem.*, **245**, 1090
26. Kueppers, F. (1971). Alpha$_1$-antitrypsin: physiology, genetics and pathology. *Humangenetik*, **11**, 177
27. Sharp, H. L. (1971). Alpha$_1$-antitrypsin deficiency. *Hosp. Pract.*, **6**, 83
28. Feldmann, G., Bignon, J., Chahinian, R., Degott, C. and Benhamou, J. P. (1974). Hepatocyte ultrastructural changes in α_1-antitrypsin deficiency. *Gastroenterology*, **67**, 1214
29. Jeppsson, J. O., Larsson, C. and Eriksson, S. (1975). Characterization of α_1-antitrypsin in the inclusion bodies from the liver in α_1-antitrypsin deficiency. *N. Engl. J. Med.*, **293**, 576
30. Chan, S. K. and Rees, P. C. (1975). Molecular basis of the α_1-protease inhibitor deficiency. *Nature* (*London*), **255**, 240
31. Le Marchand, Y., Singh, A., Assimacopoulos-Jeannet, F., Orci, L., Rouiller, R. and Jeanrenaud, B. (1973). A role for the microtubular system with release of very low density lipoproteins by perfused mouse livers. *J. Biol. Chem.*, **248**, 6862
32. Redman, C. M., Banerjee, D., Howell, K. and Palade, G. E. (1975). Colchicine inhibition of plasma protein release from rat hepatocytes. *J. Cell Biol.*, **66**, 42
33. Grätzl, M. and Schwab, D. (1976). The effect of microtubular inhibitors on secretion from liver into blood plasma and bile. *Cytobiologie*, **13**, 199
34. Ashford, A., Prass, J. W. and Southgate, P. (1968). Pharmacology and toxicology of a defibrinating substance from Malayan pit viper venom. *Lancet*, **i**, 486
35. Olmsted, J. B. and Borisy, G. G. (1973). Microtubules. *Ann. Rev. Biochem.*, **42**, 507
36. Feldmann, G. and Maurice, M. (1975). Microtubules, microfilaments et sécrétion cellulaire. *Biol. Gastroenterol.* (*Paris*), **8**, 269
37. Stadler, J. and Franke, W. W. (1972). Colchicine-binding proteins in chromatin and membranes. *Nature New Biology*, **237**, 237
38. Rothschild, M. A., Oratz, M., Zimmon, D., Schreiber, S. S., Weiner, I. and Van Ceneghem, A. (1969). Albumin synthesis in cirrhotic subjects with ascites studied with carbonate[^{14}C]. *J. Clin. Invest.*, **48**, 344
39. Rothschild, M. A., Oratz, M., Mongelli, J. and Schreiber, S. S. (1971). Alcohol-induced depression of albumin synthesis: reversal by tryptophan. *J. Clin. Invest.*, **50**, 1812
40. Jeejeebhoy, K. N., Bruce-Robertson, A., Ho, J. and Sodtke, U. (1973). The comparative effects of nutritional and hormonal factors on the synthesis of albumin, fibrinogen and transferrin. In: *Protein Turnover, Ciba Foundation Symposium 9*, pp. 217–247. (Amsterdam: Elsevier)
41. Baraona, E., Leo, M. E., Borowsky, S. A. and Lieber, C. S. (1975). Alcoholic hepatomegaly: accumulation of protein in the liver. *Science*, **130**, 794
42. Guillouzo, A., Feldmann, G., Maurice, M., Sapin, C. and Benhamou, J. P. (1976). Ultrastructural distribution of albumin in rat hepatocytes during post-natal development. *J. Microsc. Biol. Cell.*, **26**, 35
43. Guillouzo, A., Feldmann, G., Boisnard, M., Sapin, C. and Benhamou, J. P. (1975). Ultrastructural location of albumin and fibrinogen in primary cultures of new-born rat liver tissue. *Exp. Cell Res.*, **96**, 239
44. Sternlieb, I. (1972). Evolution of the hepatic lesion in Wilson's disease (hepatolenticular degeneration) In: H. Popper and F. Schaffner (eds.), *Progress in Liver Disease, Vol. IV*, pp. 511–525. (New York: Grune and Stratton)

DISCUSSION

Dr R. Schmid, San Francisco (USA): The presented observations raise the question why only a fraction of all hepatocytes, seemingly randomly distributed throughout the liver lobule, contain demonstrable albumin or fibrinogen? Does this mean that hepatocytes are heterogeneous and specialized in regard to function or that protein synthesis is cyclic, and the albumin or fibrinogen-positive cells are those in the synthetic phase at the time of investigation? The presence of fibrinogen in the endoplasmic reticulum of 90% of all hepatocytes treated with colchicine favours the second explanation, but albumin appears to behave differently.

Dr Feldmann, Clichy (France): Some observations suggest that the hepatocytes are not specialized in protein synthesis. Normally, only part of the cells produce the plasma proteins, probably in a cyclic process to be demonstrated. On stress, most and probably all cells are able to synthesize every protein. After turpentine administration the plasma concentration of several proteins produced by the liver is increased ('acute phase proteins')[1]. We found with Dr Kushner of Cleveland that when the plasma level of C-reactive protein (one of these proteins) was high after turpentine administration, most of the hepatocytes contained the protein. When the level declined, only few hepatocytes reacted. After colchicine administration, however, the percentage of albumin-containing hepatocytes was not obviously increased as it was in the case of fibrinogen. Albumin was mainly located in small vesicles of the Golgi apparatus. They were well seen on electron microscopy but were difficult to identify on light microscopy. Consequently, it was not easy to determine the percentage of albumin-containing cells. Possibly the albumin-containing cells were more numerous normally. This suggests also that the hepatocytes are not specialized in albumin production.

Dr D. S. Zimmon, New York (USA): Your anatomical findings of altered intracellular albumin distribution in alcoholic cirrhosis are particularly fascinating in view of the high rate of albumin synthesis (see reference[38] in text) and leak of albumin from liver directly into ascites[2] demonstrated in man. Is there anatomical evidence for a short or unusual albumin pathway in alcoholic cirrhosis for escape of albumin into blood or ascites?

Dr Feldmann, Clichy (France): In alcoholic cirrhosis, the cellular localization of albumin was similar to the normal liver. The protein was always located in the endoplasmic reticulum and in the Golgi apparatus and never in the cytosol of the hepatocyte. A cytoplasmic localization would be an anatomical support for a short passage of albumin from the endoplasmic reticulum towards the plasma membrane; however, this localization was never observed for albumin. Recently, some investigators have claimed that albumin was in the cytoplasm of the hepatocytes and not in the Golgi apparatus by immunoenzymatic techniques[3]. This cytoplasmic localization must be interpreted cautiously because of the problems of fixation in immunoenzymatic techniques[4]. With poor fixation, the protein is not fixed to its cellular structure; a diffusion throughout the cell may produce an abnormal localization.

Dr R. D. Soloway, Philadelphia (USA): Were any gradients between the centre and the periphery of the hepatic lobule observed for any of the four proteins measured? The gradient could reflect either cell distribution or amount of protein per cell.

Dr Feldmann, Clichy (France): The distribution of the four proteins showed no such gradient but apparently a random distribution. But the intensity of the reaction varied from cell to cell. The distribution of the protein-containing cells in the hepatic lobule differs from that of reaction products of some hepatic enzymes demonstrated with histoenzymatic techniques.

References

1. Gordon, A. H. (1970). In: M. A. Rothschild and T. Waldmann (eds.). *Plasma Protein Metabolism, Regulation of Synthesis, Distribution and Degradation*, p. 351. (New York: Academic Press)
2. Zimmon, D. S., Oratz, M., Kessler, R., Schreiber, S. S. and Rothschild, M. A. (1969). Albumin ascites: demonstration of a direct pathway bypassing the systemic circulation. *J. Clin. Invest.*, **48,** 2074
3. Lin, C.-T. and Chang, J. P. (1975). Electron microscopy of albumin synthesis. *Science*, **190,** 465
4. Feldmann, G. (1974). Localisation des antigens cellulaires par anticorps couplés à la peroxydase: techniques histologiques et ultrastructurales. *J. Microsc.*, **21,** 293

6
Biochemical Mechanisms of Glycosylation

F. W. HEMMING

Most hepatic membranes contain enzymes that catalyse the transfer of monosaccharides to glycoproteins and glycolipids. Many of the products remain as membrane components while some of the glycoproteins are secreted. A major group of glycoproteins formed in this way is that consisting of oligosaccharides linked to protein through a β-*N*-glycosidic linkage between a residue of *N*-acetylglucosamine and the amide nitrogen of an asparagine residue of the polypeptide chain. There are basically two general types of oligosaccharide units involved: the simple type (Figures 1 and 2) and the complex type (Figures 1 and 3)[1]. The former group usually contain two *N*-acetylglucosamine residues proximal to the asparagine. The second residue usually carries a mannose linked to further mannose residues to give a branched oligosaccharide of six to twelve sugar residues, the precise structure varying in different glycoproteins. Probably, in some cases, two glucose residues are present terminally[2]. The complex type have a similar, though generally smaller, core of mannose and *N*-acetylglucosamine residues and attached to this are small side-chains made up of *N*-acetylglucosamine, galactose and either sialic acid or fucose residues.

The function of glycoproteins of this kind in the plasma membrane as receptors for various biologically important effectors and for asialoglycoproteins in general is discussed elsewhere in this symposium. Some also have antigenic and immunoprotective activities while others probably play some role in cell–cell interactions or have enzymic activities. The secreted glycoproteins include most of the serum globulins with their various biological functions. The oligosaccharide portion of these molecules may play an important role in their localization in the cell, their secretion and/or in their turnover. Other membrane-bound glycoproteins may have a structural role. There is thus good reason for considering the biosynthesis of glycoproteins and ways of controlling the process.

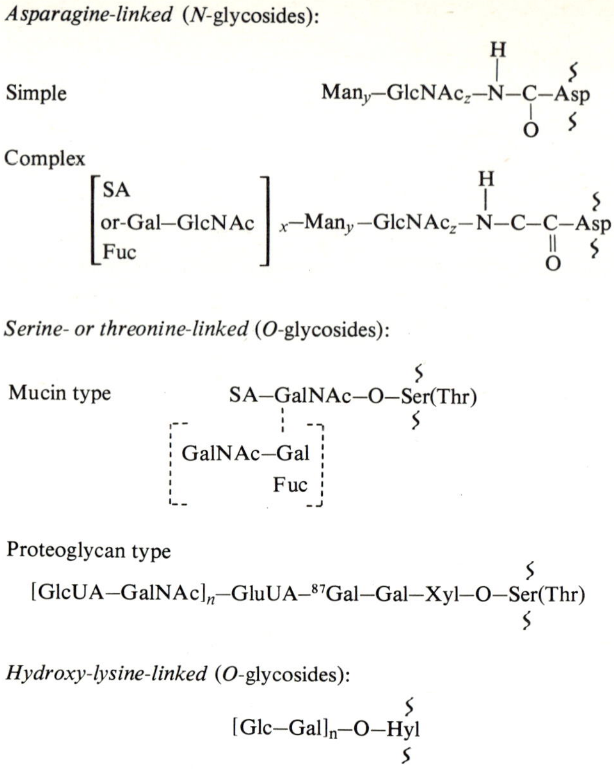

Asparagine-linked (*N*-glycosides):

Simple Man_y–$GlcNAc_z$–N(H)–C(=O)–Asp

Complex [SA or-Gal–GlcNAc Fuc]$_x$–Man_y–$GlcNAc_z$–N(H)–C–C(=O)–Asp

Serine- or threonine-linked (*O*-glycosides):

Mucin type SA–GalNAc–O–Ser(Thr)
[GalNAc–Gal, Fuc]

Proteoglycan type
[GlcUA–GalNAc]$_n$–GluUA–^{87}Gal–Gal–Xyl–O–Ser(Thr)

Hydroxy-lysine-linked (*O*-glycosides):

[Glc–Gal]$_n$–O–Hyl

Figure 1 Generalized structures of oligosaccharide portions of glycoproteins

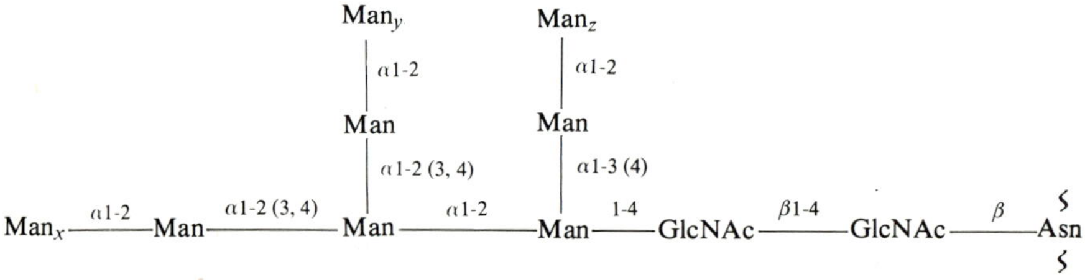

Man_y —α1-2— Man —α1-2 (3, 4)— Man
Man_z —α1-2— Man —α1-3 (4)— Man
Man_x —α1-2— Man —α1-2 (3, 4)— Man —α1-2— Man —1-4— GlcNAc —β1-4— GlcNAc —β— Asn

Figure 2 Structure of the oligosaccharide portion of thyroglobulin. An example of a simple *N*-glycosidically linked oligosaccharide (after Spiro[1])

TYPES OF GLYCOPROTEINS

There are basically two types of gene products directly concerned in protein glycosylation: those proteins acting as sugar acceptors and those that make up or assist in the formation of the components of the glycosyltransferase system. The biosynthetic pathways for different glycoproteins carrying common sugar residues are interdependent as shown at its simplest in Figure 4. Here,

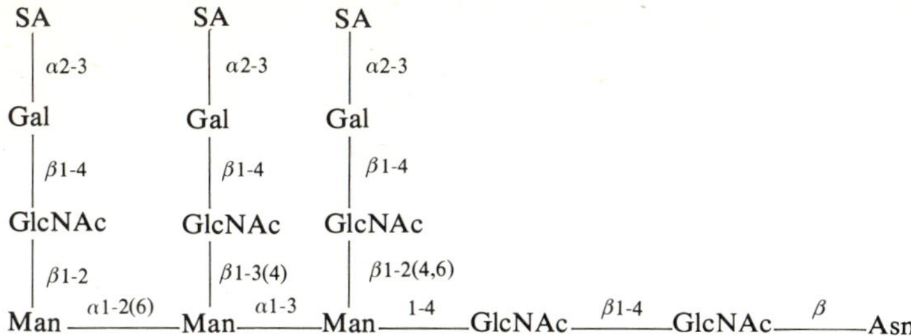

Figure 3 Structure of the oligosaccharide portion of fetuin. An example of a complex *N*-glycosidically linked oligosaccharide (after Spiro[1])

both products are dependent upon an adequate supply of the common sugar donor, the nucleotide diphosphate sugar. If the pool size of this donor is limited, overproduction of glycoprotein A may lead to underproduction of glycoprotein B.

It is generally accepted that the protein moiety of these glycoproteins is synthesized in the ribosomes of the rough endoplasmic reticulum. The ribosomes are on the cytoplasmic side of the membrane of the endoplasmic reticulum and in some way as yet not understood, the proteins formed there that are to be secreted permeate the membrane to arrive in the lumen. During progress through the lumen of the smooth endoplasmic reticulum and through the cisternae of the Golgi apparatus and secretory vesicles, progressive glycosylation occurs (Figure 5).

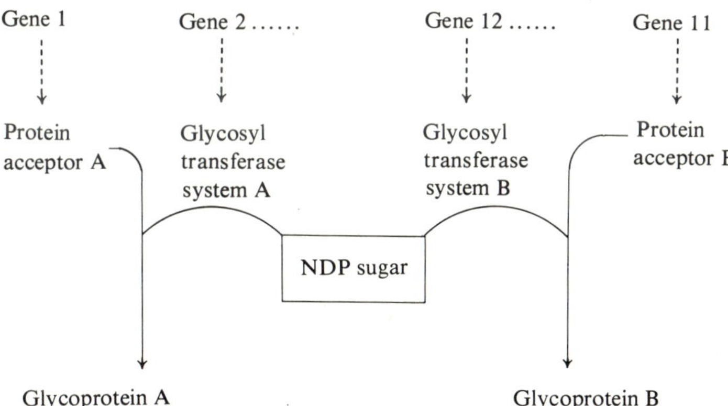

Figure 4 Simple illustration of the interdependence of glycoprotein synthesis at the glycosylation stage (NDP sugar = nucleotide diphosphate sugar)

GLYCOSYLATION STEPS

The most fully studied of these glycosylation steps are those resulting in addition of the side-chain sugars of the complex type[3] (Figures 1 and 3). The en-

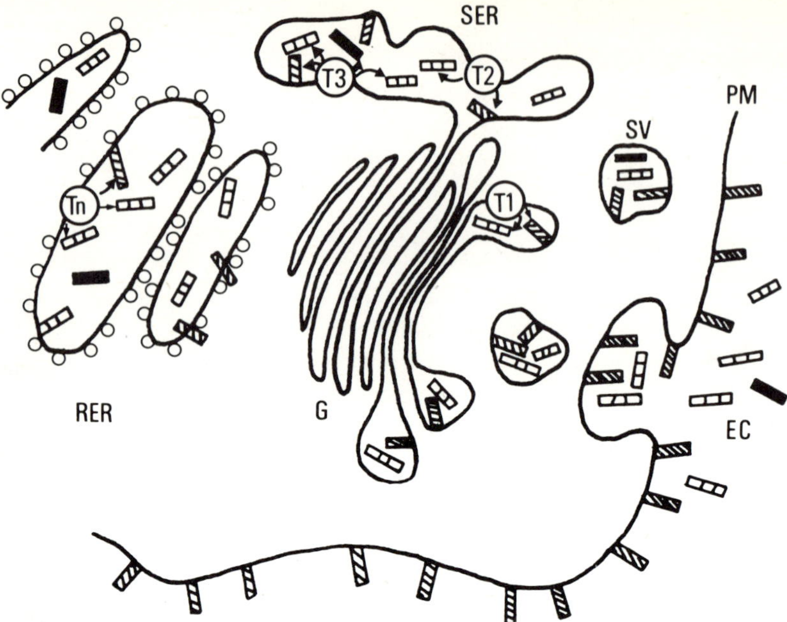

Figure 5 Diagrammatic representation of glycoprotein synthesis in a cell such as liver which secretes glycoprotein but does not have storage granules. RER, rough endoplasmic reticulum; G, Golgi apparatus; SER, smooth endoplasmic reticulum; SV, secretory vesicles; PM, plasma membrane; EC, extracellular pace; T_1 to T_n, membrane-bound glycosyltransferases

zymes responsible (transferases) occur in the membranes of the Golgi and secretory vesicles and can be solubilized and purified relatively easily. They exhibit a high degree of specificity for sugar donor and glycoprotein acceptor. The preferred donor forms of *N*-acetylglucosamine and galactose are the uridine diphosphate sugars whereas guanosine diphosphate fucose and cytidine monophosphate sialic acid donate the terminal sugars. The precise specificity of the transferases to acceptors has not been checked exhaustively but it appears that the *N*-acetylglucosaminyltransferases require a glycoprotein carrying an oligosaccharide group with a terminal mannose residue to which the *N*-acetylglucosamine is attached by a β-glycosidic linkage (Figure 6). The appropriate galactosyl transferases catalyse the formation of a β-glycosidic linkage between galactose and an *N*-acetylglucosamine residue already linked to mannose of an oligosaccharide. The sialyl and fucosyl transferases function most efficiently with a galactose acceptor already attached to an *N*-acetylglucosamine and mannose as shown in Figure 6. The enantiomeric configuration of the newly attached sugar is usually opposite to that in the donor, an inversion of configuration occurring during the transfer. Galactose is usually linked 1–4, and occasionally 1–6, sialic acid becomes linked to position 3 of galactose and fucose to position 3 or 4 of galactose. As indicated in Figure 3 *N*-acetylglucosamine may be linked to positions 2, 3, 4 or 6 of

UDPGlcNAc + GPO-Man ⟶ UDP + GPO-Man-GlcNAc

UDPGal + GPO-Man-GlcNAc ⟶ UDP + GPO-Man-GlcNAc-Gal

GDPFuc + GPO-Man-GlcNAc-Gal ⟶ GDP + GPO-Man-GlcNAc-Gal-Fuc

CMPSA + GPO-Man-GlcNAc-Gal ⟶ CMP + GPO-Man-GlcNAc-Gal-SA

Figure 6 The sequential glycosylation of the side-chains of *N*-glycosidically linked oligosaccharides of glycoproteins. GPO, glycoprotein oligosaccharide; UDP, uridine diphosphate; GDP, guanosine diphosphate; CMP, cytidine monophosphate; GlcNAc, *N*-acetylglucosamine; Man, mannose; Gal, galactose; Fuc, fucose; SA, sialic acid

mannose. It is not yet certain if the same enzyme will catalyse bonding in all four positions or if a different enzyme is needed for each position. Analogy with the specificity of the glycosyltransferases involved in the formation of blood group substances would favour the latter[4].

Thus, these transferases operate sequentially in a specific manner during the movement of protein from smooth endoplasmic reticulum to secretory vesicles and plasma membrane via the Golgi apparatus. Occasionally the oligosaccharide sequence shown in Figure 3 is incomplete leading to the phenomenon known as microheterogeneity. This is usually seen as the result partly of steric hindrance of enzymes by the increasing size of other oligosaccharide chains and partly of the acceptor moving out of the active site of the glycosyltransferase before all oligosaccharide branches are completely glycosylated. The extent of the microheterogeneity probably varies with the degree of perturbation of the membrane system by changes in physiological and pharmacological parameters.

The transfer of proximal *N*-acetylglucosaminyl and mannosyl residues to glycoproteins is less completely understood. Radioautographic studies on tissues and biochemical studies on subcellular fractions show that these sugars are added mainly in the rough and smooth endoplasmic reticulum. The evidence is uncertain regarding the sequential addition of monosaccharides to most of the glycoproteins studied. Clarification of biochemical aspects is complicated by the difficulty of solubilizing and purifying the transferase activity. Whereas direct transfer of *N*-acetylglucosamine and mannose to proximal positions on endogenous glycoprotein from appropriate nucleotide donors has not been observed, transfer from lipid-soluble donors has been demonstrated[2,5] (Figure 7). The endogenous proteins accepting these sugars have not yet been characterized but it is probable that they include various membrane-bound glycoproteins and possibly that some are subsequently further glycosylated as described above to form complex glycoproteins (Figures 1 and 3).

LIPID SOLUBLE DONORS

The type of lipid soluble donors involved are summarized in Figure 7. The lipophilic portion of the molecule is the polyisoprenoid alcohol dolichol. This

Dol P Man Dol P Glc Dol P Gal

Dol PP GlcNAc Dol PP $GlcNAc_2$

Dol PP $GlcNAc_2Man_x$ Dol PP $GlcNAc_2Man_xGlc_2$

Figure 7 Lipid soluble donors of sugars. Dol P, dolichol monophosphate (see Figure 11); Dol PP, dolichol diphosphate

effectively anchors the sugar moieties of the donors on the membrane presumably in a manner better suited to act as substrate for the next transferase. Each of these dolichol phosphate donors is formed directly from the usual nucleotide donor using the appropriate transferase (Figure 8). (The galactosyltransferase to dolichol monophosphate is much less active than the others (Figure 8) and its significance is not yet clear.) These transferases are highly specific for nucleotide and sugar structure and are tightly bound to the rough and smooth endoplasmic reticulum. They are less specific for lipid-phosphate acceptor and *in vitro* several other long-chain alcohols can substitute for dolichol[7,8]. It is relevant that the sugar-1-phosphate linkage, and hence the transfer potential of the donor is retained. The compounds are well suited to formation of glycosides in a hydrophobic environment.

In fact, most of the new glycosidic linkages so formed are as part of an oligosaccharide linked to dolichol diphosphate (Figure 9). The oligosaccharide is then transferred to an asparagine residue *en bloc*. Whether or not glucose is added to all oligosaccharides prior to transfer from dolichol diphosphate to protein is not known. These transferases have not been studied as well as the others but it does appear that a fairly high specificity to dolichol phosphate sugar as donor and to the acceptor exists.

The concentration of dolichol monophosphate is very low and is probably rate limiting. Thus variation of the concentration of dolichol monophosphate presents a possible way of controlling the rate of glycosylation of at least some glycoproteins. Possible methods of altering the concentration of dolichol monophosphate are shown in Figure 10. It is of interest that the activity of hepatic HMG-CoA reductase, an enzyme common to the biosynthetic pathway of cholesterol and dolichol, is subject to diurnal variation. Other parameters affecting the rate of biosynthesis of sterols may well affect also the concentration of dolichol and its derivatives.

$$\text{GDPMan} + \text{Dol P} \rightleftharpoons \text{Dol P Man} + \text{GDP}$$

$$\text{UDPGlc} + \text{Dol P} \rightleftharpoons \text{Dol P Glc} + \text{UDP}$$

$$\text{UDPGal} + \text{Dol P} \rightleftharpoons \text{Dol P Gal} + \text{UDP}$$

$$\text{UDPGlcNAc} + \text{Dol P} \rightleftharpoons \text{Dol PP GlcNAc} + \text{UMP}$$

Figure 8 Formation of dolichol phosphate monosaccharides from nucleotide sugar donors in pig liver

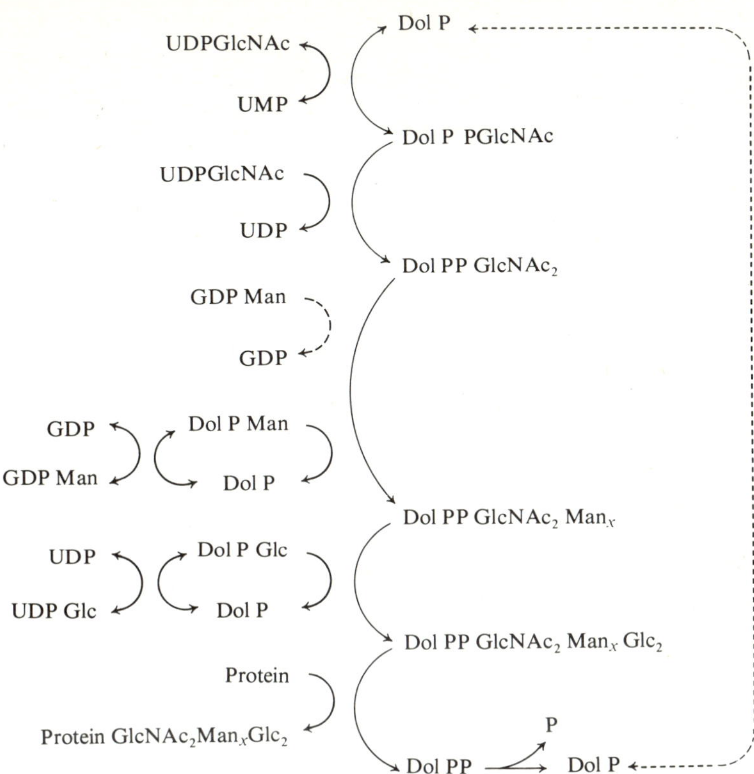

Figure 9 The role of dolichol monophosphate in glycosyl transfer to glycoprotein in rat liver[2]

Retinol (vitamin A, see Figure 11) monophosphate has also been implicated as a carrier of mannose and galactose in a manner analogous to that of dolichol monophosphate[9]. It follows that changes in retinol concentration due to dietary factors may also be important here.

VARIATIONS IN GLYCOSYLATION

The rate of glycosylation may also be affected by altering the activity of the enzyme or by changing the pool size of nucleotide diphosphate sugar. In substrate-limiting conditions the latter may affect a change in rate directly. Some glycosyltransferases respond allosterically to concentration changes of nucleotide sugars that are not substrates and recently, for example, GDP mannose has been shown to be a powerful inhibitor of UDP glucose: dolichol monophosphate glucosyltransferase.

Tunicamycin presents an interesting example of selective inhibition of a glycosyltransferase and its consequences. This is an antiviral compound that halts the formation of the envelope of several animal viruses. A major compo-

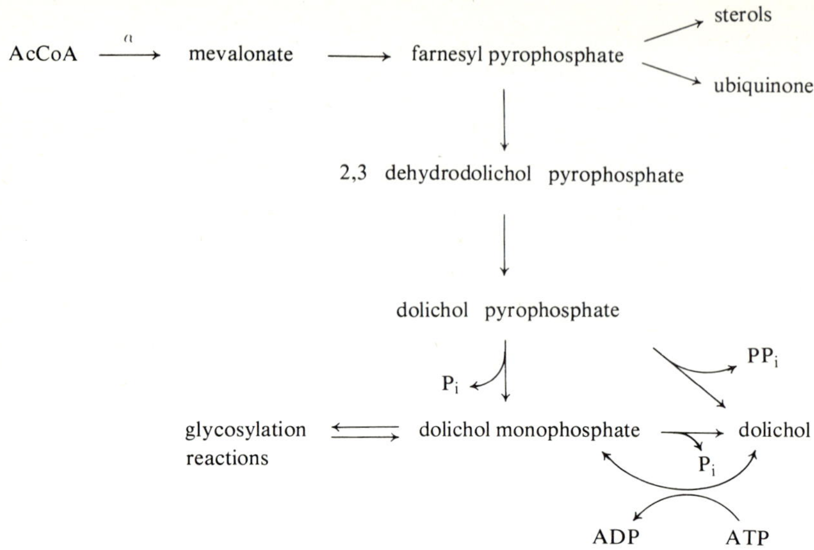

Figure 10 Steps that could possibly affect the concentration of dolichol monophosphate in mammalian cells

nent of the envelope is a glycoprotein, probably of the complex type (Figures 1 and 3). The block is achieved by inhibition of UDP *N*-acetylglucosamine: dolichol monophosphate *N*-acetylglucosaminyl phosphate transferase of the host cell (see step 1, Figure 9)[10,11] which in turn stops the glycosylation of the glycoprotein destined for the viral envelope. It has yet to be shown if tunicamycin will inhibit glycosylation of other proteins in liver.

The use of substrate analogues may also offer a way of inhibiting glycosylations. Thus deoxyglucose inhibits the formation of some glycoproteins and in yeast it appears to do this by substituting for mannose resulting in the formation of dolichol monophosphate deoxyglucose[12]. The deoxyglucose is then transferred to protein directly (in this case an *O*-glycosidic linkage with serine

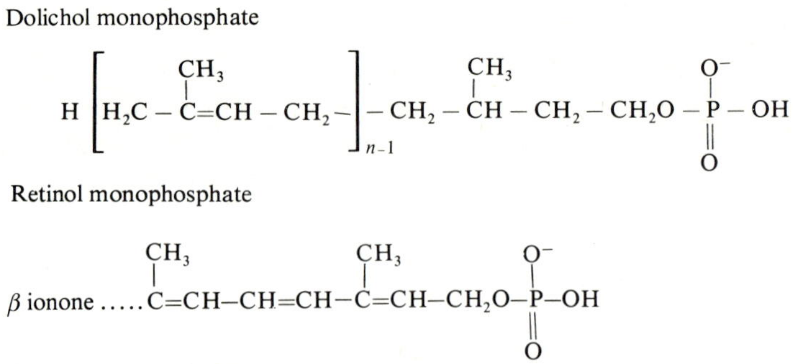

Dolichol monophosphate

$$\mathrm{H}\left[\mathrm{H_2C-\overset{\displaystyle CH_3}{\overset{|}{C}}{=}CH-CH_2-}\right]_{n-1}\mathrm{-CH_2-\overset{\displaystyle CH_3}{\overset{|}{C}H}-CH_2-CH_2O-\underset{\underset{\displaystyle O}{\|}}{\overset{\overset{\displaystyle O^-}{|}}{P}}-OH}$$

Retinol monophosphate

$$\beta\ \text{ionone}\ldots\ldots\mathrm{\overset{\displaystyle CH_3}{\overset{|}{C}}{=}CH{-}CH{=}CH{-}\overset{\displaystyle CH_3}{\overset{|}{C}}{=}CH{-}CH_2O{-}\underset{\underset{\displaystyle O}{\|}}{\overset{\overset{\displaystyle O^-}{|}}{P}}{-}OH}$$

Figure 11 The structures of dolichol and retinol phosphates. In mammalian cells (n = 14–21)

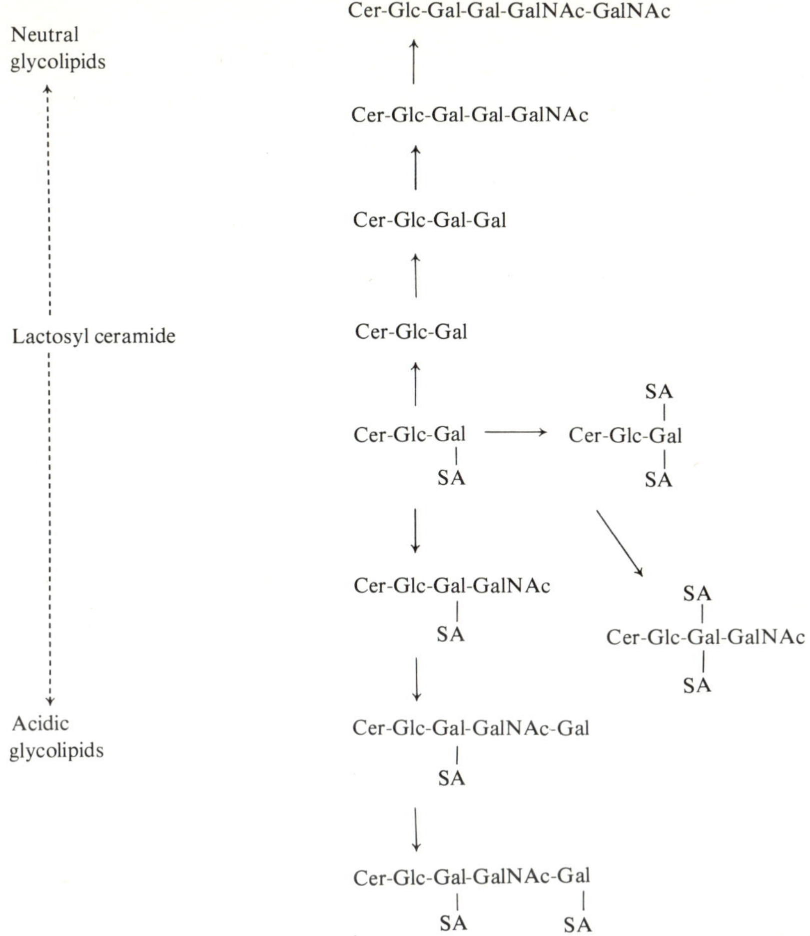

Figure 12 The biosynthetic relationship of neutral and acidic glycolipids to lactosyl ceramide. Cer, ceramide

or threonine) but further addition of sugar residues to form an oligosaccharide is blocked. Possibly an analogous block operates in mammalian tissues and other sugar derivatives will form useful inhibitors.

It is also conceivable that analogues of dolichol phosphate inhibit glycosylation, for while they can possibly act as acceptors of sugar residues the products may be inefficient as donors in the further transfer to oligosaccharide and protein. Work with cell-free preparations supports this view[7].

Since the glycosyltransferases are bound to lipid-rich membranes it is not surprising that modification of the phospholipids therein has been observed to affect their enzymic activity. For example, the transfer of galactose from UDP galactose to protein and of *N*-acetylglucosamine to protein in rat liver microsomes can be stimulated markedly by cytidine diphosphate choline in the presence of Triton X-100[3]. It was suggested that the regeneration of lecithin (at the expense of the cytidine diphosphate choline) initiated conformational changes in the enzyme that resulted in activation. Phospholipases have also been reported to stimulate both galactosyl transfer to protein and glucuronyl transfer to *p*-nitrophenol.

It is important to note that the nucleotide diphosphate sugar donors are formed in the cytoplasm which is the opposite side (outside) of the membrane from the location of the final product (inside) (Figure 5). The location of the transferases within the membrane is not yet clear. If the enzyme is also on the inside of the membrane the donor has to permeate the membrane. The rate of this process will possibly change with variation in the nature of the membrane. Alternatively if the enzyme bridges the membrane the enzymic reaction itself may result in the transfer of sugars across the membrane and a separate permeability will not be concerned.

GLYCOLIPIDS AND CONTRIBUTION OF GOLGI COMPLEX

The discussion above has been restricted primarily to the *N*-glycosidically linked glycoproteins and has allowed features common to all glycosylation processes to be introduced. However, it is necessary to consider briefly glycosyl transfer specifically to *O*-glycosidically linked glycoproteins (Figure 1) and to glycolipids (Figure 12). The major site for these enzyme activities is the Golgi apparatus. In these cases glycosyl transfer appears to be sequential and highly specific for nucleotide donor and acceptor. Dolichol phosphate or similar lipid-soluble donors are not involved. The process resembles closely that described above for addition of the branch chain sugars to complex *N*-glycosidically linked glycoproteins.

The glycolipids involved are glycosylated forms of ceramide. There are two series of compounds, one neutral and one acidic. The members of each series are related biosynthetically (Figure 12). Each step requires a specific glycosyltransferase and nucleotide sugar. Usually the sequence is not completed in all molecules and a mixture of glycolipids results. In many cultured cells the nature of the mixture is growth and cell-density dependent. At dilute suspensions a component early in the sequence predominates due to a block at one of the later steps in each sequence but as the concentration of cells increases with growth this block disappears and components later in the sequences appear. If the cells are transformed by oncogenic viruses or chemically this change with increasing cell density does not occur and the block continues. The precise step blocked is host species-specific and is not, for example, related to the type of virus used for transformation. Thus transformed cells tend to have shorter oligosaccharide chains of the glycolipids than do normal cells[14].

In contrast to this, membrane glycoproteins of transformed cells tend to contain a larger oligosaccharide unit than those of normal cells and an increased sialyltransferase activity to glycoproteins has been observed in the former[15].

COMMENT

It has been suggested that changes in plasma membrane glycoproteins and glycolipids have an important influence on cell–cell interactions and on receptor properties. Some mechanisms proposed to explain changes in these parameters require the presence of glycosyltransferases on the surface of the cell. While most glycosyltransferase activity is found associated with internal membranes there have been several reports of the presence of this activity in the plasma membrane[5,16].

It will be clear from this short review that knowledge of glycosylation mechanisms is an important aspect of a greater understanding of the relationship between membrane alterations and cell or tissue malfunction. Shortage of time and space has not allowed discussion of the two other major glycosylation processes — the formation of glycogen and glucuronidation of toxic or potentially toxic compounds. Both are important, and detailed studies of the mechanism of glycosylation involved have been fruitful and fascinating. However, they are less directly concerned with membrane phenomena and therefore slightly less relevant to this symposium.

References

1. Spiro, R. G. (1973). Glycoproteins. *Adv. Protein Chem.*, **27,** 349
2. Behrens, N. H. (1974). In: E. Y. C. Lee and E. E. Smith (eds.). *Biology and Chemistry of Eucaryotic Cell Surfaces*, pp. 159–180 (New York: Academic Press)
3. Roseman, S. (1970). The synthesis of complex carbohydrates by multiglycosyl transferase systems and their potential function in intercellular adhesion. *Chem. Phys. Lipids,* **5,** 270
4. Ginsburg, V. (1972). Biochemical basis for blood groups in man. In: R. Piras and H. G. Pontis (eds.). *Biochemistry of the Glycosidic Linkage*, pp. 387–393 (New York: Academic Press)
5. Waechter, C. J. and Lennarz, W. J. (1976). The role of polyprenol-linked sugars in glycoprotein synthesis. *Ann. Rev. Biochem.*, **45,** 95
6. Hemming, F. W. (1974). Lipids in glycan biosynthesis. In: T. W. Goodwin (ed.). *Biochemistry of Lipids, Biochem. Ser. One*, **4,** 39–98 (London: Butterworths)
7. Richards, J. B. and Hemming, F. W. (1972). The transfer of mannose from guanosine diphosphate mannose to dolichol phosphate and protein by pig liver endoplasmic reticulum. *Biochem. J.*, **130,** 77
8. Mankowski, T., Sasak, W. and Chojnachi, T. (1975). Hydrogenated polyprenol phosphates—exogenous lipid acceptors of glucose from UDP glucose in rat liver microsomes. *Biochem. Biophys. Res. Commun.*, **65,** 1292
9. Rosso, G. C., de Luca, L., Warren, C. D. and Wolf, G. (1975). Enzymatic synthesis of mannosyl retinyl phosphate from retinyl phosphate and guanosine diphosphate mannose. *J. Biol. Chem.*, **16,** 235
10. Tkacz, J. S. and Lampen, J. O. (1975). Tunicamycin inhibition of polyisoprenyl *N*-acetylglucosaminyl pyrophosphate formation in calf liver microsomes. *Biochem. Biophys. Res. Commun.*, **65,** 248

11. Takatsuki, A., Khono, K. and Tamura, G. (1975). Inhibition of biosynthesis of polyisoprenol sugars in chick embryo microsomes by tunicamycin. *Agr. Biol. Chem.*, **39,** 2089
12. Lehle, L. and Schwarz, R. T. (1976). Formation of dolichol monophosphate 2-deoxy-D-glucose and its interference with the glycosylation of mannoproteins in yeast. *Eur. J. Biochem.*, **67,** 239
13. Mookerjea, S., Cole, D. E. C. and Chow, A. (1971). Glycoprotein biosynthesis: stimulation of galactose transfer from UDP-[^{14}C]galactose into microsomal protein by cytidine 5′-diphosphocholine. *FEBS Lett.*, **23,** 257
14. Critchley, D. R. (1973). Glycolipids and cancer. In: P. W. Kent (ed.). *Membrane Mediated Information*, **1,** pp. 20–47 (Lancaster: MTP Press)
15. Warren, L., Fuhrer, J. P. and Buck, C. A. (1973). Comparison of the surface structures of virus-transformed and control cells in tissue culture. In: P. W. Kent (ed.). *Membrane Mediated Information*, **1,** pp. 142–157 (Lancaster: MTP Press)
16. Hakomori, S. I. (1975). Structures and organization of cell surface glycolipids. Dependency on cell growth and malignant transformation. *Biochim. Biophys. Acta*, **417,** 55

Recognition, Binding and Transport of Glycoproteins by the Hepatocyte Plasma Membrane

I. STERNLIEB

The ability of cellular plasma membranes to recognize and bind specific substrates while rejecting many others provides evidence for the existence of specific receptors. Some of these receptors, such as those for insulin or glucagon, play pivotal roles in the chains of events which lead to the ultimate physiological effects of these hormones on the hepatocyte. Another function of the hepatocyte, namely its catabolic function, also requires recognition of molecules of plasma proteins destined for removal from the circulation since considerable circumstantial evidence indicates that the liver is the principal organ in which plasma proteins are catabolized. The discovery of the existence of components of plasma membranes endowed with remarkably specific properties of recognition and binding of modified plasma glycoproteins has provided new insight into the mechanism of uptake and transport of macromolecules by the hepatocytes.

THE ROLE OF THE TERMINAL SIALIC ACID RESIDUE IN THE SURVIVAL OF PLASMA GLYCOPROTEINS

The work* covered in this review started in 1966 when we were studying the physiology of serum ceruloplasmin. This is a blue plasma protein which

* In collaboration with Anatol G. Morell, Richard Stockert, I. Herbert Scheinberg, Gilbert Ashwell and Gregory Gregoriadis

possesses seven copper atoms and eight to ten oligosaccharide chains composed principally of glucosamine, mannose and, in most chains, galactose terminated by a sialic acid residue. Our goal was to label this glycoprotein with tritium. By applying a method developed by Gilbert Ashwell, Anatol Morell used sequentially neuraminidase and galactose oxidase before reducing an aldehyde function on the exposed galactose by tritiated borohydride[1]. Except for a 2% diminution in molecular weight and a loss of ten electric charges, the product of this procedure possessed many of the physicochemical

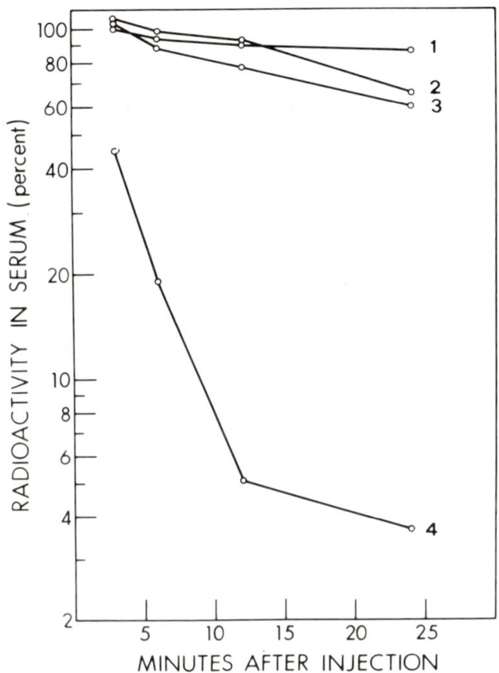

Figure 1 Disappearance from the serum of ^{64}Cu-labelled, native, and modified rabbit ceruloplasmin. Each point is the average value obtained from two animals. 1, ceruloplasmin; 2, oxidized asialoceruloplasmin; 3, asialoagalactoceruloplasmin; 4, asialoceruloplasmin

characteristics of the native protein. Moreover, the desialylated ceruloplasmin, or asialoceruloplasmin (ASCPN), could be reconstituted into native ceruloplasmin[2] and even crystallized[3].

An astonishing observation followed the i.v. injection of a preparation of asialoceruloplasmin into a rabbit: instead of the physiological half-life of 56 hours of intact ceruloplasmin we found that ASCPN disappeared from the circulation within minutes[4] (Figure 1). The phenomenon observed suggests a subtle denaturation of the protein, not detected by the physicochemical and immunochemical parameters tested for *in vitro*, but recognizable by the remarkably efficient reticuloendothelial system. No radioactivity was recovered in either spleen or lymph nodes of the rabbit though close to 90% of the injected radioactivity was recovered in the rabbit's liver. Furthermore, radioautographs of sections of this liver demonstrated the presence of silver grains exclusively in hepatocytes and none in Kupffer cells[4] (Figure 2). This was

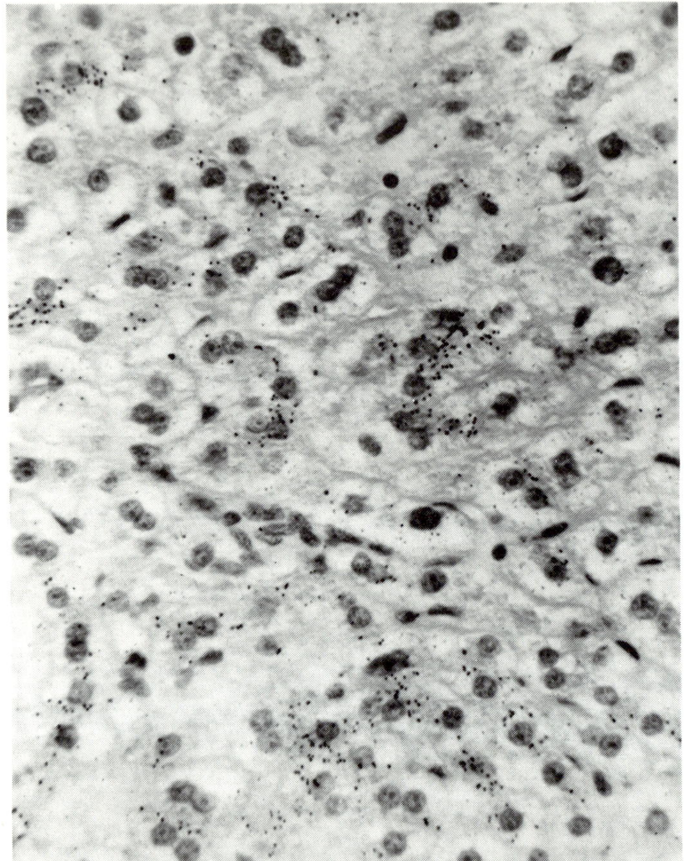

Figure 2 Radioautograph of histological sections of liver 3 min following injection of tritium-labelled asialoceruloplasmin into a rabbit. Formaldehyde fixation: NTB emulsion No. 3 (Eastman Kodak Company); Kodak Developer D 19; haematoxylin and eosin. Silver granules overlie the parenchymal cells but are absent in the Kupffer cells

confirmed when blockade of the reticuloendothelial system with the aid of heat-denatured bovine serum albumin or india ink did not affect the rate of disappearance of the desialylated ceruloplasmin.

Further studies demonstrated that the factor determining recognition and uptake by the liver was not the absence of the terminal sialic acid residue but the exposure of intact galactosyl residues[5,6]. This became evident because either oxidation of the galactosyl residue with galactose oxidase or its enzymatic removal with β-galactosidase reversed the phenomenon of rapid uptake of the glycoprotein by the liver[4] (Figure 1).

The original observations on ASCPN were extended to a number of other plasma glycoproteins (orosomucoid, fetuin, haptoglobin, α-2 macroglobulin and others), several hormones (follicle stimulating hormone, human chorionic gonadotropin and others) and even glycopeptides[7]. In all, with the exception of transferrin, the exposed galactose moiety served as a cryptic determinant of recognition and uptake by the hepatocellular plasma membrane[8].

The capacity for the transfer of desialylated glycoproteins by the hepatocellular membrane is limited however. This is shown when a trace amount of ASCPN is injected simultaneously with an excess of asialoorosomucoid: there is delayed clearance of ASCPN for as long as the excess of circulating asialoorosomucoid exceeds 0.1 mg/ml[7]. Evidently there is competitive inhibition of uptake by various desialylated glycoproteins.

ISOLATION OF THE BINDING PROTEIN FOR DESIALYLATED GLYCOPROTEINS

The specific interaction of asialoglycoproteins with the hepatocellular plasma membranes provided a basis for the isolation and purification of the binding protein specific for asialoglycoproteins, which was named HBP[9–11]. By using affinity chromatography on Sepharose to which asialoorosomucoid was covalently bound, the specific binding activity of HBP was enhanced from 0.17 in the homogenate to 1900 in the purified preparation[9]. This protein consists of a series of oligomeric proteins, the smallest unit of which has a molecular weight of approximately 500 000 daltons with two subunits of 48 000 and 40 000 molecular weight, respectively[12]. It should be noted that successful binding by the solubilized receptor required the presence of covalently linked sialic acid residues[13].

DEMONSTRATION OF A SECOND RECEPTOR SPECIFIC FOR HEXOSAMINYL CHAINS

More recently, Stockert, Morell and Scheinberg found another route of transfer in mammals for glycoproteins terminating in hexosaminyl chains which does not involve HBP. When asialoorosomucoid is injected simultaneously with agalactoorosomucoid (which terminates in *N*-acetylglucosaminyl residues) neither the rate of disappearance of this modified glycoprotein from the plasma nor the extent of its recovery from the liver are affected[14].

ULTRASTRUCTURAL STUDY OF UPTAKE AND INTRACELLULAR TRANSPORT OF PEROXIDASE-LABELLED ASIALOGLYCOPROTEIN

Recently, in collaboration with Drs Phyllis Novikoff and Michio Mori we have attempted to follow the route of uptake of asialoglycoproteins at the ultrastructural level. In preliminary experiments we have conjugated horseradish peroxidase to asialofetuin by the method of Avrameas and Ternynck[15]. One minute after intraportal injection of an asialofetuin–peroxidase conjugate we found electron-dense peroxidase reaction product in pinocytotic vesicles near the sinusoidal border of hepatocytes (Figure 3). Larger, membrane-bound, irregularly shaped vesicles were also present in the cytoplasm near the

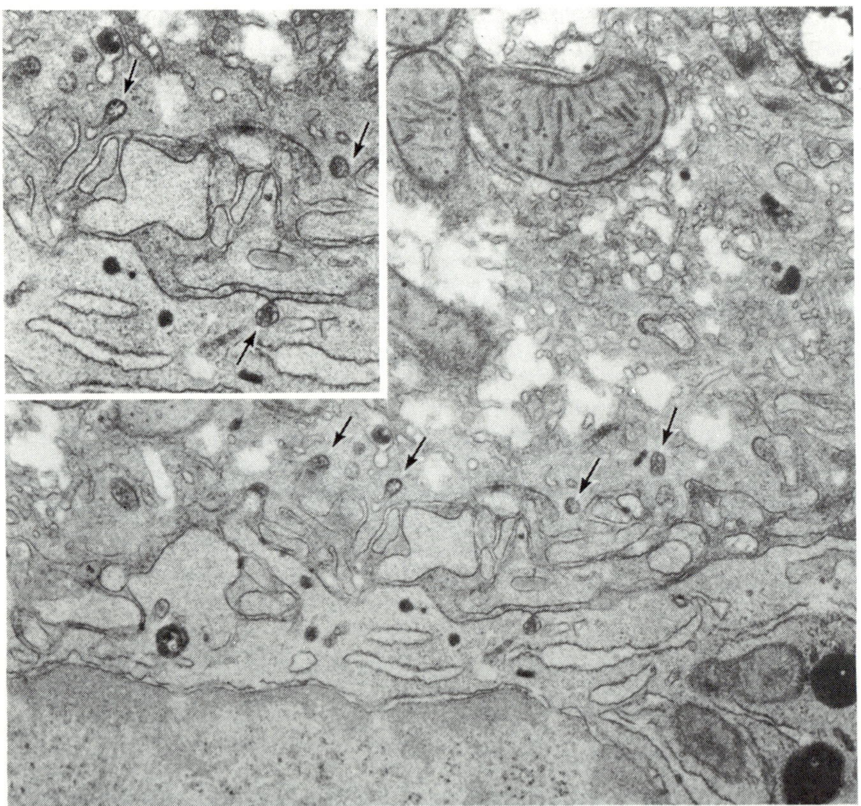

Figure 3 Electron micrograph of thin section of rat liver 1 min after intraportal injection of asialofetuin–peroxidase conjugate. Note presence of electron-dense peroxidase reaction product (arrows) in pinocytotic vesicles near the sinusoidal border of a hepatocyte adjacent to a Kupffer cell. Inset: higher magnification of central portion. Osmium tetroxide; epon; uranyl acetate and lead citrate; RCA EMU 3H (× 18 500; inset × 26 000)

sinusoidal border (Figure 4). At 10 min reaction product was visible close to the biliary pole of the hepatocyte. More detailed studies are required before we are able to map out the pathways followed by asialoglycoproteins from the hepatocellular membrane to their catabolic site, the lysosome[16].

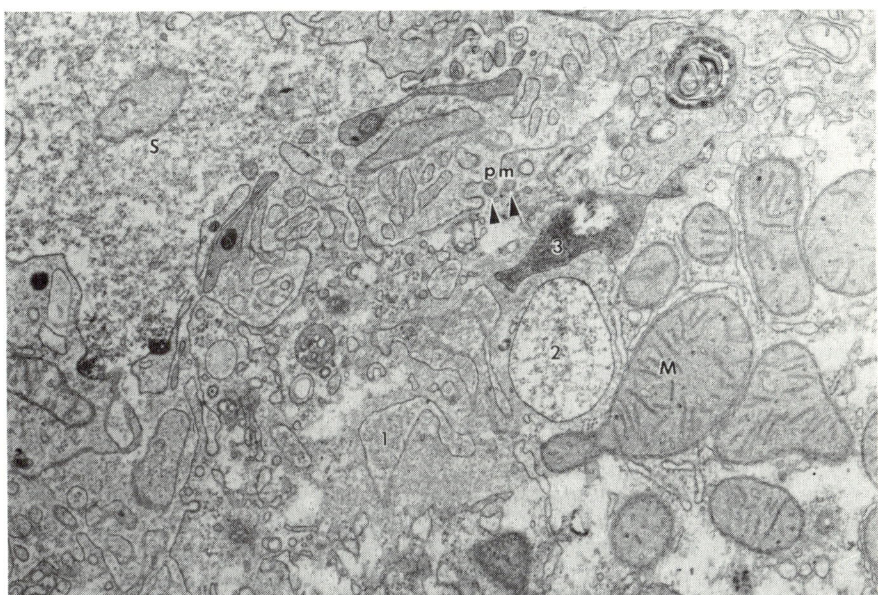

Figure 4 Same experiment as in Figure 3. Portion of hepatocyte and sinusoid showing large membrane bound, irregularly shaped vesicles (1–3) containing peroxidase reaction product near the sinusoidal border at 1 min after injection as well as pinocytotic vesicles (arrowheads). M, mitochondrion; pm, plasma membrane; S, sinusoid containing peroxidase reaction product (×15 500)

Evidently the hepatocellular plasma membranes contain binding proteins which are specific for different carbohydrate moieties of glycoproteins. One of these binding proteins, specific for galactosyl residues, named HBP, was isolated and purified from mammalian liver. There are binding proteins specific for other carbohydrate residues of which at least one specific for hexosaminyl residues has been identified. These binding proteins seem to initiate the uptake and transport of plasma glycoproteins towards their catabolic sites. Practical applications of these observations may emerge through the delivery of missing enzymes or the targeting of drugs to hepatocytes.

SUMMARY

The hepatocytic plasma membrane shows selectivity in recognition and uptake of modified glycoproteins from the plasma. Specific binding proteins responsible for these phenomena have been demonstrated. Although their physiological role is still uncertain, the specificity of the interaction of the modified glycoproteins with the intact receptor provides a tool for the study of factors which affect the recognition, uptake, transport and intracellular catabolisms of macromolecules by hepatocytes.

Acknowledgements

The original work referred to in this review is supported by US Public Health Service Grant No. AM–1059 from the National Institute of Arthritis, Metabolism and Digestive Diseases and CA-6576 from the National Cancer Institute.

References

1. Morell, A. G., van den Hamer, C. J. A., Scheinberg, I. H. and Ashwell, G. (1966). Physical and chemical studies on ceruloplasmin. IV. Preparation of radioactive, sialic acid-free ceruloplasmin, labeled with tritium on terminal D-galactose residues. *J. Biol. Chem.,* **241,**
2. Hickman, J., Ashwell, G., Morell, A. G., van den Hamer, C. J. A. and Scheinberg, I. H. (1970). Physical and chemical studies on ceruloplasmin. VIII. Preparation of *N*-acetylneuraminic acid-1-[^{14}C]labeled ceruloplasmin. *J. Biol. Chem.*, **245,** 759
3. Morell, A. G., Sternlieb, I. and Scheinberg, I. H. (1969). Physical and chemical studies on ceruloplasmin: crystallization of desialized human ceruloplasmin asialoceruloplasmin. *Science*, **166,** 1293
4. Morell, A. G., Irvine, R. A., Sternlieb, I., Scheinberg, I. H. and Ashwell, G. (1968). Physical and chemical studies on ceruloplasmin. V. Metabolic studies on sialic acid-free ceruloplasmin *in vivo*. *J. Biol. Chem.*, **243,** 155
5. van den Hamer, C. J. A., Morell, A. G., Scheinberg, I. H., Hickman, J. and Ashwell, G. (1970). Physical and chemical studies on ceruloplasmin. IX. The role of galactosyl residues in the clearance of ceruloplasmin from the circulation. *J. Biol. Chem.*, **245,** 4397
6. Sternlieb, I., Morell, A. G. and Scheinberg, I. H. (1973). The remarkable selectivity of hepatocytes in the uptake of glycoproteins. *Gastroenterology*, **64,** 1049
7. Morell, A. G., Gregoriadis, G., Scheinberg, I. H., Hickman, J. and Ashwell, G. (1971). The role of sialic acid in determining the survival of glycoproteins in the circulation. *J. Biol. Chem.*, **246,** 1461
8. Ashwell, G. and Morell, A. G. (1974). The role of surface carbohydrates in the hepatic recognition and transport of circulating glycoproteins. In: A. Meister (ed.). *Advances in Enzymology, Vol. 41*, pp. 99–128
9. Hudgin, R. L., Pricer, W. E., Ashwell, G., Stockert, R. J. and Morell, A. G. (1974). The isolation and properties of a rabbit liver binding protein specific for asialoglycoproteins. *J. Biol. Chem.*, **249,** 5536
10. Morell, A. G. and Scheinberg, I. H. (1972). Solubilization of hepatic binding sites for asialoglycoproteins. *Biochem. Biophys. Res. Commun.*, **48,** 808
11. Stockert, R. J., Morell, A. G. and Scheinberg, I. H. (1974). Mammalian hepatic lectin. *Science*, **186,** 365
12. Kawasaki, T. and Ashwell, G. (1976). Chemical and physical properties of a hepatic membrane protein that specifically binds asialoglycoproteins. *J. Biol. Chem.*, **251,** 1296

13. Pricer, W. E. and Ashwell, G. (1971). The binding of desialylated glycoproteins by plasma membranes of rat liver. *J. Biol. Chem.*, **246,** 4825
14. Stockert, R. J., Morell, A. G. and Scheinberg, I. H. (1976). The existence of a second route for the transfer of certain glycoproteins from the circulation into the liver. *Biochem. Biophys. Res. Commun.*, **68,** 988
15. Avrameas, S. and Ternynck, T. (1971). Peroxidase labeled antibody and F_{ab} conjugates with enhanced intracellular penetration. *Immunochemistry*, **8,** 1175
16. Gregoriadis, G., Morell, A. G., Sternlieb, I. and Scheinberg, I. H. (1970). Catabolism of desialylated ceruloplasmin in the liver. *J. Biol. Chem.*, **245,** 5833

7
Plasma Membrane Protein Biogenesis and Turnover in Hepatoma Tissue Culture Cells

D. DOYLE, BARBARA ENGLAND, ELSE FRIEDMAN, ESTHER HOU and J. TWETO

Despite the important role played by the mammalian plasma membrane in many normal cell phenomena and the role proposed for this cell organelle in abnormal cell growth[1], little is known about the pathway of biogenesis of this membrane. One reason for this has been the difficulty of isolating a homogeneous plasma membrane fraction such that the polypeptides can be identified and characterized. However, the introduction in recent years of chemical reagents that cannot cross the limiting membrane of intact cells but which react with polypeptides exposed on the surface of these cells offers the possibility of easily and unequivocably identifying the polypeptides of the surface membrane.

LABELLING OF MEMBRANE

We use lactoperoxidase-catalysed iodination[2,3] for labelling the surface proteins of hepatoma tissue culture (HTC) cells. We have shown[4] that this procedure labels specifically the polypeptides of the plasma membrane of this cell. To summarize these studies, light microscopic autoradiography of sectioned cells showed the incorporated label to be localized primarily at the periphery of the cell. Most of this label could be released from the cell by trypsin, but not by collagenase or hyaluronidase. The label was recovered from the cells as either monoiodotyrosine or diiodotyrosine after hydrolysis of cell extracts with a mixture of proteolytic enzymes. Finally, the label copurified during cell fractionation with 5′-nucleotidase, an enzyme confined primarily to the plasma membrane of hepatocytes.

Iodination of HTC cells *in situ* does not affect cell viability, cell growth rate or cell function as measured by the induction of tyrosine aminotransferase in these cells by corticosteroids, suggesting that the turnover of the plasma membrane could be studied using the incorporated iodide as a specific marker for the polypeptides in this membrane.

However, before examining the turnover of the iodinated membrane polypeptides, it was necessary to determine whether iodination itself affected the turnover of cell protein and whether the iodinated polypeptides in the membrane turn over in a way similar to that of unmodified plasma membrane proteins. A double isotope procedure was used to assess the effect of iodination on the rate constant of degradation of both total protein and plasma membrane protein of HTC cells (Table 1). In this experiment, cells were

Table 1 Effect of iodination on turnover of total protein and plasma membrane protein of HTC cells

Conditions	*Cell fraction*					
	Homogenate			*Plasma membrane*		
	dpm × 10^{-6}/mg of protein					
	^3H	^{14}C	$^3H/^{14}C$	^3H	^{14}C	$^3H/^{14}C$
Iodinated cells 24 hours in culture	1.8	0.86	2.1	1.2	0.44	2.7
Uniodinated control cells 24 hours in culture	1.7	0.92	1.9	1.7	0.66	2.6

(See reference [5] for experimental details)
Reproduced from reference [5] with permission

labelled for 1 hour with [^3H]tyrosine. Then, one aliquot of the ^3H-labelled cells was immediately given a second pulse of tyrosine, but labelled with ^{14}C. This culture was the zero time control. A second aliquot of the ^3H-labelled cells was incubated at 37 °C for 24 hours before administration of [^{14}C]tyrosine. A third aliquot of the ^3H-labelled cells was iodinated with unlabelled iodide before being incubated for 24 hours at 37 °C. Then, the second [^{14}C]tyrosine pulse was given. The zero time control cells had $^3H/^{14}C$ ratios of about three for each of several cell fractions analysed. This value is the ratio to be expected with no turnover of protein. The control cells, not iodinated, had isotope ratios of about 2.0 and 2.5 for total protein and plasma membrane protein respectively. Thus, some of the ^3H-labelled protein was degraded during the 24-hour incubation. The iodinated cells also had $^3H/^{14}C$ ratios of about 2.0 and 2.5 for total protein and plasma membrane protein, respectively. The higher isotope ratio of the plasma membrane fraction relative to total cell protein suggests that plasma membrane proteins are turning over at a slower rate than total cell protein. The fact that the isotope ratios for the iodinated cells were essentially identical with those for the uniodinated control cells indicates that iodination does not affect significantly the turnover of either total cell protein or plasma membrane protein. This being the case, it is possible to examine the fate of the iodinated polypeptides with some confidence that their turnover reflects adequately the normal turnover of unmodified plasma membrane proteins. We believe and will try to show here that an understanding of the general features of membrane protein degradation can give important insight into the complex process of membrane biogenesis.

DEGRADATION OF MEMBRANE PROTEINS

We first compared the degradation rates of iodinated membrane proteins in both exponentially growing HTC cells and in non-growing density inhibited cells by following the loss with time of ^{125}I in trichloracetic acid-insoluble material[5]. In both exponentially growing cells and in density-inhibited cultures the iodinated proteins are degraded with a half-life of about 100 hours. The label lost from protein was released into the medium at a continuous rate. Most, 90% or more, of the label in the medium was soluble in 10% trichloracetic acid and was identified by chromatography as either free iodide or monoiodotyrosine and diiodotyrosine.

The degradation rate of total protein of HTC cells also was measured by monitoring the loss of acid-insoluble radioactivity from cells that were grown in the presence of [^{3}H]leucine for 4 hours. In both growing and density-inhibited cells, the half-life for the degradation of total protein was approximately 60 hours, confirming the results of Table 1; the half-life of total cell protein is significantly shorter than that of plasma membrane proteins. The value for the half-life of total cell protein is probably overestimated somewhat, since the leucine, in contrast to iodotyrosine, is certainly reutilized to some extent. Similar values for the half-life of total cell protein were obtained in both exponentially growing and in density-inhibited cells, indicating that similar to the turnover of plasma membrane proteins, the overall turnover of cell protein is not a function of cell density[5,6].

COMPLEXITY OF PLASMA MEMBRANE

We next examined the complexity of the HTC cell plasma membrane with respect to those polypeptides having tyrosine residues externally oriented and accessible to iodination. The results are presented in Figure 1. Lane A shows the Coomassie staining pattern of the polypeptides in a homogenate of HTC cells labelled *in situ* with ^{125}I and separated by electrophoresis in a polyacrylamide gel containing dodecyl sulphate. Despite the extreme complexity of this pattern clear differences can be seen between it and the autoradiogram of the homogenate displayed in lane B. Some of the Coomassie bands are not

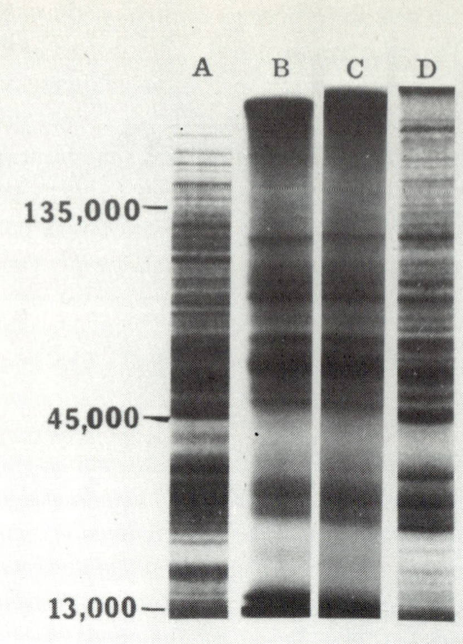

Figure 1 Dodecyl sulphate–polyacrylamide gel electrophoresis of the polypeptides in HTC cell fractions. Polypeptides were separated by electrophoresis in slab gels of 9% acrylamide. A, homogenate stained for protein; B, autoradiogram of the homogenate; C, autoradiogram of the purified plasma membrane; D, plasma membrane stained for protein (see reference[4] for details; reproduced with permission)

labelled, and the intensity of labelling of different bands does not correlate well with the staining intensity. Lane C is an autoradiogram of a plasma membrane fraction of HTC cells purified by a step sucrose gradient method[4]. Comparison of this pattern with the autoradiogram of the homogenate in B shows that all of the labelled proteins in the homogenate are present in the purified membrane. Thus, none of these proteins was selectively lost during cell fractionation. Lane D is a Coomassie-stained pattern of the plasma membrane preparation. In C and D, there is a remarkable similarity between the patterns, with many of the labelled bands corresponding in a roughly quantitative fashion with the Coomassie staining bands. This should be contrasted with the homogenate staining and labelling patterns in lanes A and B. Several stained bands in D do not have a corresponding labelled band, suggesting that these polypeptides are not accessible to the lactoperoxidase probe, either because they represent contamination of the plasma membrane preparation or because they are proteins which are located on the cytoplasmic side of the membrane. One of the major polypeptides associated with the membrane has an apparent molecular weight of about 45 000 daltons. This polypeptide is not iodinated and probably is the actin-like protein which is associated with the interior surface of many eukaryotic cells[7]. The fact that it is not labelled with iodide is additional evidence that only externally disposed membrane proteins are labelled. The degree of coincidence between the Coomassie-stained pattern and the autoradiographic pattern of the membrane polypeptides suggests that polypeptides containing tyrosines accessible to iodination comprise a significant proportion of the protein mass of the membrane. A higher resolution dodecyl sulphate–acrylamide gel system was used to resolve some of the high molecular weight polypeptides and some of the low molecular weight polypeptides which were not resolved in Figure 1. Autoradiography of this gel (Figure 2) shows that iodination with different amounts of carrier iodide or

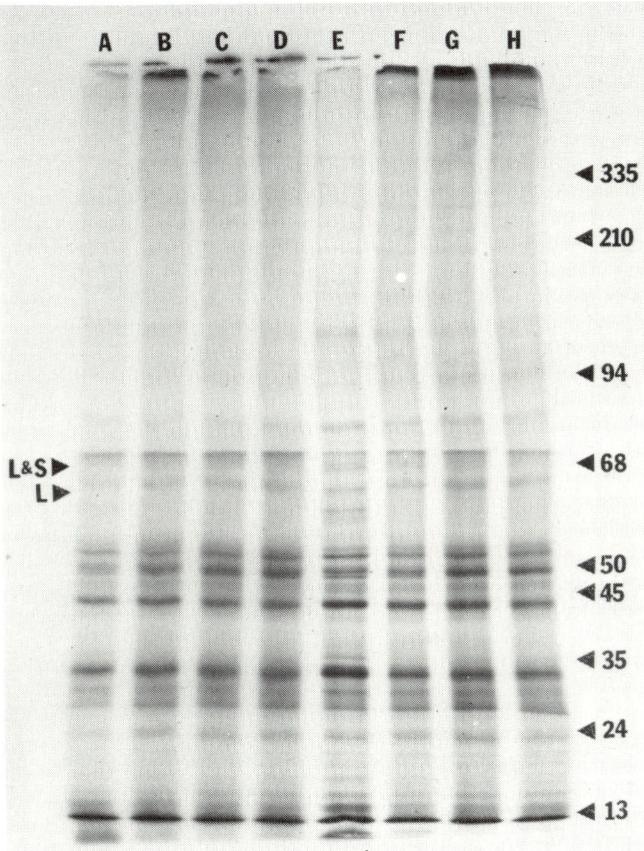

Figure 2 Autoradiogram of plasma membrane polypeptides from iodinated HTC cells. HTC cells were iodinated with ^{125}I (A–D) or ^{131}I (F–H) in the presence of varying amounts of unlabelled iodide; A and E, no unlabelled iodide; B and F, 50 μM unlabelled iodide; C and G, 200 μM unlabelled iodide; D and H 250 μM unlabelled iodide. A constant amount of radioactivity was applied to each slot of the slab gel. The gel was composed of 7.5% acrylamide on top of 10% acrylamide. The numbers indicate the approximate mobilities of several standard proteins of known molecular weight (expressed as mol. wt. $\times$ 10^{-3}); L: the mobility of the major component in the preparation of lactoperoxidase used for iodination; S: the mobility of the major protein component in the medium in which the cells are grown. The marker proteins were iodinated and localized by autoradiography. Each marker was run in the presence of unlabelled, crude HTC cell membranes (from Tweto, Friedman and Doyle[4], reproduced with permission)

with different isotopes does not affect significantly the pattern of labelled polypeptides. Differences exist between the ^{131}I-labelling pattern and the ^{125}I pattern, but these are due to the greater sensitivity and resolution when ^{131}I is used. Between 50 and 60 distinct bands are evident in the ^{131}I autoradiogram.

Since ^{125}I and ^{131}I label the same membrane proteins to the same relative

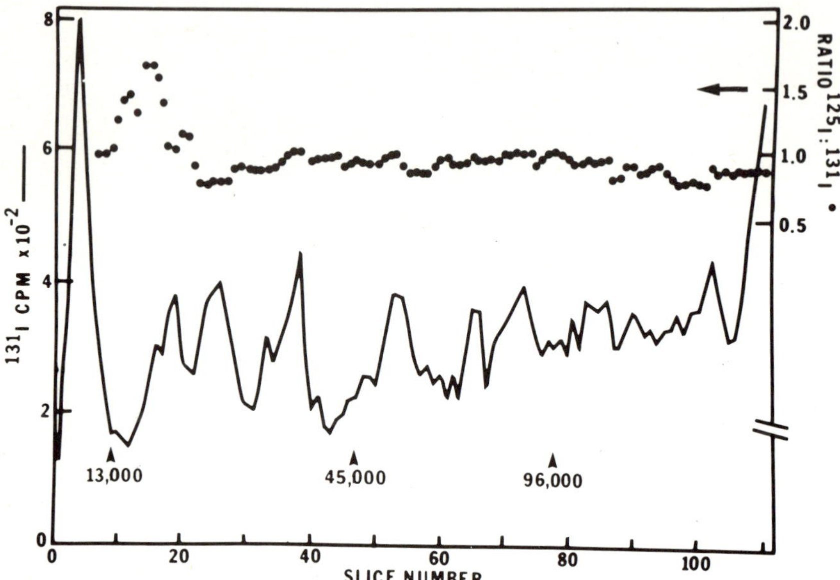

Figure 3 ^{125}I/^{131}I Ratios of polypeptides in the plasma membrane fraction of HTC cells as separated on dodecyl sulphate–polyacrylamide gels. A suspension of washed HTC cells (10^8 cells) was divided into two parts. Half of the cells were iodinated with ^{131}I, while the other half were iodinated with ^{125}I as described in reference[5]. Immediately after iodination, the ^{131}I-labelled cells were homogenized and frozen while the ^{125}I-labelled cells were placed in culture for 48 hours at a density of 1×10^6/ml. A homogenate of these cells was prepared and frozen with the frozen homogenate of the ^{131}I-labelled cells which were not in culture. A plasma membrane fraction was then prepared from the combined homogenates. The polypeptides in the membrane fraction were dissociated in dodecyl sulphate and dithiothreitol and separated on disc gels of 9% acrylamide. Approximate mobility of standard proteins in this system is indicated. ●, ratio of ^{125}I/^{131}I in the acrylamide fractions; —, ^{131}I-labelled polypeptides in the membrane fraction. The arrow indicates the ratio of ^{125}I/^{131}I in cells, neither of which was in culture (zero time control) (reproduced from Tweto and Doyle[5] with permission)

extent, a double isotope[5,8] procedure could be used to examine the relative rates of degradation of these proteins in the plasma membrane. This experiment, presented in Figure 3, was done with non-growing cells. The details of the design of the experiment are described in the legends to the figure. Essentially, the ^{131}I radioactivity represents the polypeptides in the membrane accessible for iodination, while the ^{125}I radioactivity shows what happens to these polypeptides as the cells are in culture. The ratio of ^{125}I/^{131}I in the membrane fraction, then, is a measure of protein degradation. Figure 3 shows the results from an experiment in which the ^{125}I-labelled cells were in culture for 48 hours before mixing with the ^{131}I-labelled cells which were not cultured. The mean isotope ratio for the separated polypeptides in this membrane preparation was 0.9 with a standard deviation of 0.1. Thus, turnover did occur relative to a control experiment in which cells labelled with ^{125}I were mixed at zero time with ^{131}I-labelled cells. In the control experiment there is no turnover of the plasma membrane proteins of the ^{125}I-labelled cells. These cells had a mean ^{125}I/^{131}I ratio for the separated polypeptides of 1.5 with a standard deviation of 0.2. There was no significant deviation from the mean ratio of 0.9 among the polypeptides separated from the cells which underwent membrane turnover except at the low molecular weight region of the gel. This was also found in the control experiment and, therefore, cannot be due to degradation. In experiments in which the ^{125}I-labelled cells were in culture for shorter (5 hours) or longer (90 hours) periods of time there was no indication of heterogeneity of degradation rates among the dissociated membrane polypeptides. Similarly, there was no indication of heterogeneity of degradation rates among plasma membrane polypeptides in exponentially growing cells[5].

Since some membrane proteins do not have tyrosine residues exposed on the outside of the cell, and cannot be labelled by lactoperoxidase-catalysed iodination, we employed a more conventional double isotope procedure using labelled leucine as precursor to examine the degradation of all the proteins associated with the plasma membrane. These results are presented in Figure 4. It is obvious that all of the leucine-labelled polypeptides in this preparation of plasma membrane had similar ^{3}H/^{14}C ratios, indicating similar rates of degradation. All of these results support a model for the turnover of the plasma membrane in which the proteins are removed in large structural units, implying then some sort of coordinated regulation of either the synthesis and/or the insertion into the membrane of a large number of different polypeptides.

REPLACEMENT OF POLYPEPTIDES

An indication of how the cell might accomplish the simultaneous replacement of this large number of membrane polypeptides was obtained by examining the turnover of the glycoproteins in the plasma membrane. Membrane glycoproteins were labelled by growing HTC cells in the presence of [^{14}C] fucose, while externally disposed plasma membrane proteins in the same cells were labelled with ^{125}I by lactoperoxidase-catalysed iodination. The plasma membrane contains a set of fucose-labelled glycoproteins which are externally disposed. However, in HTC cells, in contrast to other cell types, the fucose incorporated into membrane glycoproteins does not fractionate solely with the plasma membrane. Rather, as shown in Table 2, a significant amount of the fucose-labelled glycoprotein fractionates with an internal membrane system(s) of this cell.

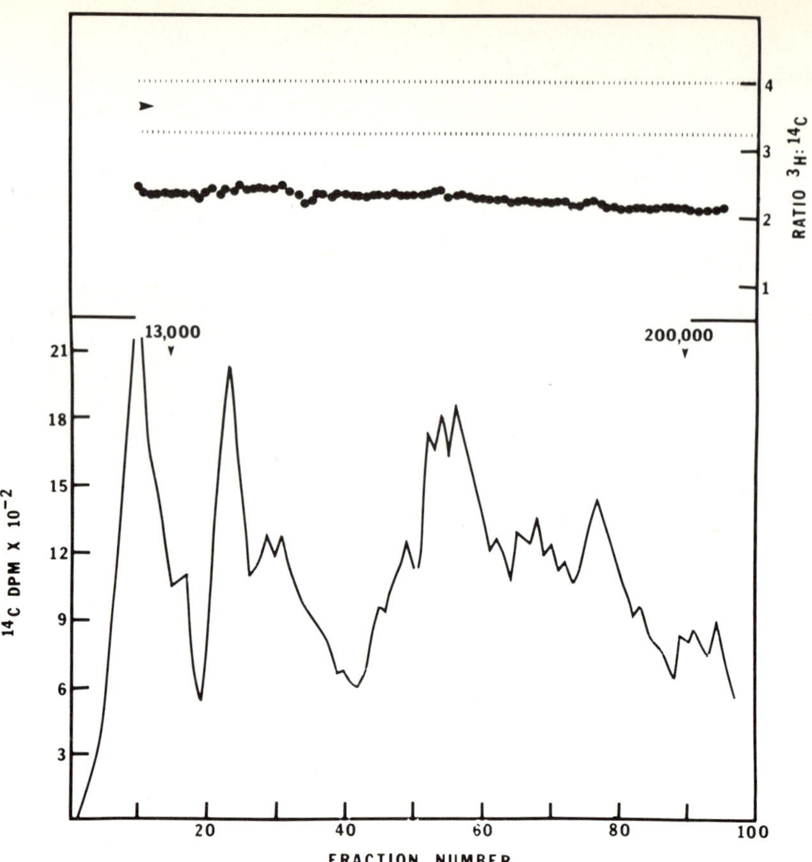

Figure 4 ^{3}H/^{14}C Ratios of the leucine-labelled polypeptides in the plasma membrane fraction of HTC cells. A culture of HTC cells was divided into two equal parts. Half of the cells were incubated in complete growth medium containing [^{3}H]leucine for 24 hours, while the other half were incubated for the same interval of time in medium containing [^{14}C]leucine. At this time, an aliquot of the ^{14}C-labelled cells and an aliquot of the ^{3}H-labelled cells were homogenized and frozen. A plasma membrane fraction then was prepared from a mixture of the homogenates (zero time control). The other ^{3}H-labelled cells were incubated in growth medium for 72 hours at a density of 1.2×10^6/ml. A homogenate of these cells then was prepared and frozen with the frozen homogenate of the ^{14}C-labelled cells. Closed circles represent ^{3}H/^{14}C ratios of the polypeptides in the membrane fraction obtained from the latter mixture of cells. The arrows and dotted lines represent the mean ^{3}H/^{14}C ratio and one standard deviation of the mean for the polypeptides in the plasma membrane prepared from the zero time-control cells; solid lines denote ^{14}C-labelled polypeptides in the membrane preparation separated in the 9% dodecyl sulphate–acrylamide disc gel system (see reference [5] for details; reproduced with permission)

GLYCOPROTEINS IN THE MEMBRANE

We then examined the glycoprotein composition of these different HTC cell fractions and compared it to the glycoprotein composition of the plasma membrane. As shown in the fluorogram of Figure 5, each cell fraction contains the

Table 2 Biochemical analysis of microsomal and plasma membrane fractions of HTC cells

	Cell fraction			
	Crude microsomes	*Light microsomes*	*Heavy microsomes*	*Plasma membrane*
[^{125}I]Acid insoluble radioactivity	0.4	0.6	0.4	15
[^{14}C]Fucose—acid insoluble radioactivity	3.0	6.0	1.5	2.7
[^{3}H]Uridine—acid insoluble radioactivity	—	1.9	2.7	0.7
5′-Nucleotidase	2.0	5.0	0.5	14
Glucose-6-phosphatase	3.1	5.4	6.4	<1
CMP-*N*-Acetylneuraminic acid: glycoprotein transferase	2.5	3.0	0.9	1.2

Details for the cell fractionation are given in references[4] and[5]. The crude microsomal fraction was separated on a step sucrose gradient into light and heavy subfractions. The numbers represent specific activities relative to the homogenate. The specific activities of the homogenate were 21 000 ^{125}I acid-insoluble CPM/mg, 30 000 [^{14}C]fucose acid-insoluble CPM/mg; 540 000 [^{3}H]uridine acid-insoluble CPM/mg, 6 units/mg of 5′-nucleotidase; 3.1 units/mg of glucose-6-phosphatase, and 656 CPM per h/mg of CMP-*N*-acetylneuraminic acid : glycoprotein transferase activity

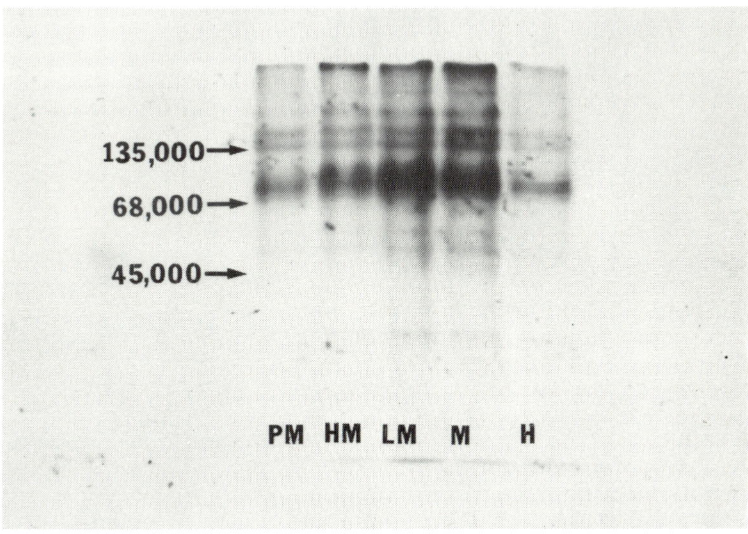

Figure 5 Dodecyl sulphate–polyacrylamide gel electrophoresis of the fucose-labelled glycoproteins in HTC cell fractions. Cells were grown in the presence of [^{3}H]fucose for 48 hours and the following fractions were prepared from an homogenate (H) of these cells, as described in references[4] and[5]: crude microsomes (M), light microsomes (LM), heavy microsomes (HM) and plasma membrane (PM). Aliquots of each fraction were dissociated by heating at 100 °C for 1 min in 1% dodecyl sulphate, 1% 2-mercaptoethanol. Polypeptides were separated on 9% polyacrylamide slab gels and visualized by fluorography

same set of fucose-containing glycoproteins as resolved by electrophoresis in the sodium dodecyl sulphate–polyacrylamide gels in about the same relative proportion to each other. Separation of these membrane glycoproteins in a more sensitive two-dimensional electrophoretic system also indicates that a similar and probably identical set of glycoproteins exists in both the plasma membrane fraction and an internal membrane fraction of HTC cells.

If the same fucose-containing glycoproteins are present in the plasma membrane and in an internal membrane system(s) of the cell, then the question arises as to whether the glycoproteins in the microsomal fraction of HTC cells are intrinsic to the membrane systems in this fraction or whether they are secretory proteins sequestered within the lumen of the microsomes. As originally shown by Kreibich and Sabatini[9,10] low concentrations, i.e. less than 0.05%, of deoxycholate will release rapidly labelled, presumably secretory, proteins

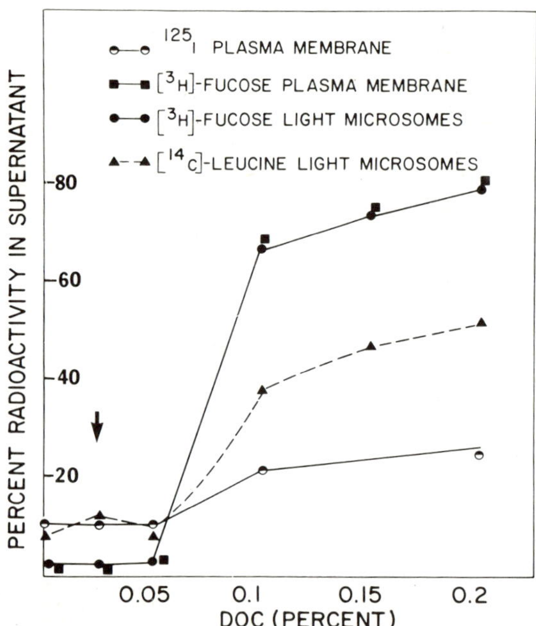

Figure 6 Release of labelled proteins from the light microsomal and the plasma membrane fractions of HTC cells by deoxycholate. One aliquot of cells was grown in the presence of [^{3}H]fucose, 1 μCi/ml for 72 hours. These cells were collected, washed, and resuspended in medium containing [^{14}C]leucine, 1 μCi/ml, for an additional 24 hours. A second aliquot of HTC cells was iodinated with ^{125}I. Light microsomal and plasma membrane fractions were prepared from both groups of cells. The membrane fractions were treated with the concentration of detergent indicated in the figure. After 30 min incubation at 4 °C, supernatant and pellet fractions were obtained by centrifugation for the determination of acid-insoluble radioactivity. The arrow indicates the lowest deoxycholate concentration at which radioactivity is released from a liver microsomal fraction prepared from mice that had been injected with [^{3}H]leucine 1 hour prior to sacrifice. Mice were used as a source of rapidly labelled secretory proteins rather than HTC cells because the latter cells do not secrete proteins into the medium

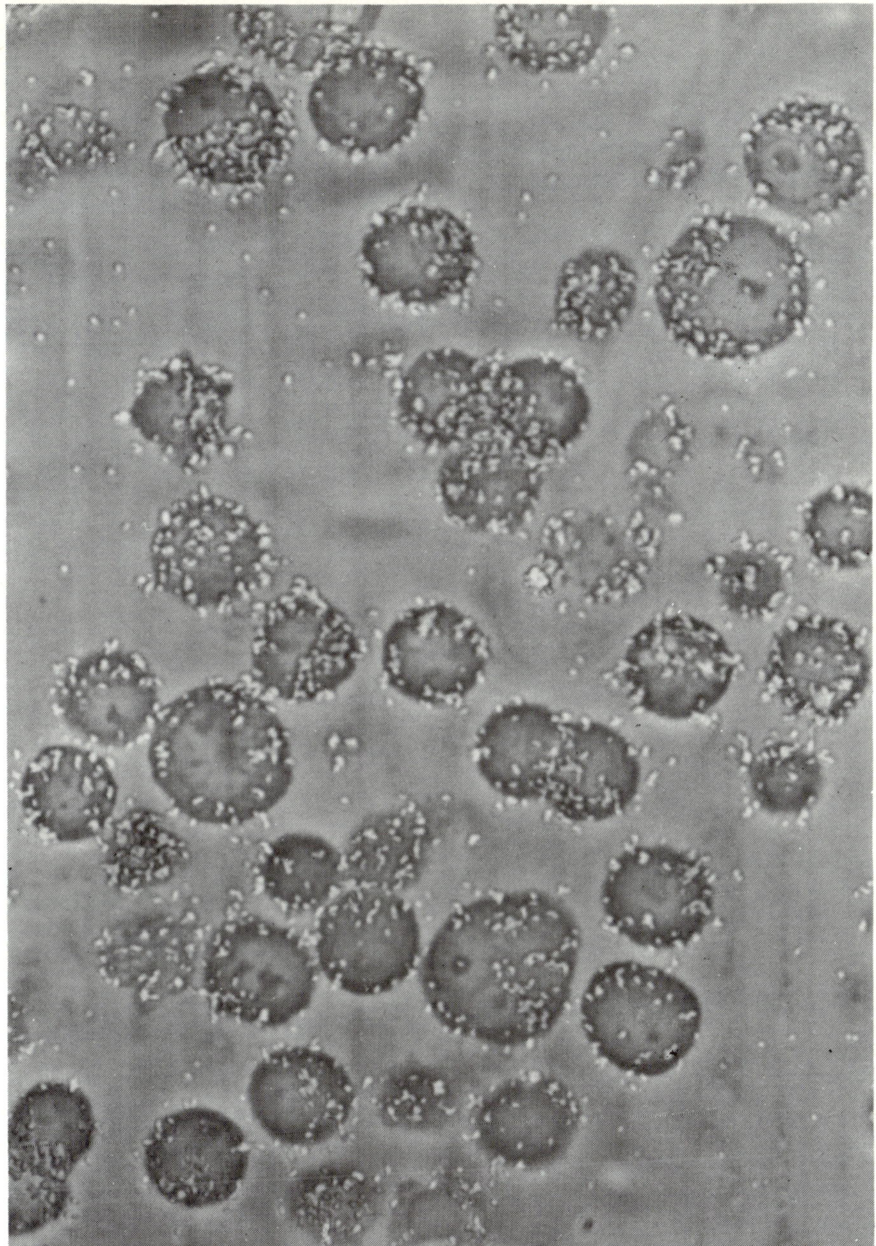

Figure 7a

Figure 7 Autoradiographs of thin sections of HTC cells that were (7a) grown in the presence of [^{3}H]fucose for 48 hours, or (7b) iodinated with ^{125}I, or (7c) grown in the presence of [^{3}H]fucose for 48 hours followed by 72 hours of growth in unlabelled medium. Figure 7b is reproduced from reference [4] with permission

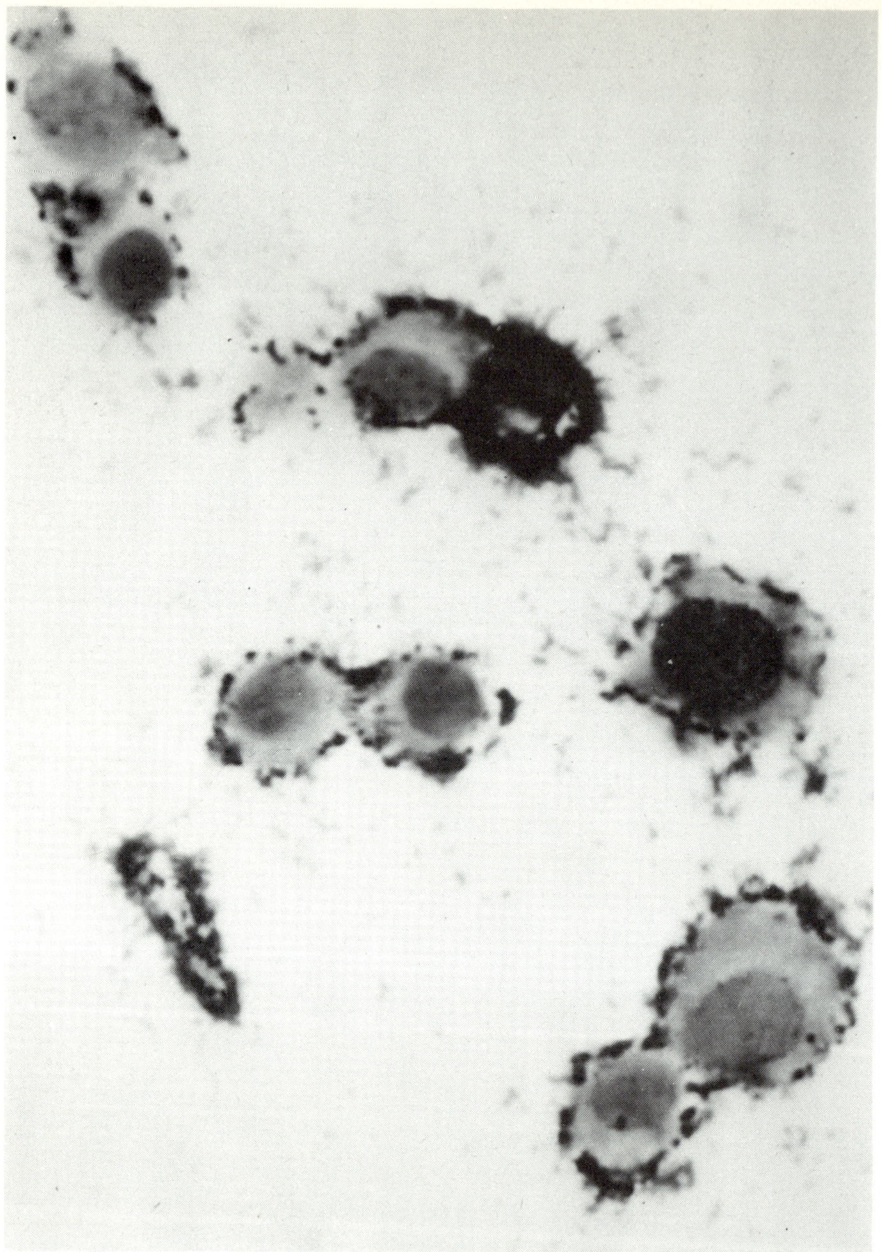

Figure 7b

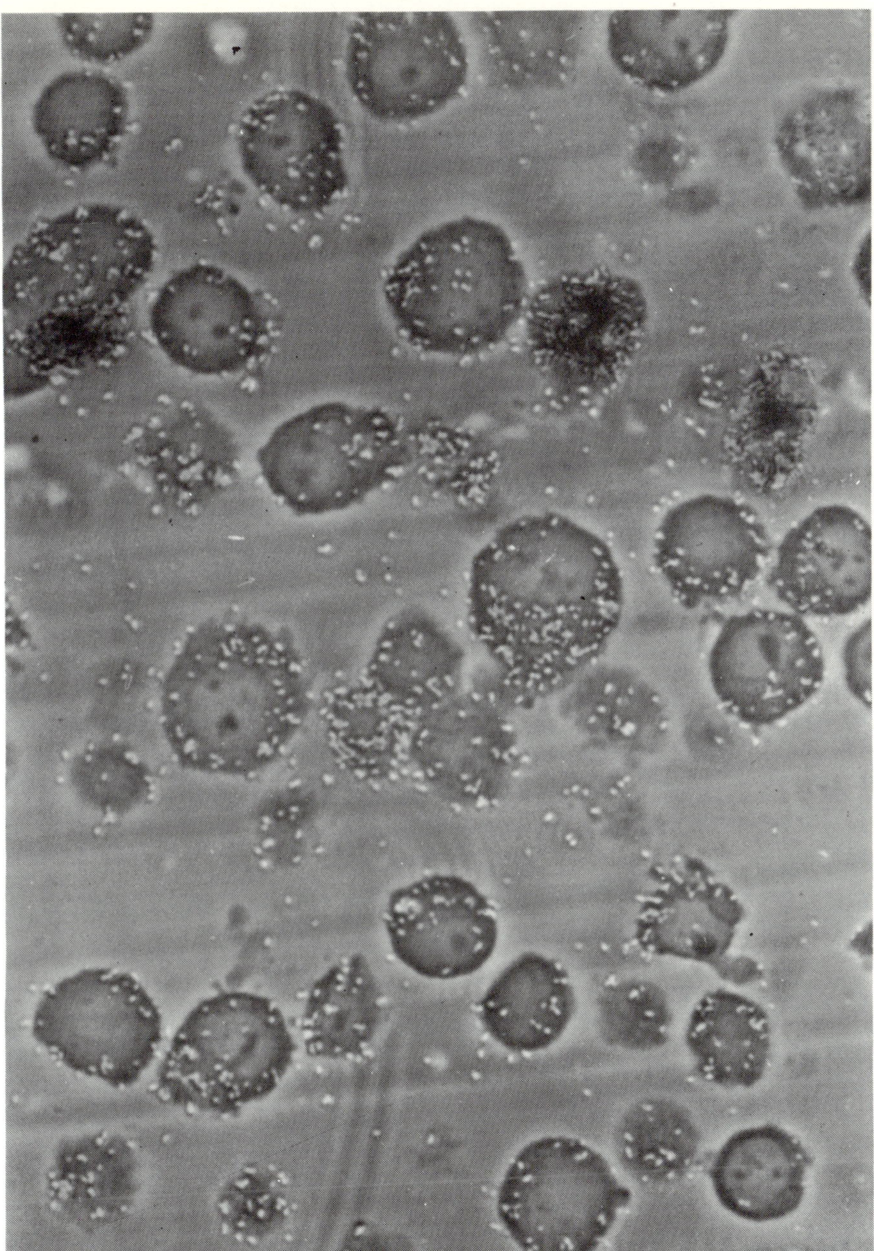

Figure 7c

from liver microsomes, while higher concentrations are required to 'solubilize' intrinsic membrane proteins.

The fucose-labelled glycoproteins of the plasma membrane fraction are released from the lipid bilayer at a concentration of deoxycholate higher than that required to release secretory proteins from either rat or mouse liver microsomes (Figure 6). Indeed, these glycoproteins appear to be intrinsic proteins of the plasma membrane of HTC cells[4,5], even though they are more soluble in deoxycholate than most of the other plasma membrane proteins which can be iodinated. The concentration-dependent release of the fucose-labelled glycoproteins from the microsomal fraction by deoxycholate is identical to that from the plasma membrane, suggesting by analogy that these glycoproteins are intrinsic also to the membrane system(s) present in the microsomal fraction.

Autoradiography of thin sections from HTC cells that were grown for 48 hours in the presence of [^{3}H]fucose also shows that there is an internal membrane compartment(s) containing fucose-labelled glycoproteins. That is, autoradiographic grains representing incorporated fucose are not confined to the cell periphery but are present over most of the cell section, with the exception of the nucleus (Figure 7a). The localization of the incorporated fucose should be contrasted with the localization of iodide incorporated via lactoperoxidase-catalysed iodination. In the latter case, the grains are confined almost exclusively to the periphery of the cell (Figure 7b).

The fucose-labelled glycoproteins of the internal membrane system are turning over at about twice the rate of those in the plasma membrane fraction (50 hours versus 100 hours). Furthermore, in contrast to the rather homogeneous rates of turnover of the iodinated polypeptides and the leucine-labelled polypeptides of the plasma membrane (Figures 3 and 4), the fucose-labelled constituents show some heterogeneity in turnover (Figure 8), even more so than those of the microsomal fraction (Figure 9). Thus, while the microsomal and plasma membrane fractions of HTC cells have many of the major fucose-labelled glycoproteins in common, those of the microsomes are turning over faster and with more homogeneity than those of the plasma membrane.

We have recently shown that normal rat and mouse liver as well as other tissue culture cells derived from mouse and rat hepatomas have a similar set of fucose-labelled glycoproteins as HTC cells. However, in rodent liver[11], and most likely in other tissue culture cells[12], the fucose-labelled glycoproteins are either secreted or confined primarily to the plasma membrane. That is, labelled fucose when administered to these cells, 'chases' to the plasma membrane or out of the cell. The fucose-labelled glycoproteins of the internal membrane system(s) of HTC cells are not secretory proteins and cannot be easily 'chased' to the plasma membrane. This is illustrated most directly in autoradiographs of thin sections from HTC cells that were grown in labelled fucose for 48 hours followed by a 72-hour period of growth in unlabelled fucose. During the latter period of growth the specific activity of the incorporated fucose decreased

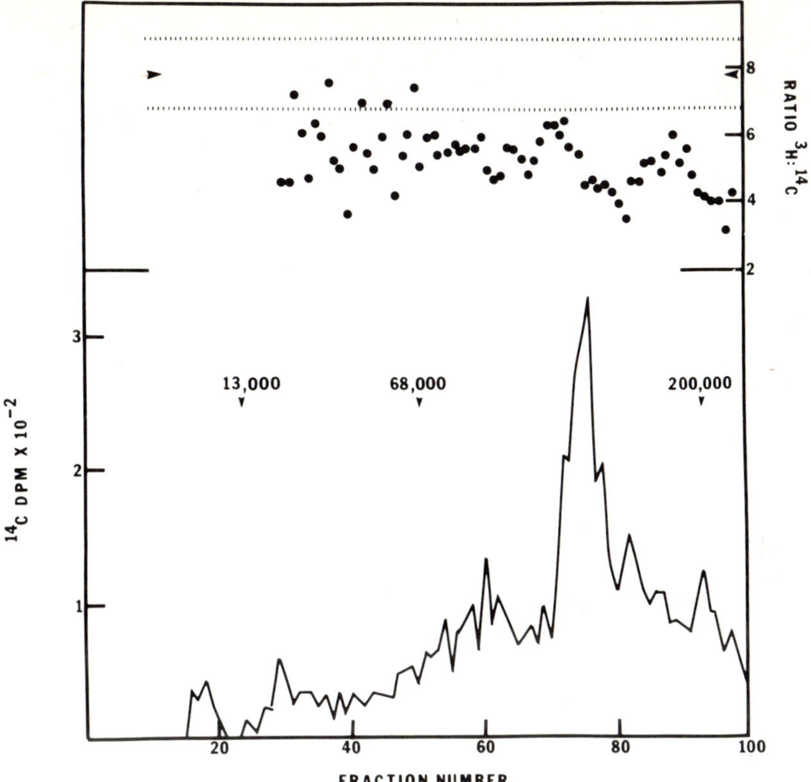

Figure 8 $^3H/^{14}C$ Ratios of fucose-labelled polypeptides in the plasma membrane fraction of HTC cells. A culture of HTC cells was divided into two parts. One part was grown in complete growth medium containing L-[6-^{3}H]fucose. The other cells were grown in complete growth medium containing L-[1-^{14}C]fucose. At the end of 24 hours at 37 °C, the cells were collected and washed twice with Earle's balanced salt solution. At this time, aliquots of the ^{14}C-labelled cells and the ^{3}H-labelled cells were homogenized and mixed together. This mixture, representing the zero time control, as well as an homogenate of the other ^{14}C-labelled cells, was immediately frozen. The other ^{3}H-labelled cells were suspended in complete growth medium at a concentration of 1.2 × 10^6/ml. These cells were incubated at 37 °C for 72 hours. At the end of the incubation, the cell density was still 1.2 × 10^6/ml. These ^{3}H-labelled cells were homogenized. The homogenate was mixed and frozen with the frozen homogenate of the ^{14}C-labelled cells. A plasma membrane fraction and a microsomal fraction were prepared from this mixture of cells and from the zero time control cells. These latter cells had $^3H/^{14}C$ isotope ratios of 7.8 for protein in each of the cell fractions analysed. The arrow enclosed by dotted lines denotes the mean ratio and one standard deviation of the mean for the polypeptides in the plasma membrane fraction of the zero time control cells. The polypeptides were dissociated in 1% dodecyl sulphate and 1% 2-mercaptoethanol, and separated on a 9% polyacrylamide disc gel containing 0.1% dodecyl sulphate. ●, $^3H/^{14}C$ ratios for the fucose-labelled polypeptides in the plasma membrane fraction when the ^{3}H-labelled cells were in culture for 72 hours. ——, [^{14}C]fucose radioactivity of the polypeptides in this fraction (see reference[5]; reproduced with permission)

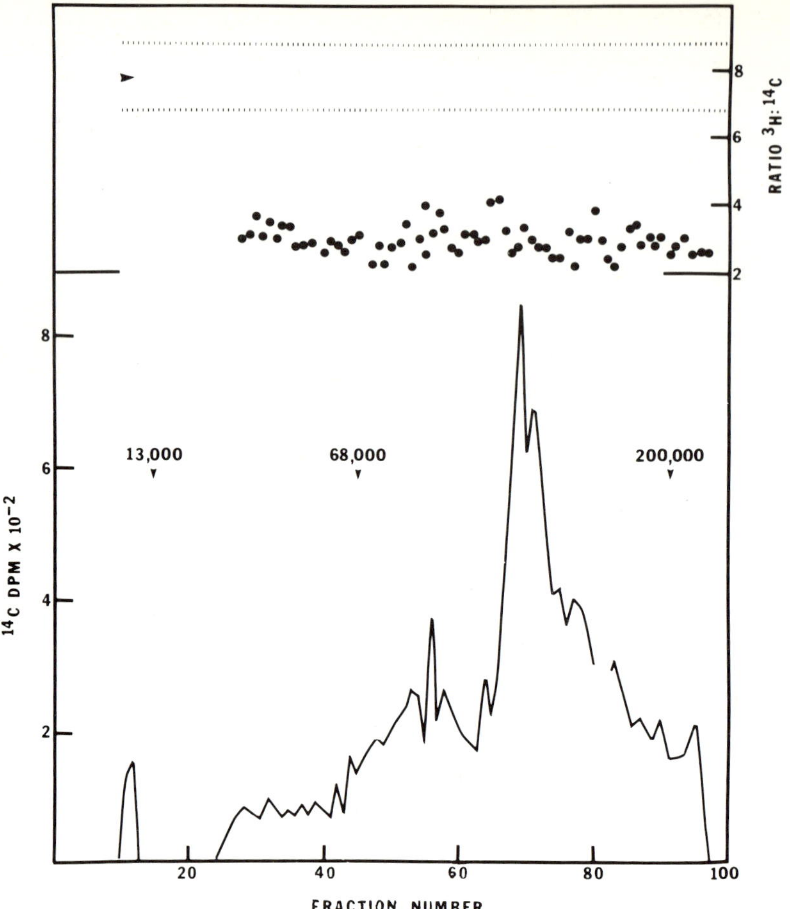

Figure 9 $^3H/^{14}C$ Ratios of fucose-labelled polypeptides in the microsomal fraction of HTC cells. Experimental details are the same as in Chapter 5, Figure 8, except that the polypeptides in the microsomal fraction of the cell were separated on the 9% acrylamide–dodecyl sulphate disc gel (see reference[5]; reproduced with permission)

about 10-fold. Yet, the distribution of grains is very similar to that in cells not grown in unlabelled medium (compare Figures 7a and 7c). These results indicate that the fucose-labelled glycoproteins of the internal membrane compartment(s) in HTC cells reside in this compartment for extended intervals of time.

To reconcile the differences in the kinetics of fucose labelling and the ultimate intracellular localization of incorporated fucose between HTC cells and other cells of liver origin, we propose that HTC cells are defective in one or more steps in the synthesis, assembly or turnover of plasma membrane polypeptides. That is, HTC cells, in contrast to normal hepatocytes and other cells in culture, accumulate formed membrane structure in an internal cell compartment. This structure(s), we propose, is the normal precursor to the plasma membrane and is part of the mechanism used by the cell to replace the relatively large area of membrane removed as a unit during normal turnover.

Supporting the existence of a formed plasma membrane precursor are the equivalent relative concentrations of each of the fucose-labelled glycoproteins in each of the cell membrane fractions, the solubility properties of these glycoproteins in deoxycholate — suggesting a similar organization of the glycoprotein in the different cell fractions with respect to the lipid bilayer, and the turnover properties of the fucose-labelled glycoproteins in the internal membrane compartment — they all turn over at similar rates although with a half-life somewhat faster than that of the same glycoproteins in the plasma membrane fraction (50–60 hours versus 90–100 hours).

The slow rate of turnover of the polypeptides forming the plasma membrane of HTC cells (half-life = 100 hours) compared to either hepatocytes or other cells in culture — for example, a half-life of about 24 hours for the plasma membrane proteins of both HeLa[13] and mouse L cells[14] — is consistent with a defect in the degradation of the plasma membrane in HTC cells leading to an accumulation of the membrane precursor inside the cell. In this context, it is interesting that HeLa cells have about the same generation time as HTC cells, 24–30 hours, a much faster rate of membrane turnover, 24 versus 100 hours, and little or no internal compartment of fucose-labelled glycoproteins[12].

A working hypothesis is proposed suggesting that during biogenesis proteins destined for the plasma membrane are assembled first into a precursor membrane system which in most cells becomes rapidly incorporated into the limiting cell membrane but which accumulates in HTC cells because of a defect in the degradation of organized units of plasma membrane interiorized via normal pinocytosis. It should be mentioned, in the light of our model, that Steinman *et al.*[15] have suggested recently that in cultured cells the rate of plasma membrane protein interiorization is much faster than the rate of membrane protein degradation. Their results imply that much of the interiorized membrane protein is recycled to the cell surface. If the proteins of the plasma membrane are first assembled in a precursor form, it would be important to determine whether the precursor vesicle is assembled completely *de novo* or whether new membrane proteins are added to an already existing membrane structure that is being recycled through the interior compartment of the cell. The latter mechanism, if it exists, could have important implications for the control of specific membrane protein biogenesis and the way a cell responds to a change in the environment.

Acknowledgements

This work was supported by grants from the US Public Health Service.

References

1. Oseroff, A. R., Robbins, P. W. and Burger, M. M. (1973). The cell surface membrane: biochemical aspects and biophysical probes. *Ann. Rev. Biochem.*, **42,** 647
2. Phillips, D. R. and Morrison, M. (1971). Exposed protein on the intact human erothrocyte. *Biochemistry*, **10,** 1766
3. Hubbard, A. L. and Cohn, Z. A. (1972). The enzymatic iodination of the red cell membrane. *J. Cell Biol.*, **55,** 390
4. Tweto, J., Friedman, E. and Doyle, D. (1976). Protein of the hepatoma tissue culture cell plasma membrane. *J. Supramolec. Struct.*, **4,** 141
5. Tweto, J. and Doyle, D. (1976). Turnover of the plasma membrane proteins of hepatoma tissue culture cells. *J. Biol. Chem.*, **251,** 872
6. Eagle, H., Piez, K. A., Fleischman, R. and Oyama, V. I. (1959). Protein turnover in mammalian cell cultures. *J. Biol. Chem.*, **234,** 592
7. Pollard, T. D. and Weighing, R. R. (1973). *Crit. Rev. Biochem.*, **2,** 1
8. Doyle, D. and Tweto, J. (1975). In: D. M. Prescott (ed.). *Methods in Cell Biology, Vol. 10*, pp. 235–260 (New York: Academic Press)
9. Kreibich, G., Debey, P. and Sabatini, D. D. (1973). Selective release of content from microsomal vesicles without membrane disassembly. I. Permeability changes induced by low detergent concentrations. *J. Cell Biol.*, **58,** 436
10. Kreibich, G. and Sabatini, D. D. (1974). Selective release of content from microsomal vesicles without membrane disassembly. II. Electrophoretic and immunological characterisation of microsomal subfabrications. *J. Cell Biol.*, **61,** 789
11. Bennett, G., Leblond, C. P. and Haddad, A. (1974). Migration of glycoprotein from the Golgi apparatus to the surface of various cell types as shown by radioautography after labeled fucose injection into rats. *J. Cell Biol.*, **60,** 258
12. Atkinson, P. H. (1975). Synthesis and assembly of HeLa cell plasma membrane glycoproteins and proteins. *J. Biol. Chem.*, **250,** 2123
13. Cook, J. S., Will, P. C., Proctor, W. R. and Brake, E. T. (1976). In: J. S. Cook (ed.). *Biogenesis and Turnover of Membrane Macromolecules* (New York: Raven Press)
14. Hubbard, A. L. and Cohn, Z. A. (1975). Externally disposed plasma membrane proteins. II. Metabolic fate of iodinated polypeptides of mouse L cells. *J. Cell Biol.*, **64,** 461
15. Steinman, R. M., Brodie, S. E. and Cohn, M. A. (1976). Membrane flow during pinocytosis. A stereologic analysis. *J. Cell Biol.*, **68,** 665

An Electron Microscopic, Histochemical and Cytochemical Study of Membrane Alterations in the Duckling Liver Caused by Vitamin E Deficiency and X-Irradiation

W. A. R. HUIJBERS, M. J. HARDONK and I. MOLENAAR

Vitamin E deficiency causes changes in cellular membranes. In epithelial cells of the human and duckling intestine and in duckling hepatocytes a decrease in contrast of intracellular membranes has been found[1–4]. Biochemical investigation of liver cellular membranes of vitamin E-deficient ducklings showed that this loss of contrast could be correlated to a decrease of polyunsaturated fatty acids[4, 5]. Histochemical and biochemical studies demonstrated changes of membrane proteins of liver parenchymal cells; in intracellular membranes as well as in the plasma membrane changes in activity of some membrane-bound enzymes could be demonstrated[6].

In view of these results the following hypothesis for the function of vitamin E has been proposed[7]. Vitamin E protects unsaturated fatty acids at those sites in the membrane where they can be damaged by free radicals from electron transfer chains[8–12]. The protecting effect of the tocopherol molecule is exerted by the redox function of its chromanol ring[13–15]. The composition of the different cellular membranes, mainly their content of cholesterol and unsaturated fatty acids, is probably of great importance with regard to their sensitivity to an excess of free radicals occurring in vitamin E deficiency.

In vitamin E deficiency mitochondrial and endoplasmic reticulum membranes mainly are involved. However, it might be possible to learn about the sensitivity of all cellular membranes to peroxidative processes by equally charging them with free radicals. This could be carried out by X-irradiation of the liver of normal and vitamin E-deficient ducklings. Furthermore, information about the differences in sensitivity can be elaborated by studying the action of free radicals in a gradient.

METHODS

Vitamin E deficiency in ducklings was induced as described previously[6]. The gradient in X-ray damage to liver cellular membranes had been produced by an irradiation with 1500 rad (29 kV, 25 mÅ) of a field of 2 cm diameter on the

Table 1 **Electron microscopic, histochemical and cytochemical alterations in duckling liver membranes in vitamin E deficiency and X-irradiation**

		Vitamin E-deficient liver	*X-irradiation of control liver*			*X-irradiation of vitamin E-deficient liver*		
			I	*II*	*III*	*I*	*II*	*III*
Plasma membrane								
Membrane contrast	sin.	–	↓↓	↓	–	↓↓	↓	–
	interc.	–	↓↓	↓	–	↓↓	↓	–
	b.c.	–	–	–	–	–	–	–
Membrane contrast after ruthenium red staining	sin.	–	–	–	–	–	–	–
	interc.	–	–	–	–	–	–	–
	b.c.	–	–	–	–	–	–	–
Glycoprotein staining	sin.	–	–	–	–	↓↓	↓	–
	interc.	–	–	–	–	↓↓	↓	–
	b.c.	–	–	–	–	↓↓	↓	–
Alkaline phosphatase		↓	↓↓	↓	–	↓↓	↓	–
ATP-ase		–	–	–	–	–	–	–
5′-Nucleotidase		↑↑50%*	↓	–	–	↑	–	–
Leucyl-β-naphthylamidase	sin.	–	↓	–	–	↓↓	↓	–
	b.c.	↑	–	–	–	↑	–	–
Lysosomes								
Number		↑	↑	–	–	↑↑	↑	–
Uptake of ruthenium red		↑	↑	↑	–	↓↓	↓	–
Acid phosphatase		↑	↑	–	–	a	a	a

Mitochondria							
Membrane contrast	↓60%†	↓	↓	↓	↓	↓	↓
Membrane contrast after ruthenium red staining	–	–	–	–	–	–	–
Configurational changes	present	present	present	present	present	present	present
Succinate dehydrogenase	–	↑	↑	↑	↑	↑	↑
Endoplasmic reticulum							
Membrane contrast	↓	↓	↓	↓	↓	↓	↓
Membrane contrast after ruthenium red staining	–	–	–	–	–	–	–
Glucose-6-phosphatase	↑45%*	↑	↑	↑	↑	↑	↑

The changes after X-irradiation of vitamin E-deficient liver are observed with respect to a non-irradiated vitamin E-deficient liver.
I = 0–1 mm penetration depth of the X-rays.
II = 1–2 mm penetration depth of the X-rays.
III = 2–3 mm penetration depth of the X-rays.
sin. = sinusoidal wall; interc. = intercellular wall; b.c. = bile canalicular wall.
– = unchanged; ↑ = increased; ↑↑ = strongly increased; ↓ = decreased; ↓↓ = strongly decreased.
a = histochemically increased and cytochemically decreased, both in a gradient.
* Hulstaert *et al.*[6]. † Molenaar *et al.*[4].

surface of the liver. The intensity grade of the X-irradiation causes a corresponding decrease in the amount of free radicals, resulting in a zone of detectable alterations of 3 mm under the liver capsule.

These changes have been studied by electron microscopic, histochemical and cytochemical techniques. Two hours after the irradiation procedure the livers of the animals were perfused with a fixative containing 1.5% glutaraldehyde, 0.075 M cacodylate buffer (pH 7.4) and 1% sucrose. Vibratome® sections (100 μm) were postfixed in 2% OsO_4 in 0.1 M cacodylate buffer. Ruthenium red staining of vibratome sections was performed according to Luft[16]. The tissue was embedded in epon. Ultrathin sections for electron microscopy were stained with uranyl acetate and lead citrate. Cryostat sections (10 μm) were used for enzyme histochemical reactions, as described by Pearse[17, 18]. The reaction products of four plasma membrane enzymes (alkaline phosphatase, ATP-ase, 5′-nucleotidase and leucyl-β-naphthylamidase), one lysosomal enzyme (acid phosphatase), one mitochondrial enzyme (succinate dehydrogenase) and one endoplasmic reticulum enzyme (glucose-6-phosphatase) were studied. After perfusion fixation 30 μm vibratome sections were incubated for enzyme cytochemical studies of the following enzymes: ATP-ase[19], 5′-nucleotidase[19], acid phosphatase[20] and alkaline phosphatase[21].

RESULTS

The results are summarized in Table 1. Osmium tetroxide (OsO_4) reacts with the double bonds of the unsaturated fatty acids of the cellular membranes. Therefore the membrane contrast after OsO_4 staining will provide information about the degree of unsaturation of the membrane. The staining with ruthenium red[22] not only indicates the presence of phospholipids in membranes, but also provides information about alterations in the glycoprotein fraction of the

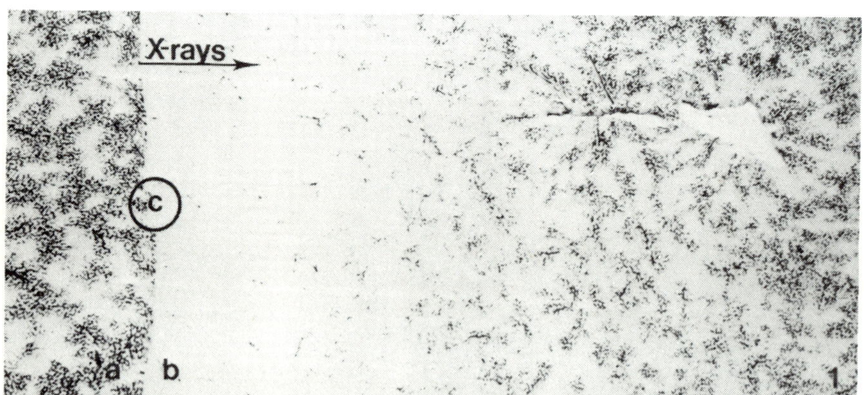

Figure 1 Alkaline phosphatase in control duckling liver: a = non-irradiated; b = after X-irradiation. In the irradiated liver the decrease in enzyme activity shows a gradient in the direction of the X-rays. c = liver capsule (× 20)

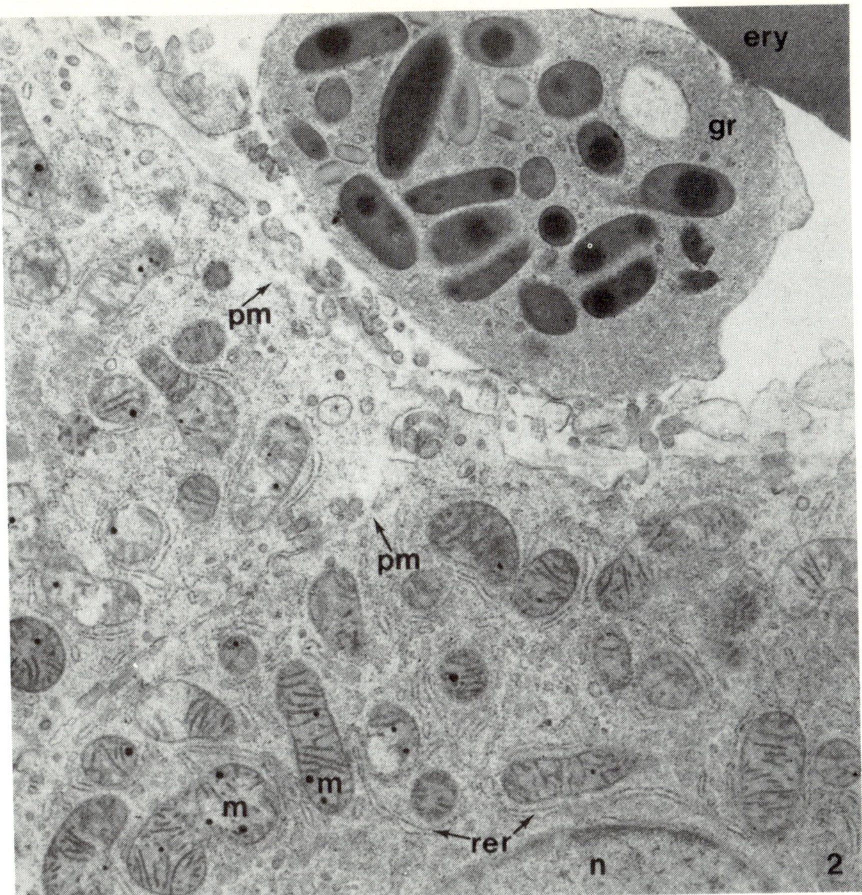

Figure 2 Liver parenchymal cells of control duckling after irradiation, just beneath the liver capsule. In several places the plasma membrane (pm) cannot be observed in positive contrast (compare with Figure 3). gr = granulocyte; ery = erythrocyte; m = mitochondrion; rer = rough endoplasmic reticulum; n = nucleus (× 15 000)

membrane[22–24]. Alterations in enzyme activities can be regarded as indicators for damage of the protein fraction of the membranes.

In vitamin E deficiency striking histochemical differences are observed in some membrane-bound enzymes: of the plasma membrane-bound enzymes alkaline phosphatase has decreased in activity, while the activity of 5′-nucleotidase has increased; the lysosomal enzyme acid phosphatase shows an increase in activity, and so does glucose-6-phosphatase, localized in the endoplasmic reticulum. In electron micrographs an increase in the number of lysosomes around the bile capillaries in the hepatocytes was found. These lysosomes contain electron-dense material after staining with ruthenium red. Contrary to the plasma membrane the membrane contrast of mitochondria

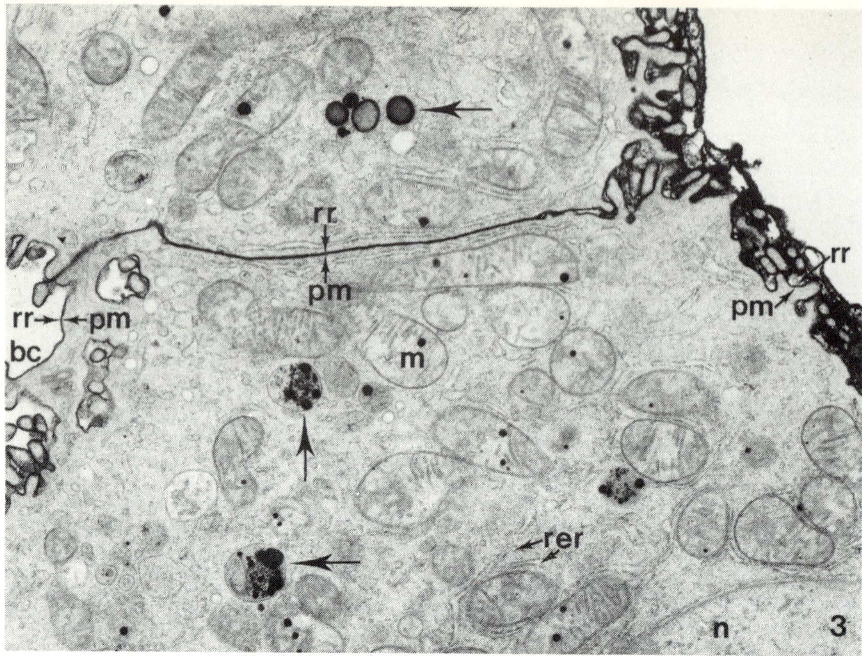

Figure 3 Liver parenchymal cells of control duckling after irradiation. After staining with ruthenium red (rr) the plasma membrane (pm) is clearly observable (compare Figure 2). A number of lysosomes (→) contain electron-dense material. m = mitochondrion; rer = rough endoplasmic reticulum; bc = bile canaliculus; n = nucleus (× 15 000)

and endoplasmic reticulum has decreased. However, no changes were visible after staining with ruthenium red. In vitamin E deficiency some mitochondria had abnormal shapes, i.e. they were larger and showed ramifications.

After irradiation of the liver of the control duckling the activity of the plasma membrane-bound enzymes decreased (Figure 1). Enzymes bound to the lysosomes, mitochondria and endoplasmic reticulum, however, showed an increase in activity. A gradient can be seen only in the changes that occurred in the plasma membrane and in the lysosomes. In the plasma membrane and in the membranes of mitochondria and endoplasmic reticulum the positive contrast, as normally obtained after OsO_4 fixation, has decreased (Figure 2), but it is present after staining with ruthenium red. The number of lysosomes has increased and they appeared to contain ruthenium red (Figure 3). The mitochondria show swelling (Figure 4) and rupture, changes in shape and longitudinal cristae (Figure 5).

After irradiation of the liver of the deficient duckling the activity of the enzymes that are situated in the sinusoidal plasma membrane has decreased (alkaline phosphatase and leucyl-β-naphthylamidase). On the other hand, the activity of the enzymes that are localized on the plasma membrane along the bile capillaries has increased (5′-nucleotidase (Figure 6) and leucyl-β-

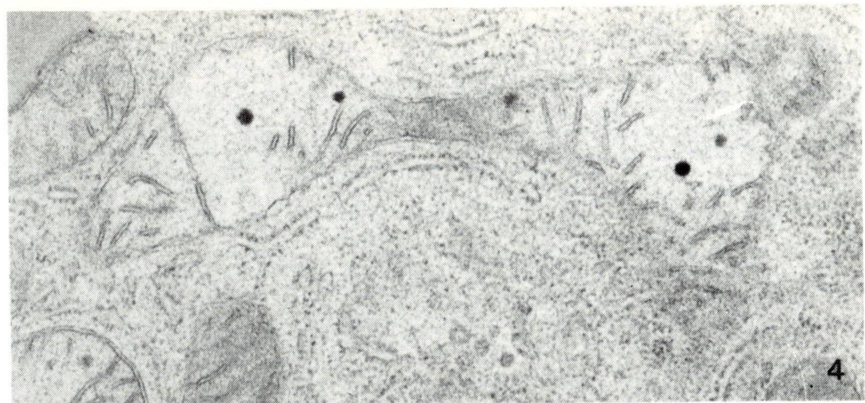

Figure 4 Swollen mitochondrion in hepatocyte of control duckling after irradiation (× 35 000)

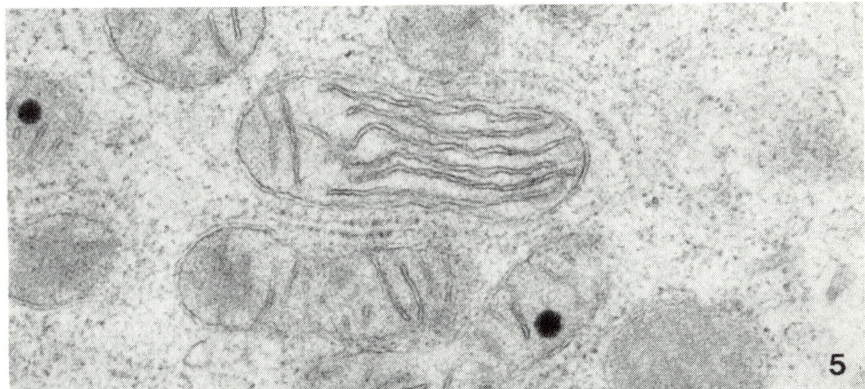

Figure 5 Mitochondrion with longitudinal cristae in hepatocyte of control duckling after irradiation (× 35 000)

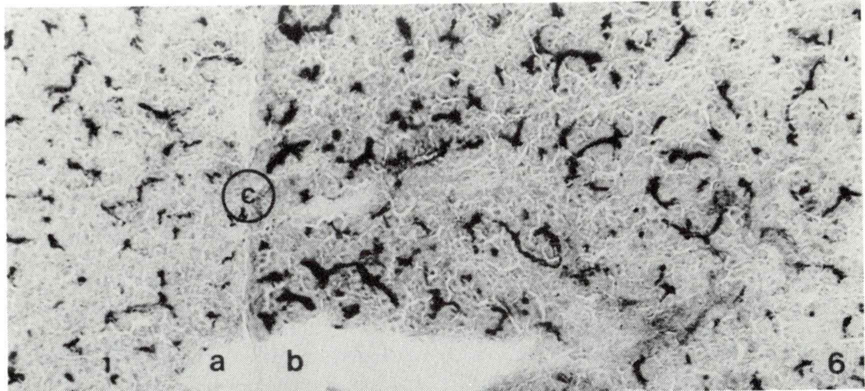

Figure 6 5′-Nucleotidase in vitamin E-deficient duckling: a = non-irradiated; b = after X-irradiation. After irradiation the enzyme activity in the bile capillaries has increased. c = liver capsule (× 150)

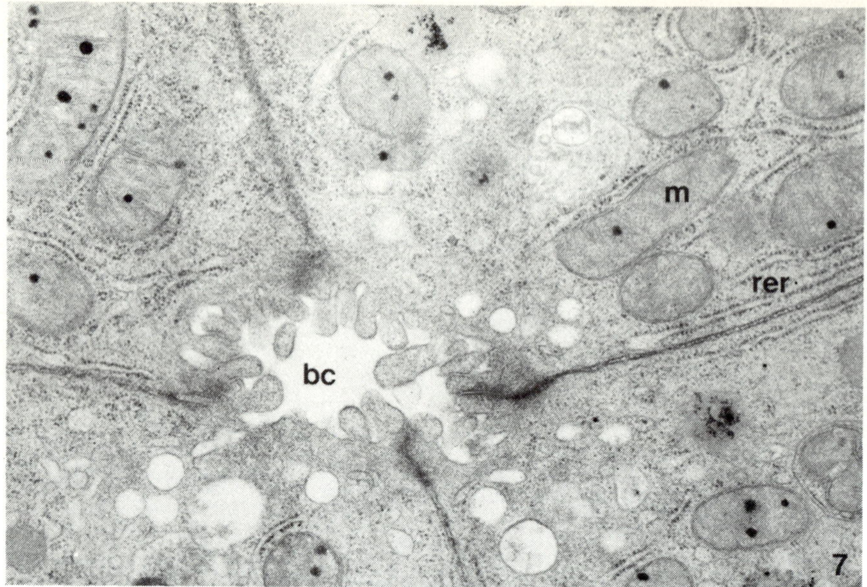

Figure 7 Bile capillary (bc) in liver of vitamin E-deficient duckling after irradiation, just beneath the liver capsule. Staining with ruthenium red; there is no electron-dense material observable along the extracellular side of the membrane of the bile capillary. After staining with ruthenium red the membranes of mitochondria (m) and endoplasmic reticulum (rer) show a normal positive contrast. Compare Figure 3 (× 20 000)

naphthylamidase). ATP-ase showed an intercellular activity. The enzymes of lysosomes, mitochondria and endoplasmic reticulum showed an increase in activity as has been found in the control animals. A gradient could be found only in the changes observed in the lysosomes and in the plasma membrane. The ultrastructural changes that occurred after irradiation of the deficient animals are for the greater part identical to those that occur after irradiation of the control animal, particularly as far as the mitochondria and the endoplasmic reticulum are concerned. In the case of the plasma membrane and the lysosomes, however, there is a striking difference: after staining with ruthenium red no electron-dense material adhering to the plasma membrane was observed, especially in the area that has been exposed to a large dose of irradiation (Figure 7). In such areas the lysosomes do not stain with ruthenium red (Figure 7), nor could acid phosphatase be demonstrated enzymecytochemically.

CONCLUSIONS

From the point of view of their composition cellular membranes can be divided into two groups. The plasma membrane and the lysosomal membrane with a protein/lipid ratio of about 1[25], a cholesterol/phospholipid ratio of 0.5–1.0[26–28], and characterized by presence of carbohydrates[26], low content of vitamin E[29,30] and absence of electron transfer chains, form the first group.

The endoplasmic reticulum membrane and mitochondrial inner membrane with a protein/lipid ratio of 2–3[25], a cholesterol/phospholipid ratio of 0.015–0.16[31–34], and characterized by absence of carbohydrates[35], high content of vitamin E[36,37] and presence of electron transfer chains, form the second group. The mitochondrial outer membrane occupies a position in between these two groups: a protein/lipid ratio of about 1[25] and a cholesterol content two to three times that of the inner membrane[32].

How does this division contribute to the insight into the cause of alterations in cellular membranes due to peroxidative processes originated by vitamin E deficiency and/or X-irradiation?

Under normal conditions free radicals are generated in endoplasmic reticulum membranes and mitochondrial inner membranes, in which vitamin E is present to inactivate the radicals. In vitamin E deficiency the mitochondrial membrane, particularly the outer membrane and the endoplasmic reticulum membrane are much more sensitive to free radical damage than the lysosomal membrane and the plasma membrane in which hardly any damage can be observed. The last-mentioned membrane systems show a low sensitivity to vitamin E deficiency because of the absence of the source of free radicals (the electron transfer chains). Furthermore, peroxides produced in endoplasmic reticulum and in mitochondria, escaping into the cytoplasm, can be captured by the cytoplasmic glutathione peroxidase[38,39]. Only with simultaneous deficiency of vitamin E and selenium (an essential constituent of glutathione peroxidase[40]) cytoplasmic peroxides can no longer be inactivated. In that situation the plasma membrane and the lysosomes are damaged relatively soon, as was also observed by Machado *et al.*[41]

After X-irradiation free radicals and peroxides originate in every area of the cell. In this case plasma membranes and lysosomal membranes are damaged most seriously. These membranes contain much lipid and cholesterol, being points of attack of free radicals. This results in the production of a considerable amount of peroxides. Besides, these membranes contain hardly any or no vitamin E. In the mitochondrial outer membrane lipids and cholesterol are points of attack of the free radicals, whereas the tocopherol molecule is missing as a protector. Membranes from the second group contain far less lipids and cholesterol, and moreover vitamin E protects these membranes against free radicals.

The combination of X-irradiation and vitamin E deficiency resulted in the same sequences of sensitivity as with X-irradiation only. The changes in the plasma membrane and in the lysosomal membrane are more pronounced than after X-irradiation of a control animal.

SUMMARY

The morphological, histochemical and cytochemical effects of free radical damage, caused by vitamin E deficiency and X-irradiation, on the ultrastructure and enzyme activities of several cellular membranes have been described, particularly involving the plasma membrane and the membranes of lysosomes, mitochondria and endoplasmic reticulum. The plasma membrane and the lysosomal membranes are highly sensitive to X-irradiation and show a low sensitivity to vitamin E deficiency. By contrast membranes of mitochondria and endoplasmic reticulum are highly sensitive to vitamin E deficiency and show a low sensitivity to X-irradiation.

References

1. Molenaar, I., Hommes, F. A., Braams, W. G. and Polman, H. A. (1968). Effect of vitamin E on membranes of the intestinal cell. *Proc. Natl. Acad. Sci. USA*, **61,** 982
2. Molenaar, I., Vos, J., Jager, F. C. and Hommes, F. A. (1970). The influence of vitamin E deficiency on biological membranes. An ultrastructural study on the intestinal epithelial cells of ducklings. *Nutr. Metab.*, **12,** 358
3. Molenaar, I., Vos, J., Jager, F. C. and Hommes, F. A. (1970). The effect of vitamin E deficiency on the ultrastructure of intestinal epithelial cells and their membranes in particular. In: N. Shimazono and Y. Takagi (eds.). *International Symposium on Vitamin E: Recent Advances in Physiology and Clinical Use*, pp. 76–101. (Tokyo: Kyoritsu Shuppan)
4. Molenaar, I., Vos, J. and Hommes, F. A. (1972). Effect of vitamin E deficiency on cellular membranes. *Vitam. Horm.*, **30,** 45
5. Vos, J., Molenaar, I., Searle-van Leeuwen, M. and Hommes, F. A. (1972). Mitochondrial and microsomal membranes from livers of vitamin E-deficient ducklings. *Ann. N.Y. Acad. Sci.*, **203,** 74
6. Hulstaert, C. E., Gijzel, W. P., Hardonk, M. J., Kroon, A. M. and Molenaar, I. (1975). Cellular membranes and membrane-bound enzymes in vitamin E deficiency. A histochemical, cytochemical, biochemical, and morphologic study of the liver of the Pekin duckling. *Lab. Invest.*, **33,** 176
7. Molenaar, I., Hulstaert, C. E. and Hardonk, M. J. (1977). Role in function and ultrastructure of cellular membranes. In: L. J. Machlin (ed.). *Vitamin E*. (New York: Marcel Dekker). (In press)
8. Hemmerich, P., Ehrenberg, A., Walker, W. H., Eriksson, L. E. G., Salach, J., Bader, P. and Singer, T. P. (1969). On the structure of succinic dehydrogenase flavocoenzyme. *FEBS Lett.*, **3,** 37
9. Fong, K., McCay, P. B., Poyer, J. L., Keele, B. B. and Mistra, H. (1973). Evidence that peroxidation of lysosomal membranes is initiated by hydroxyl free radicals produced during flavin enzyme activity. *J. Biol. Chem.*, **248,** 7792
10. Borg, D. C. (1972). Applications of electron spin resonance in biology. In: W. A. Pryor (ed.). *Free Radicals in Biology, Vol. I.*, pp. 69–147. (New York: Academic Press)
11. Mead, J. F. (1976). Free radical mechanisms of lipid damage and consequences for cellular membranes. In: W. A. Pryor (ed.). *Free Radicals in Biology, Vol. I.*, pp. 51–68. (New York: Academic Press)
12. Pryor, W. A. (1976). The role of free radical reactions in biological systems. In: W. A. Pryor (ed.). *Free Radicals in Biology, Vol. I.*, pp. 1–49. (New York: Academic Press)
13. Lucy, J. A. (1972). Functional and structural aspects of biological membranes: a suggested structural role for vitamin E in the control of membrane permeability and stability. *Ann. N.Y. Acad. Sci.*, **203,** 4
14. McCay, P. B., Poyer, J. L., Pfeifer, P. M., May, H. E. and Gilliam, J. M. (1971). A function for α-tocopherol: stabilization of the microsomal membrane from radical attack during TPNH-dependent oxidations. *Lipids*, **6,** 297

15. McCay, P. B., Pfeifer, P. M. and Stipe, W. H. (1972). Vitamin E protection of membrane lipids during electron transport functions. *Ann. N.Y. Acad. Sci.*, **203,** 62
16. Luft, J. H. (1971). Ruthenium red and violet. I: Chemistry, purification, methods of use for electron microscopy and mechanism of action. *Anat. Rec.*, **171,** 374
17. Pearse, A. G. E. (1968). *Histochemistry, Theoretical and Applied, Vol. I.* (Edinburgh, London: Churchill Livingstone)
18. Pearse, A. G. E. (1972). *Histochemistry, Theoretical and Applied, Vol. II.* (Edinburgh, London: Churchill Livingstone)
19. Wachstein, M. and Meisel, E. (1957). Histochemistry of hepatic phosphatases at a physiologic pH. With special reference to the demonstration of bile canaliculi. *Am. J. Clin. Pathol.*, **27,** 13
20. Barka, T. and Anderson, P. J. (1962). Histochemical methods for acid phosphatase using hexazonium pararosanilin as coupler. *J. Histochem. Cytochem.*, **10,** 741
21. Hugon, J. and Borgers, M. (1966). A direct lead method for the electron microscopic visualization of alkaline phosphatase activity. *J. Histochem. Cytochem.*, **14,** 429
22. Luft, J. H. (1971). Ruthenium red and violet, II: Fine structural localization in animal tissues. *Anat. Rec.*, **171,** 369
23. Blanquet, P. R. (1976). Ultrahistochemical study on the ruthenium red surface staining. I: Processes which give rise to electron-dense marker. *Histochemistry*, **47,** 63
24. Oda, M., Price, V. M., Fisher, M. M. and Philips, M. J. (1974). Ultrastructure of bile canaliculi, with special reference to the surface coat and the pericanalicular web. *Lab. Invest.*, **31,** 314
25. Guidotti, G. (1972). Membrane proteins. *Ann. Rev. Biochem.*, **41,** 731
26. Henning, R., Kaulen, H. D. and Stoffel, W. (1970). Isolation and chemical composition of the lysosomal and the plasma membrane of the rat liver cell. *Hoppe Seyler's Z. Physiol. Chem.*, **351,** 1191
27. Van Hoeven, R. P. and Emmelot, P. (1972). Studies on plasma membranes. XVIII: Lipid class composition of plasma membranes isolated from rat and mouse liver and hepatomas. *J. Membrane Biol.*, **9,** 105
28. Thinès-Sempoux, D. (1975). A comparison between the lysosomal and the plasma membrane. In: J. T. Dingle and H. B. Fell (eds.). *Lysosomes in Biology and Pathology, Vol. 3.* (Amsterdam: North-Holland)
29. Kayden, H. J. and Bjornson, L. (1972). The dynamics of vitamin E transport in the human erythrocyte. *Ann. N.Y. Acad. Sci.*, **203,** 127
30. Chow, C. K. (1975). Distribution of tocopherols in human plasma and red blood cells. *Am. J. Clin. Nutr.*, **28,** 756
31. Graham, J. M. and Green, C. (1970). The properties of mitochondria enriched *in vitro* with cholesterol. *Eur. J. Biochem.*, **12,** 58
32. Parsons, D. F. and Yano, Y. (1967). The cholesterol content of the outer and inner membranes of guinea-pig liver mitochondria. *Biochim. Biophys. Acta*, **135,** 362
33. Franke, W. W., Deumling, B., Zentgraf, H., Falk, H. and Rae, P. M. M. (1973). Nuclear membranes from mammalian liver. IV: Characterization of membrane-attached DNA. *Exp. Cell. Res.*, **81,** 365
34. Feo, F., Canuto, R. A., Bertone, G., Garcea, R. and Pani P. (1973). Cholesterol and phospholipid composition of mitochondria and microsomes isolated from Morris hepatoma 5123 and rat liver. *FEBS Lett.*, **33,** 229
35. Renkonen, O., Gahmberg, C. G., Simons, K. and Kääriänen, L. (1970). Enrichment of gangliosides in plasma membranes of hamster kidney fibroblasts. *Acta Chem. Scand.*, **24,** 733
36. Wiss, O., Bunnel, R. H. and Gloor, U. (1962). Absorption and distribution of vitamin E in the tissues. *Vitam. Horm.*, **20,** 441
37. Oliviera, M. M., Weglicki, W. B., Nason, A. and Nair, P. P. (1969). Distribution of α-tocopherol in beef heart mitochondria. *Biochim. Biophys. Acta*, **180,** 98

38. Hoekstra, W. G. (1975). Biochemical function of selenium and its relation to vitamin E. *Fed. Proc.*, **34,** 2083

39. Diplock, A. T. (1976). Metabolic aspects of selenium action and toxicity. *CRC Crit. Rev. Toxicol.*, **6,** 271

40. Rotruck, J. T., Pope, A. L., Ganther, H. E., Swanson, A. B., Hafeman, D. G. and Hoekstra, W. G. (1973). Selenium: biochemical role as a component of glutathione peroxidase. *Science*, **179,** 588

41. Machado, E. A., Porta, E. A., Hartroft, W. S. and Hamilton, F. (1971). Studies on dietary hepatic necrosis. II: Ultrastructural and enzymatic alterations of the hepatocytic plasma membrane. *Lab. Invest.*, **24,** 13

8
Enzymes of the Hepatocyte Plasma Membrane

M. H. WISHER and W. H. EVANS

The metabolic complexity of mammalian cells, especially those organized into tissues and organs is accompanied by the differentiation of the plasma membrane into morphologically identifiable areas. These domains are specialized for absorption, secretion, intercellular communication and attachment to a substratum or neighbouring cells. At the biochemical level this functional differentiation is a corollary of an uneven regional distribution in the plasma membrane of enzymes, receptors and structural components. Knowledge of the organization and biosynthesis of topographically specialized plasma membrane domains constitutes a major goal in explaining the biochemical basis of intracellular metabolic compartmentation, cellular behaviour and cell–

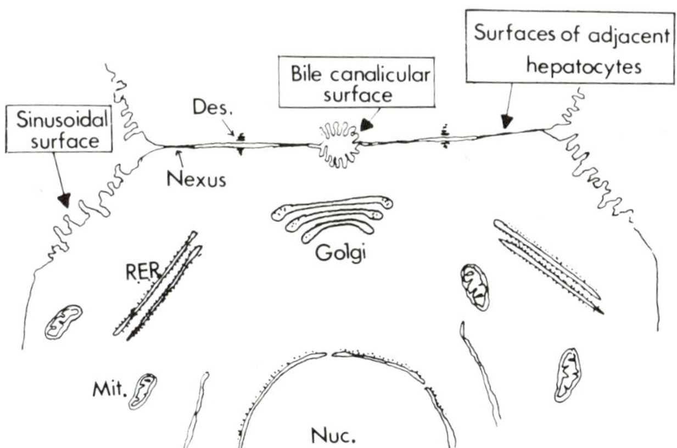

Figure 1 Schematic diagram of the hepatocyte. Nuc, nucleus; Mit, mitochondria; RER, rough endoplasmic reticulum; Des, desmosome

cell interactions. The present account describes approaches towards explaining the biochemical basis of the differentiation of the hepatocytic plasma membrane into its three major functional areas, namely the blood sinusoidal, bile canalicular and contiguous (lateral) areas (shown schematically in Figure 1).

HEPATOCYTE PLASMA MEMBRANE FUNCTIONS

The three major surfaces may be delineated on the basis of their participation in the wide range of hepatic functions which are summarized in Table 1. The sinusoidal plasma membrane region constitutes the major metabolic interface

Table 1 Physiological functions of the hepatocyte plasma membrane regions

Region	*Per cent of total surface area*	*Physiological function*
Sinusoidal	40–50 Microvillar	(1) Absorption or uptake from serum of amino acids, sugars, bile acids, desialylated glycoproteins and drugs. Gaseous exchange of O_2 and CO_2 (2) Involvement in secretion into serum of proteins, glycoproteins and lipoproteins (e.g. albumin, transferrin, α_1-glycoprotein, very low density lipoproteins) (3) Circulating hormones (e.g. glucagon, insulin, adrenaline) react with hormone receptors thus controlling hepatic metabolism
Contiguous	30–40 'Flat'	(1) Tight junctions maintain blood–bile barrier (2) Desmosomes function in intercellular adhesion and maintenance of cell shape (3) Gap junctions involved in intercellular communication and metabolic cooperation (4) Absorption of blood components?
Canalicular	10 Microvillar	(1) Bile formation (secretion of conjugated bile acids, bile pigments, cholesterol, phospholipids and protein) (2) Excretion of drugs and other compounds absorbed at the sinusoidal surface and modified or detoxified during transit across the hepatocyte

of the hepatocyte with the space of Disse to which blood is brought by the hepatic artery and portal vein. A wide range of small and large molecules are absorbed from and secreted into the blood at this microvillar surface which accounts for approximately half the hepatocyte's surface area. Secretory vesicles derived from the Golgi apparatus dictyosomes appear to move predominantly towards and fuse with the blood sinusoidal plasma membrane region. The bile canalicular plasma membrane region is also microvillar and accounts for about 10% of the hepatocyte's surface area. It is involved primarily in the transfer into the canaliculi of bile, that may also contain foreign compounds previously absorbed from the blood sinusoidal plasma membrane and metabolically modified or detoxified in the endoplasmic reticulum membranes during transit across the hepatocyte. At the contiguous or lateral surface region, separating the blood and bile cavities, the main morphological features of the plasma membrane are the intercellular junctions, of which the gap, tight and desmosomes are the major categories. Tight junctions and desmosomes are thought to fulfil a mainly mechanical function related to maintaining blood–bile separation and cell shape respectively, whereas the gap junctions are increasingly recognized as symmetrical structures providing direct channels for intercellular communication between hepatocytes that may underlie the metabolic cooperation of cells so necessary in tissues.

TOPOGRAPHY OF HEPATOCYTE PLASMA MEMBRANE ENZYMES

Although a wide range of plasma membrane enzymes are known and serve as markers in subcellular fraction studies[1,2], two major questions arise related to their topography on the cell surface. The first question is related to the disposition of these enzymes with respect to the transverse symmetry of the membrane. This is an important consideration since a plasma membrane enzyme's active site may be positioned on the outside (environmental) or inside (cytoplasmic) of the membrane and the protein itself may span the width of the membrane. Enzymes whose active sites are located on the outside are known as ecto-enzymes[3] (Table 2). Many plasma membrane marker enzymes are now categorized as ecto-enzymes on the basis of the following criteria: substrates are hydrolysed by intact cells[3,4], the enzyme activity is inhibited by antisera raised against plasma membranes[5,6], and enzymes are iodinated after lactoperoxidase-catalysed iodination of whole cells[7]. Many plasma membrane enzymes are also glycosylated (Table 2) and since it is hypothesized that most glycoproteins span the lipid bilayer[18], then it is probable that these plasma membrane enzymes also span the membrane. In contrast adenylate cyclase is a plasma membrane enzyme located at the cytoplasmic face (Table 2). However, this enzyme may act in concert with hormone receptors (e.g. glucagon, adrenaline, insulin receptors) which are located on the outside of the plasma membrane.

The second major topographical question is related to the regional distribution of enzymes in the lateral plane of the plasma membrane. In metabolically specialized cells, plasma membrane enzymes are localized at different concentrations at the different functional surfaces. For instance, histochemical studies have shown that there is an uneven distribution of enzymes at the sinusoidal and canalicular surfaces of the hepatocyte (Table 2). Because of this heterogeneous distribution of plasma membrane enzymes the isolation of a plasma membrane fraction showing high specific activities of marker enzymes, relative to the homogenate, does not guarantee that a representative fraction of

Table 2 Properties and regional distribution of some liver plasma membrane enzymes

Plasma membrane enzyme	*Properties*			*Regional distribution*		
	Mol. wt.	*Glycoprotein*	*Ecto-enzyme*	*Blood sinusoidal*	*Contiguous*	*Bile canalicular*
5′-Nucleotidase	70 000–140 000[8]	√ [8]	√ [6]	++	+	+++[9]
Nucleotide pyrophosphatase (alkaline phosphodiesterase[1])	130 000[10]	√ [10]	√ [7]	++	+	+++[11]
Alkaline phosphatase	154 000[12]	√ [13]	√ [3]	++	+	+++[9]
Leucine aminopeptidase	*140 000[37]	* √ [14]	√ [3]	++	+	+++[15]
Cytidine monophosphate, neuraminic acid hydrolase	?	√ [38]	√ [38]	+++	++	+[38]
Adenylate cyclase	*100 000[16]	?	endo-enzyme[3]	+++	++	—[17]

Data obtained from pig renal cortex membranes

Definitions

Ecto-enzyme — a plasma membrane enzyme whose active site is accessible from the cell's exterior

Endo-enzyme — a plasma membrane enzyme whose active site is accessible to the cell's cytoplasm

the hepatocyte surface has been isolated. Indeed, the majority of plasma membrane isolation studies have been concerned merely with the preparation of plasma membrane fractions of acceptable purity and yield and rarely address themselves to the likelihood that the presence of different functional domains may result in heterogeneity of plasma membrane fragments.

IDENTIFICATION OF PLASMA MEMBRANE FRACTIONS

To obtain plasma membrane fractions derived predominantly from each of the three functionally specialized surfaces of the hepatocyte, Wisher and Evans[19] purified six plasma membrane subfractions. These were prepared from low and high speed fractions of rat liver, dispersed in hypo- and isotonic media, using a combination of differential, rate-zonal and density-gradient centrifugation (Figure 2). The plasma membrane fractions, which varied in density, were identified by studying their morphology, histochemistry, enzymic, protein and glycoprotein composition, hormone binding and by using domain-specific labels.

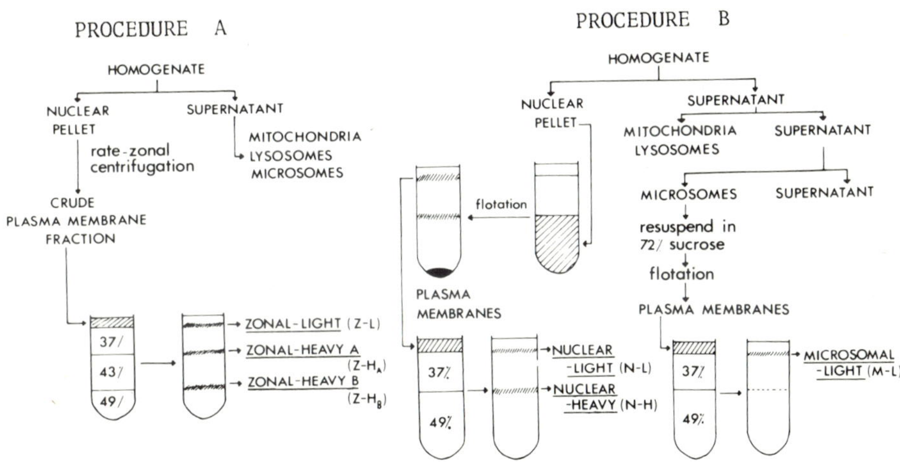

Figure 2 Origin of plasma membrane fractions

The contiguous plasma membranes, were isolated from sucrose gradients at a density of 1.16–1.18 (nuclear-heavy, zonal-heavy A and B, Figure 2) and contained mainly large vesicles and membrane strips associated with characteristic desmosomes and gap junctions. The canalicular plasma membranes were isolated as vesicles of density 1.12–1.13 (zonal-light, Figure 2) after their mechanical release from a plasma membrane fraction prepared by rate-zonal centrifugation of a low speed pellet of homogenate (procedure A[20]). This fraction contained extremely high specific activities (100- and 200-

fold relative to the homogenate) of a number of plasma membrane hydrolytic enzymes (Figure 3) that histochemical studies had shown to be mainly located at the bile canaliculus (Table 2). Furthermore although this subfraction contained basal adenylate cyclase activity (Figure 3), the enzyme was inhibited by a range of glucagon concentrations.

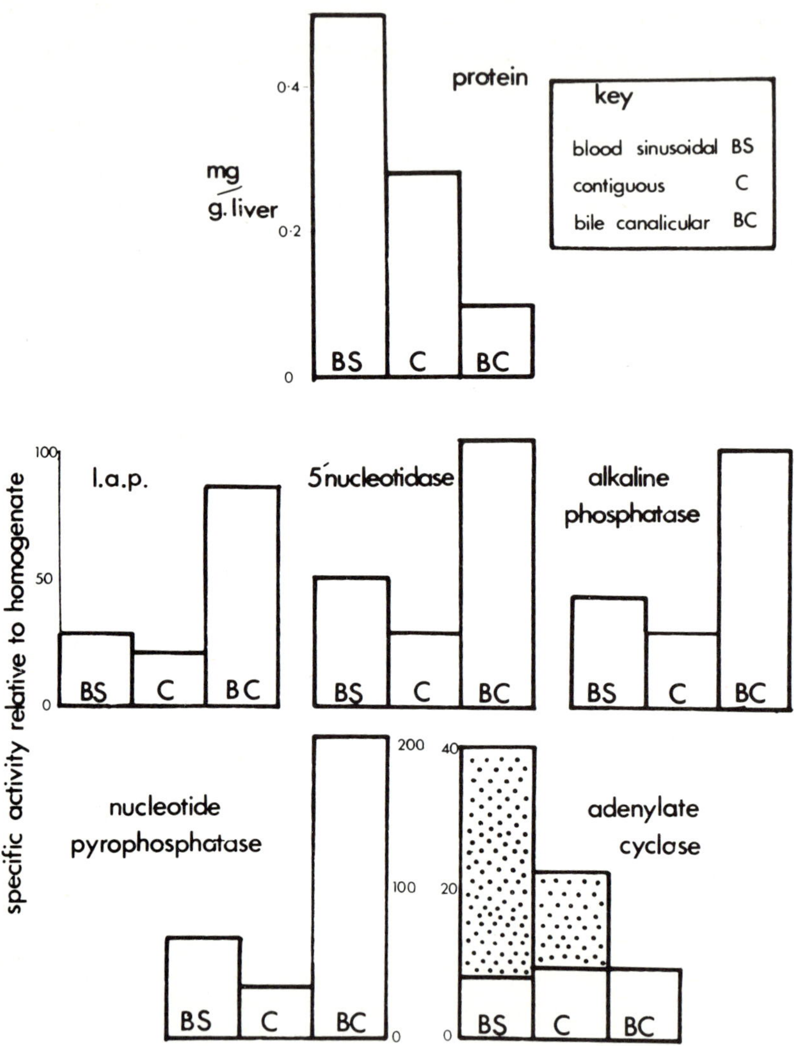

Figure 3 Distribution of plasma membrane enzymes on the hepatocyte surface. Protein recovered is expressed as mg of protein/g wet weight of liver. Enzyme-specific activities are expressed relative to that of the homogenate. Basal adenylate cyclase activity (open bars) and glucagon (2 μM) stimulated activity (shaded bars) are both expressed relative to the homogenate basal adenylate cyclase activity. BS, blood sinusoidal; C, contiguous; BC, bile canalicular; l.a.p., leucine aminopeptidase

The sinusoidal plasma membranes proved to be the most difficult to purify and characterize, since they possess similar densities in sucrose gradients and some similar enzymic properties to those described for Golgi apparatus membranes and canalicular plasma membranes. Two plasma membrane subfractions of a density of 1.12–1.13 on sucrose gradients (nuclear-light, microsomal-light, Figure 2) prepared from 'nuclear' and 'microsomal' fractions of liver homogenates dispersed in isotonic sucrose (procedure B), were highly enriched, relative to the homogenate, in glucagon-stimulated adenylate cyclase activity in addition to the usual hydrolytic plasma membrane marker enzymes (Figure 3). Histochemical studies have shown that glucagon-stimulated adenylate cyclase was located on the blood sinusoidal surface of the hepatocyte (Table 2). The highest specific binding of [^{125}I]insulin to plasma membranes was also found in these two subfractions while the lowest specific binding was present in the canalicular fraction[21,22]. The most direct evidence used to identify the sinusoidal plasma membrane fraction was obtained by the use of a domain-specific labelling technique. This involved the brief perfusion via the portal vein of liver sinusoids with radio-iodinated ligands such as wheatgerm haemagglutinin and glucagon[23]. The distribution of radio-iodinated ligands among the six plasma membrane fractions was studied and the highest amounts were found in the designated sinusoidal plasma membrane fractions.

DISTRIBUTION OF LIPIDS AND PROTEINS IN THE HEPATOCYTE PLASMA MEMBRANE

Analysis of the lipid, protein and glycoprotein composition of the sinusoidal, canalicular and contiguous plasma membrane subfractions indicated that the functionally differentiated regions of the membrane are constructed of a similar population of proteins, glycoproteins and lipids[20, 24]. Polyacrylamide-gel electrophoresis of the plasma membrane subfractions resolved the proteins into about 20 major polypeptide bands, but although the staining patterns of the subfractions were similar, differences in the intensities of a number of bands were apparent (Figure 4a). Furthermore the staining patterns of glycoproteins were also similar (Figure 4b) with the sinusoidal and canalicular fractions containing the highest amount of a number of glycoproteins. This uneven distribution of glycoproteins corresponded with the differential location of glycoenzymes (Table 2), hormone receptors[22] and glycolipids[25], in the plasma membrane fractions.

The neutral-lipid to phospholipid ratio was similar in all three plasma membrane fractions, and the lipid to protein ratios varied in agreement with the different densities of the fractions (Table 3). A more detailed examination of the neutral-lipid and phospholipid composition also indicated similarity, but the canalicular fraction contained significantly higher amounts of unesterified

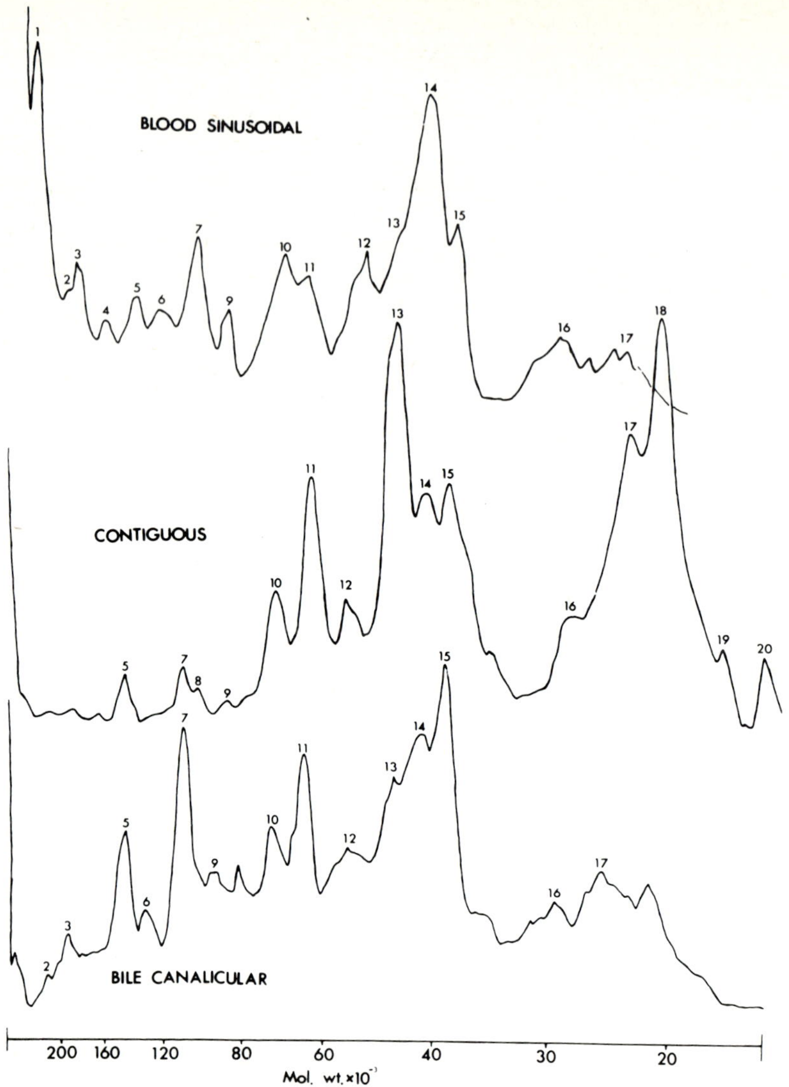

Figure 4 Polypeptide and glycoprotein composition of plasma membrane fractions. Polyacrylamide gel electrophoresis of plasma membrane fractions in sodium dodecyl sulphate was carried out and the gels were stained for (a) protein or (b) glycoprotein. Apparent molecular weights were estimated by using gels calibrated by proteins of known molecular weight

cholesterol and sphingomyelin. Since these two lipid classes may be located in the outer half of the lipid bilayer of animal cell plasma membranes[26,27] their location at higher concentrations in canalicular plasma membranes may be related to a role in the maintenance of membrane stability in an environment subject to perturbation by bile salts.

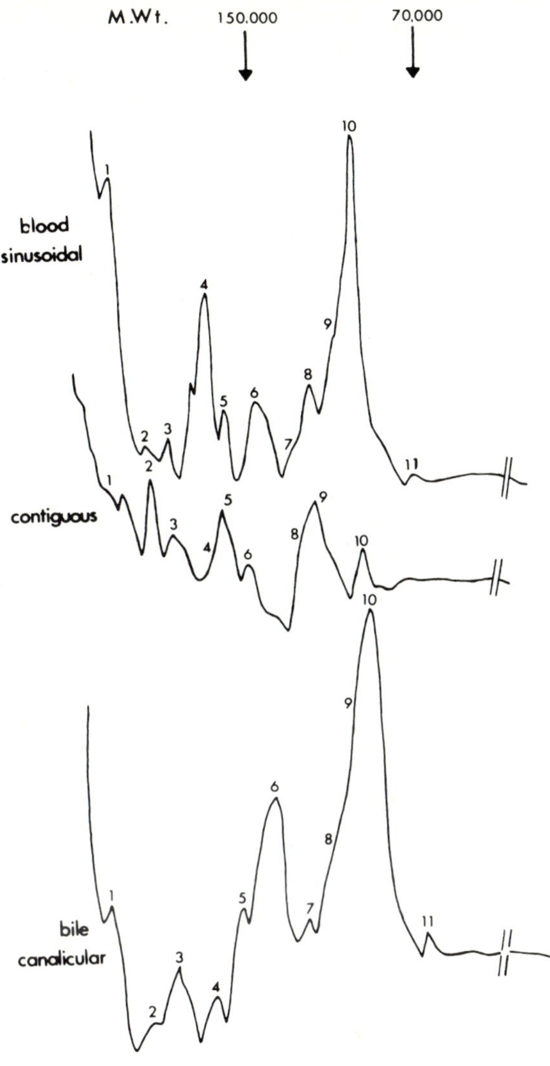

Figure 4(b)

PLASMA MEMBRANE SUBFRACTIONS FROM ISOLATED HEPATOCYTES

One cause of the heterogeneous properties of the various plasma membrane fractions prepared from liver could be their derivation from cells other than hepatocytes. Since hepatocytes account for 85–95% of the cellular weight of the liver[28] it can be argued that the contribution of non-parenchymal cell membranes (e.g. reticuloendothelial cells) to the characterized subfractions is likely

Table 3 Lipid distribution on the hepatocyte surface membrane

	Blood sinusoidal plasma membrane	*Contiguous plasma membrane*	*Bile canalicular plasma membrane*
Protein content (% by wt.)	50.6	56.8	47.4
Phospholipid content (% by wt.)	34.6	30.5	36.5
Neutral-lipid content (% by wt.)	14.8	12.7	16.1
Unesterified cholesterol (% neutral lipid)	51.0	60.7	73.4
Unesterified cholesterol/ phospholipid (molar ratio)	0.42	0.49	0.65
Sphingomyelin (mol% of phospholipid)	12.8	15.6	24.4

to be small. However, to assess experimentally the contribution of non-parenchymal cells to the heterogeneity of the plasma membrane subfractions, livers were dissociated into their constituent cells by collagenase perfusion. Isolated hepatocytes were then disrupted in isotonic sucrose and three plasma membrane fractions were isolated with similar densities and morphological features to those prepared from the whole tissue[29]. As shown in Figure 5, the distribution of hydrolytic enzymes and glucagon-stimulated adenylate cyclase was also similar in plasma membranes prepared from either liver tissue or isolated hepatocytes. Collagenases used in the perfusion procedure are known to contain proteases, and therefore the relatively lower specific activities in the isolated cell subfractions of the ecto-enzymes 5′-nucleotidase and alkaline phosphodiesterase may be attributed to proteolytic action. The inclusion of soybean trypsin inhibitor in the perfusion medium increased the specific activities and recovery of these ecto-enzymes in the isolated fractions. Polyacrylamide-gel electrophoresis showed that the polypeptide distribution was also similar in plasma membrane subfractions from whole liver and isolated hepatocytes.

ORGANIZATION OF A MULTIFUNCTIONAL PLASMA MEMBRANE

It is now thought that many plasma membrane proteins are mobile, being able to diffuse laterally in the plane of the membrane and can be redistributed around the cell surfaces forming aggregates or patches. The lateral movement of membrane proteins and glycoproteins may be restricted by their association with intra- or submembranous structures such as microfilaments or microtubules. Furthermore the binding of hormones to their corresponding receptors may trigger a conformational change so that the receptor interacts with membrane enzymes (such as adenylate cyclase) restricting lateral diffusion away

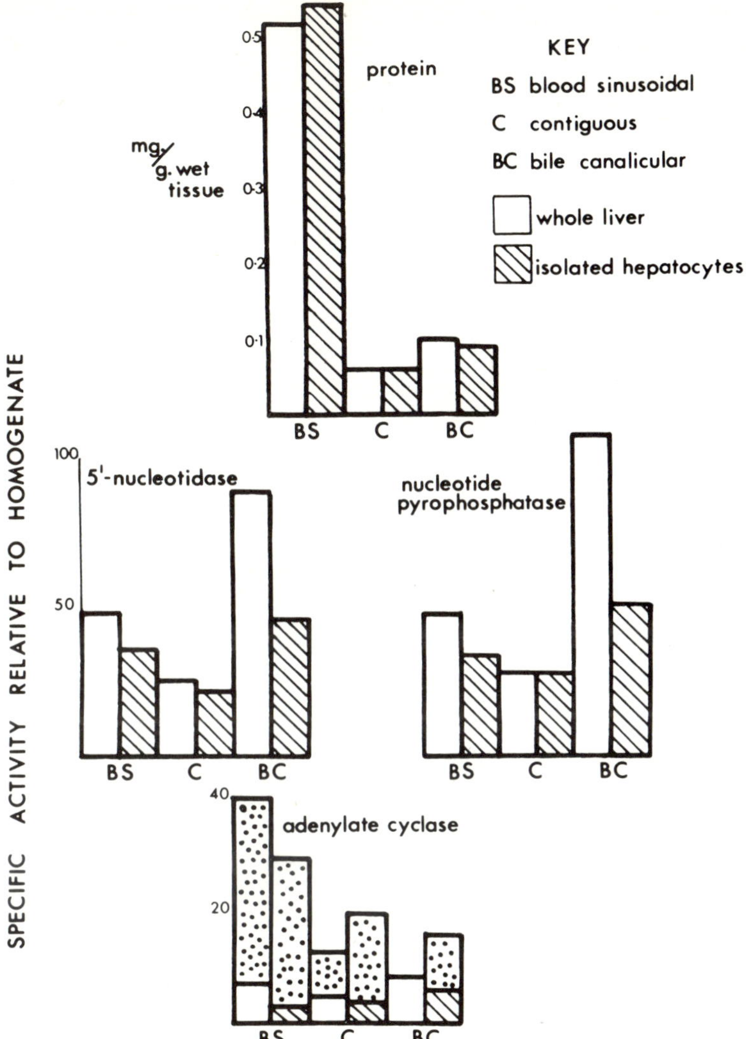

Figure 5 Distribution of enzymes in membranes from liver tissue and isolated hepatocytes. Plasma membrane fractions were isolated from whole liver (open bars) and isolated hepatocytes (hatched bars). BS, blood sinusoidal; C, contiguous; BC, bile canalicular. The designated bile canalicular fraction from isolated cells corresponds to the nuclear-light subfraction (see Figure 2) and is thought to consist of bile canalicular vesicles intermixed with some blood sinusoidal vesicles. Protein recovered is expressed as mg of protein/g wet wt. of liver. Enzyme-specific activities are expressed relative to that of the homogenate. Basal adenylate cyclase and glucagon (2 μM) stimulated activity (shaded bars) are both expressed relative to the homogenate basal adenylate cyclase activity

from the hormone-containing environment[30]. In tissues, the movement of membrane proteins and enzymes may also be restricted by short- and long-range interactions between intramembranous proteins of adjacent cells during gap and tight junction formation. Since gradients of plasma membrane components are retained in dissociated cells, the generation and maintenance of these gradients must be linked to subcellular events since the constraints imposed by interactions with adjacent cells have been removed. Indeed the uneven distribution of proteins and glycoproteins is probably a consequence of the metabolic compartmentation of the hepatic cytoplasm, as discussed in the next section in the context of plasma membrane biogenesis.

BIOGENESIS OF A MULTIFUNCTIONAL PLASMA MEMBRANE

After synthesis intracellularly, proteins and lipids may be inserted into the plasma membrane by two possible mechanisms. Membrane components may be inserted directly into the cell surface domains at which they are functionally active and from which lateral movement is restricted. A second mechanism postulates the insertion of membrane components at a single pole of the cell from which they redistribute around the cell by lateral diffusion. Supporting this second mechanism is radioautographic evidence for the insertion of glycosylated components into the plasma membrane in a range of mammalian tissues[31].

Using 5′-nucleotidase as an example, the pathway of synthesis, insertion and loss from the membrane of this glycosylated plasma membrane ecto-enzyme (Table 2) can be traced. In liver, this enzyme is synthesized predominantly by membrane-bound ribosomes[32], and was detected histochemically on the Golgi apparatus membranes[33]. Although it is not known how newly synthesized proteins may pass from the rough endoplasmic reticulum to the Golgi apparatus, the detection of activity in Golgi cisternal and secretory elements can at least be interpreted as indicating where glycosylation occurs en route to the plasma membrane. Interaction of Golgi secretory vesicles with the sinusoidal plasma membrane during the discharge of the vesicle contents can result in the selective transfer of the 5′-nucleotidase into the sinusoidal plasma membrane[34] from which lateral diffusion to the contiguous region of the plasma membrane subsequently occurs. This example indicates that blood sinusoidal domain-specific insertion mechanisms can be regarded as a consequence of hepatic secretion into the blood and may also apply to other membrane glycoproteins (*cf.* Table 2) and possibly glycolipids.

It is, however, difficult to explain, by a mechanism involving a blood sinusoidal domain-specific insertion, how the highest concentrations of hydrolytic enzyme activities occur at the bile canalicular plasma membrane. If the organization of the tight junctions results in a complete seal surrounding the canalicular area thus preventing further lateral diffusion of proteins in the membrane, then it is possible that glycosylated ecto-enzymes, such as 5′-

nucleotidase, nucleotide pyrophosphatase and leucine aminopeptidase, are inserted directly into the canalicular membrane. Although interaction of Golgi vesicles with the canalicular plasma membrane is not observed, the possible direct insertion of enzymes and lipids into the bile canalicular surface may be associated with the formation of bile. The lipid and protein composition of bile differs radically from that of the bile canalicular membrane and the small amounts of plasma membrane marker enzymes found in bile probably represent limited detergent solubilization of these components by bile salts[35].

PROSPECTS FOR FURTHER STUDIES

The isolation and characterization of fractions derived predominantly from the blood sinusoidal, bile canalicular and contiguous surfaces of the hepatocyte now makes possible the detailed study of the metabolic functions of these different domains. For instance the binding of hormones to receptors can now be studied in plasma membrane fractions originating from the blood sinusoidal surface rather than the previously used plasma membrane fractions that have now been shown to be derived predominantly from the contiguous and bile canalicular surfaces[19]. Interaction of glucagon and adrenaline with liver plasma membranes stimulates adenylate cyclase activity but it is not clear what role other enzymes, such as 5′-nucleotidase, nucleotide pyrophosphatase and adenosine triphosphatase and the products of their action play in controlling the activity of adenylate cyclase. Furthermore, although increased cyclic AMP concentrations stimulate protein phosphorylation the effect of phosphorylation of liver plasma membrane enzymes has not been studied in detail.

As indicated earlier, the bile canalicular fraction is potentially useful in studying the involvement of the plasma membrane in bile formation and changes occurring in hepatobiliary disease. Moreover, changes in the composition of the plasma membrane during neoplastic transformation may be studied by comparing strictly representative plasma membrane fractions from isolated hepatocytes and hepatomas. However, the interpretation of these differences in unsynchronized dividing cells may be complicated since the composition and structural organization of the plasma membrane may change during the cell cycle[36].

References

1. Benedetti, E. L. and Emmelot, P. (1968). Structure and function of plasma membranes isolated from liver. In: A. J. Dalton and F. Hagueman (eds.). *The Membranes*, pp. 33–120. (New York: Academic Press)
2. Solyom, A. and Trams, E. G. (1972). Plasma membrane marker enzymes. *Enzyme,* **13,** 329
3. DePierre, J. W. and Karnovsky, M. L. (1974). Ectoenzymes of the guinea pig polymorphonuclear leukocyte. *J. Biol. Chem.,* **249,** 7111

4. Trams, E. G. and Lauter, C. J. (1974). On the sidedness of plasma membrane enzymes. *Biochim. Biophys. Acta,* **345,** 180
5. Newby, A. C., Luzio, J. P. and Hales, C. N. (1975). The properties and extracellular location of 5′-nucleotidase of the rat fat-cell plasma membrane. *Biochem. J.,* **146,** 625
6. Gurd, J. W. and Evans, W. H. (1974). Distribution of liver plasma membrane 5′-nucleotidase as indicated by its reaction with anti-plasma membrane serum. *Arch. Biochem. Biophys.,* **164,** 305
7. Evans, W. H. (1974). Nucleotide pyrophosphatase, a sialoglycoprotein located on the hepatocyte surface. *Nature* (*London*), **250,** 391
8. Evans, W. H. and Gurd, J. W. (1973). Properties of a 5′-nucleotidase purified from mouse liver plasma membranes. *Biochem. J.,* **133,** 189
9. Essner, E., Novikoff, A. B. and Masek, B. (1958). Adenosine-triphosphatase and 5′-nucleotidase activities in the plasma membrane of liver cells as revealed by electron microscopy. *J. Biophys. Biochem. Cytol.,* **4,** 711
10. Evans, W. H., Hood, D. O. and Gurd, J. W. (1973). Purification and properties of a mouse liver plasma-membrane glycoprotein hydrolysing nucleotide pyrophosphate and phosphodiester bonds. *Biochem. J.,* **135,** 819
11. Sierakowska, H., Szemplinska, H. and Shugar, D. (1963). Intracellular localization of phosphodiesterase by a cytochemical method. *Acta Biochem. Polonica,* **10,** 399
12. Ohkubo, A., Langerman, N. and Kaplan, M. M. (1974). Rat liver alkaline phosphatase. *J. Biol. Chem.,* **249,** 7174
13. Emmelot, P. and Bos, C. J. (1966). On the participation of neuraminidase-sensitive sialic acid in the K^+-dependent phosphohydrolysis of *p*-nitrophenyl phosphate by isolated rat liver plasma membranes. *Biochim. Biophys. Acta,* **115,** 244
14. Wacker, H. (1974). The role of carbohydrate moieties in the activity and properties of aminopeptidase from pig kidney. *Biochim. Biophys. Acta,* **334,** 417
15. Monis, B. (1975). Isozymes of arylamidase (aminopeptidase): source of blood aminoacyl naphthylamidase in the rat. *Nature* (*London*), **208,** 587
16. Queener, S. F., Fleming, J. W. and Bell, N. H. (1975). Solubilization of calcitonin-responsive renal cortical adenylate cyclase. *J. Biol. Chem.,* **250,** 7586
17. Ong, S. H., Whitely, T. H., Stowe, N. W. and Steiner, A. L. (1975). Immunohistochemical localization of 3′,5′-cyclic AMP and 3′,5′-cyclic GMP in rat liver, intestine and testis. *Proc. Natl. Acad. Sci. USA,* **72,** 2022
18. Edelman, G. M. (1976). Surface modulation in cell recognition and cell growth. *Science,* **192,** 218
19. Wisher, M. H. and Evans, W. H. (1975). Functional polarity of the rat hepatocyte surface membrane. *Biochem. J.,* **146,** 375
20. Evans, W. H. (1970). Fractionation of liver plasma membranes prepared by zonal centrifugation. *Biochem. J.,* **116,** 833
21. Dron, D. I., Wisher, M. H. and Thomas, J. H. (1976). Are insulin-degrading enzymes located on liver plasma membranes? *Diabetologia,* **12,** 387
22. Evans, W. H., Bergeron, J. J. M. and Geschwind, I. I. (1973). Distribution of insulin receptor sites among liver plasma membrane subfractions. *FEBS Lett.,* **34,** 259
23. Carey, F. and Evans, W. H. (1977). Identification of blood sinusoidal plasma membrane fractions from rat liver homogenates by radioiodinated ligand binding. *Biochem. Soc. Trans.* (In press)
24. Kremmer, T., Wisher, M. H. and Evans, W. H. (1976). The lipid composition of plasma membrane subfractions originating from the major functional domains of the hepatocyte cell surface. *Biochim. Biophys. Acta,* **455,** 655
25. Evans, W. H. and Gurd, J. W. (1971). Biosynthesis of liver membranes. *Biochem. J.,* **125,** 615
26. Zwaal, R. F. A., Roelafsen, B. and Colley, C. M. (1973). Localization of red cell membrane constituents. *Biochim. Biophys. Acta,* **300,** 159

27. Fisher, K. A. (1976). Analysis of membrane halves: Cholesterol. *Proc. Natl. Acad. Sci. USA*, **73,** 173
28. Greengard, O., Federman, M. and Knox, W. E. (1972). Cytomorphometry of developing rat liver and its application to enzymic differentiation. *J. Cell Biol.,* **52,** 261
29. Wisher, M. H. and Evans, W. H. (1977). Preparation and characterization of plasma membrane subfractions from isolated rat hepatocytes. *Biochem. J.* (In press)
30. Cuatrecasas, P. (1974). Membrane receptors. *Ann. Rev. Biochem.,* **43,** 169
31. Bennett, G., Leblond, C. P. and Haddad, A. (1974). Migration of glycoprotein from the Golgi apparatus to the surface of various cell types as shown by radio-autography after labeled fucose injection into rats. *J. Cell Biol.,* **60,** 258
32. Bergeron, J. J. M., Berridge, M. and Evans, W. H. (1975). Biogenesis of plasmalemmal glycoproteins. *Biochim. Biophys. Acta,* **407,** 325
33. Farquhar, M. G., Bergeron, J. J. M. and Palade, G. E. (1973). Cytochemistry of Golgi fractions prepared from rat liver. *J. Cell Biol.,* **60,** 8
34. Bergeron, J. J. M., Evans, W. H. and Geschwind, I. I. (1973). Insulin binding to rat liver Golgi fractions. *J. Cell Biol.,* **59,** 771
35. Evans, W. H., Kremmer, T. and Culvenor, J. C. (1976). Role of membranes in bile formation. *Biochem. J.,* **154,** 589
36. Johnsen, S., Stokke, T. and Prydz, H. (1975). HeLa cell plasma membranes. Changes in membrane protein composition during the cell cycle. *Exp. Cell Res.,* **93,** 245
37. Wacker, H., Lehky, P., Vanderhaeghe, F. and Stein, E. A. (1976). On the subunit structure of particulate aminopeptidase from pig kidney. *Biochim. Biophys. Acta,* **429,** 546
38. Van Dijck, W., Maier, H. and Van den Eignden, D. H. (1977). CMP-NANA hydrolase, an ectoenzyme unevenly distributed over the hepatocyte surface. *Biochim. Biophys. Acta.* (In press)

9
Hepatocyte Surface Enzymes and their Appearance in the Bile

R. COLEMAN, G. HOLDSWORTH and O. S. VYVODA

Detergents are one of the principal tools of the biochemist in the chemical dissection and dissolution of membranes and in the preparation of many isolated membrane components. Mammalian liver, however, makes and secretes large amounts of detergents, the bile salts, without obvious signs of morphological damage. We were intrigued by this ability of the liver to secrete detergents without killing itself and have studied some of the membrane biochemistry which may explain this.

BILIARY ENZYME PROFILE

Our initial experiments[1] were aimed at finding the extent to which liver cells were damaged as a result of bile production. To do this we had to use a more sensitive tool than morphology and we have therefore examined bile from various mammalian species for the presence of a whole range of cellular enzymes. The choice of the particular enzymes used was governed by two factors: (i) their subcellular location in hepatic cells; enzymes were chosen to be representative of different cellular compartments and membranes, (ii) their retention or partial retention of activity in the presence of bile (and bile salts). This second point is important since the apparent absence of an enzyme from bile would not be a significant finding if that enzyme (e.g. glucose-6-phosphate, 6-phosphogluconate dehydrogenase) was strongly inhibited by biliary constituents. We have therefore chosen only those enzymes which were capable of detection in bile.

The virtual absence from bile of several intracellular enzymes representing both cytosol and organelles, suggests that bile production had not resulted in the breakage, dissolution, or shedding of cells (Table 1). Since this was observed in all species tested it may be a general feature of bile production that the hepatocytic plasma membrane remains a barrier to the non-specific loss of large molecules*.

Table 1 Enzyme profiles of mammalian bile

Derived from Holdsworth and Coleman[1] from which further details can be obtained. The values are means and are expressed as the ratio of the specific activity of bile to the specific activity of the liver homogenate from the donor animal. In the case of man, only bile was available and the values are specific activity (μmol/mg protein/h). The inhibition factor relates 1 volume of sheep liver homogenate + 9 vol of sheep bile incubated for up to 16 hours. All values are *uncorrected* for inhibition by biliary constituents.

Enzyme	*Subcellular location*	*Bile inhibition factor %*	*Specific activity ratio*				*Sp. act. man*
			Sheep	*Pig*	*Ox*	*Rat*	
Lactate dehydrogenase	Cytosol	0	0	0	0	0	(0)
Alcohol dehydrogenase	Cytosol	0	0	0	0	0	(0)
Acid phosphatase	Lysosomes	0	0	0	0.3	0.05	(0)
Succinate dehydrogenase	Mitochondria	30	0	0	0	0	(0)
Alkaline phosphatase	Plasma membrane	35	24.3	7.5	6.3	0.2	(0.5)
5′-Nucleotidase	Plasma membrane	0	6.1	0.2	14.0	1.7	(1.1)
Alkaline phosphadiesterase	Plasma membrane	30	7.2	1.8	0.3	0.1	(0.1)
Leucyl-β-naphthylamidase	Plasma membrane	80	0.4	1.6	0	0.6	(2.8)*

*OD units/mg/h

The bile of all species contained appreciable amounts of enzymes normally ascribed, in both histological and cell fractionation studies, to plasma membranes[4]. The specific activities of these enzymes in the various bile samples were often considerably higher than in the donor livers, which contrasts with the low levels of intracellular enzymes previously discussed. Both sets of observations, therefore, suggest that a more specific process than generalized cell damage should account for the presence of putative plasma membrane enzymes in bile[1].

Alkaline phosphatase, 5′-nucleotidase and L-leucyl-β-naphthylamidase have been observed by others in the bile of rat and of man[2,3,5–12]. Our studies indicate that they have wider occurrence in the bile of mammalian species and are probably therefore a general feature of biliary composition. Our studies also show that alkaline phosphodiesterase I occurs widely in animal biles; the

* Low levels of other putative intracellular enzymes have been found in bile[1–3]; the presence of these does not invalidate the general conclusion. Studies on their rate of appearance in bile suggest that this is not related to bile salt output.

occurrence of this enzyme in rat bile has recently been confirmed by Evans *et al.*[13] who have also carried out a detailed study of the enzyme content of rat and mouse bile.

NATURE AND ORIGIN OF THE BILIARY ENZYMES

When sheep or rat bile was passed through a carbohydrate affinity column, made with Concanavalin A–Sepharose, we observed that considerable amounts of all four putative plasma membrane enzymes were retained[14,15]. If the loaded column was then eluted with carbohydrate eluting agents, α-methyl glucoside or α-methylmannoside, in the presence of detergent, various amounts of each of the enzymes were eluted[14,15]. Also, if the column had been previously equilibrated with α-methylmannoside, the enzymes could not then be adsorbed to the column[15]. Such behaviour strongly indicates glycoproteins[14,15]. The graded nature of the response to adsorption and carbohydrate elution probably points to differences and heterogeneity in the carbohydrate sequences of the proteins. In other studies[9], neuraminidase treatment, which removes sialic acid residues from glycoproteins, reduced the electrophoretic mobility of human bile alkaline phosphatase.

The occurrence, in bile, of enzymes with reactions similar to enzymes located in the hepatocytic plasma membrane may indicate that the enzymes are derived from the plasma membrane during the process of bile formation. The plasma membrane forms of the enzymes discussed above have all been isolated and purified from liver plasma membranes; in the isolation of all these enzymes, detergents or n-butanol are required, suggesting that the proteins were embedded in the hydrophobic region of the membrane. In addition, the isolated enzymes have all been shown to be glycoproteins[7,13–25]. The similarities with biliary enzymes are obvious. Evans *et al.* have shown the kinetic and immunological identities of the plasma membrane and the biliary forms of 5′-nucleotidase[13].

In many types of cell, including liver cells, the plasma membrane counterparts of the biliary enzymes are 'ecto-enzymes', i.e. enzymes so located in the plasma membrane that their activity (active centre) occurs on the outer face of the plasma membrane. They are non-latent and affected by inhibitors and immunological reagents presented to the outside of the cell[4,13,18,26–33].

EFFECT OF DIFFERENT BILE SALTS UPON ISOLATED MEMBRANE PREPARATIONS

Persistent morphological and functional intactness of the membranes of liver cells, in the presence of bile salts, may be due to one of several factors: (i) the membrane-damaging ability of the bile salts secreted, (ii) the effective concentration in contact with individual membranes, (iii) the composition of the membranes in actual contact with the bile salts, (iv) the biosynthetic repair of any damage caused by bile salt action.

The level of bile salts in hepatic bile may reach about 1% and in the gall bladder may be as high as 7%[34,35]. At concentrations in the region of 1%, deoxycholate is often employed in the dissolution of membranes and in the isolation of many membrane components[36–38]. We have now compared the effects of other bile salts on membranes to those of deoxycholate.

Treatment of rat liver plasma membrane preparations, in isotonic solution at 37 °C and at pH 7.5, with high levels of deoxycholate, cholate, taurocholate or glycocholate, brought about almost total removal of phospholipids. Protein removal was less complete than phospholipid removal, but was greatest for deoxycholate. Among the proteins removed by all the bile salts were: alkaline phosphatase, alkaline phosphodiesterase, 5′-nucleotidase and L-leucyl-β-naphthylamidase. The residue after deoxycholate treatment had lost its membrane-like appearance and had become granular and disorganized with only occasional desmosomes and tight junctions remaining. Cholate and its conjugates, however, left behind residues which still contained large numbers of vesicular membrane-like structures which possessed recognizable trilamellar

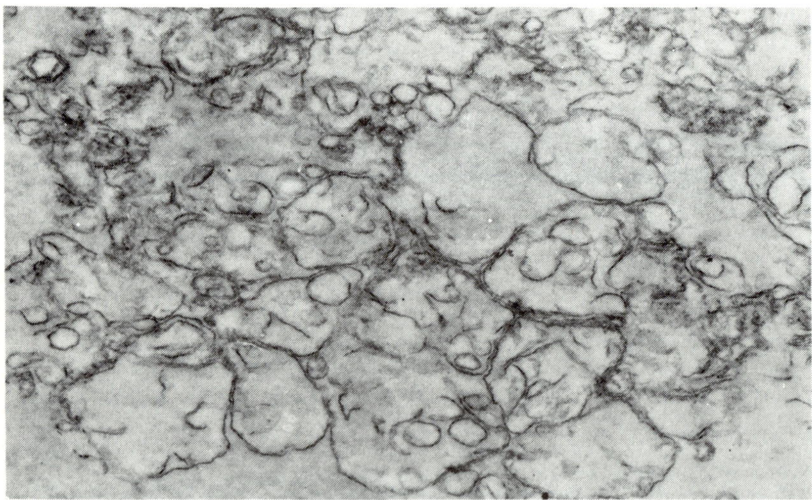

Figure 1 Membrane-like material remaining after cholate treatment of rat liver plasma membranes. Incubation: 37 °C, 10 min, pH 7.5, in isotonic buffered saline + 20 mM cholate, Centrifugation: 100 000 g × 60 min. The pellet had lost approximately 50% protein, 95% phospholipid, 50% L-leucyl-β-naphthylamidase, 80% alkaline phosphatase, 95% 5′-nucleotidase, 100% alkaline phosphodiesterase compared to a control (cholate-omitted) pellet (× 26 000)

profiles, even though depleted in phospholipids and to a considerable degree of protein[38,39] (Figure 1).

Vesicular structures, possessing trilamellar profiles, also remained after treatment of erythrocyte membranes with high levels of cholate, taurocholate or glycocholate at 37 °C. These 'membrane' preparations were depleted in lipid and had lost a proportion of the protein content of the original membrane preparations; this protein loss was selective and included the ecto-enzyme, acetylcholinesterase, and the peripheral proteins (i.e. polypeptide bands 1, 2, 4.1/4.2, 5, 6, 7). The lipid-depleted 'membranes' contained largely those polypeptides which spanned the original membrane (i.e. bands 3 and 4.5 together with some of band 4.1/4.2)[40,41]. Deoxycholate was extremely damaging and caused a complete breakdown of the membrane structure. Dehydrocholate, a synthetic bile salt which does not form micelles, had no effect upon the erythrocyte membranes.

It seems therefore, that bile salts differ in their ability to perturb and damage membranes; deoxycholate, a dihydroxy bile salt, is much more damaging than are cholate and its conjugates, which are trihydroxy bile salts. Synthetic detergents also show a range of solubilizing abilities[37,38].

EFFECTS OF DIFFERENT BILE SALTS UPON INTACT CELLS

Direct analyses of bile salt concentrations in liver tissue or subcellular fractions indicate that the levels of total bile salts inside liver cells are probably low, i.e. 0.01% or 0.2 mM[42,43]. The high, potentially membrane-damaging, levels of bile salts occur, however, in hepatic and gall-bladder bile, both of which are extracellular. Thus, it may be that any gross membrane-damaging effects of bile salts might be exerted from the outside, rather than from the inside of the cell. We have, therefore, studied the effects of different bile salts when presented to the outsides of intact cells (maintained in isotonic media at physiological temperature and pH).

Although intact human erythrocytes were lysed by high concentrations of all the bile salts tested[44] (with the exception of the non-micelle-forming dehydrocholate), they were much more resistant to the lytic action of concentrations of cholate, glycocholate and taurocholate, than to deoxycholate and glycodeoxycholate. Moreover, we observed that cholate and more particularly its conjugates, were able to remove appreciable amounts of membrane components without causing the cells to lyse[44].

Prelytic extracts, made with glycocholate at a concentration of 0.6%, contained proteins, including a proportion of the ecto-enzyme acetylcholinesterase[44,45], but were essentially devoid of intracellular proteins, e.,g. haemoglobin[44], and peripheral proteins characteristic of the inner face of the membrane. About 10–15% of the phospholipids, and some of the cholesterol could also be removed before the onset of lysis[44]. The composition of the phospholipid extracted did not reflect the composition of the intact membrane, or that extracted at comparable detergent levels from the permeable membrane preparations[44]. Its composition more closely resembled that suggested for the outer leaflet of the phospholipid bilayer[46,47], being largely composed of phosphatidylcholine and sphingomyelin and only having small amounts of PE, PS and PI[44]. When these experiments were repeated with sheep erythrocytes, in which sphingomyelin replaces phosphatidylcholine, the prelytic extract made with glycocholate is composed of almost 80% sphingomyelin[48].

It therefore seems that the milder type of bile salt can remove a proportion of the components of the outer leaflet of the membrane bilayer and possibly proteins embedded in the outer leaflet, but not spanning the membrane, without causing sufficient membrane damage to result in the loss of intracellular components[33,38,39,44]. Analogous studies by Pearlstein and Seaver[49], using a range of synthetic non-ionic detergents, have achieved a non-lytic release of outwardly disposed plasma membrane components from intact fibroblast and BHK cells.

When pig lymphocytes were treated with glycocholate, plasma membrane components were released before intracellular enzymes could be detected in the medium[33]. Among the plasma membrane constituents detected were several ecto-enzymes, e.g. 5′-nucleotidase, alkaline phosphatase and alkaline phosphodiesterase. These enzymes are analogous to the ecto-enzymes of the hepatocyte and their shedding during glycocholate treatment, without the loss of intracellular enzymes, may assist in understanding some of the events leading up to the secretion of bile and its effect upon hepatic cells[33,38,39].

WORKING HYPOTHESIS FOR THE RELATION BETWEEN BILE SALT SECRETION AND COMPOSITION OF BILE AND EFFECTS ON MEMBRANES

The appearance of plasma membrane enzymes in bile may be part of a wider series of phenomena connected with the output of detergents by hepatic cells. Our experiments on the composition of bile and the effects of different bile salts on cells and membranes have encouraged us to develop[14,33,38–41,44,45] a working hypothesis relating to bile salt secretion and its effects on membranes (Figure 2). This hypothesis seeks not only to explain the specific enzyme pattern of bile, but also its specific phospholipid pattern. (Bile contains large amounts of phosphatidylcholine, present as mixed micelles with the bile salts; the output of this phospholipid is dependent upon the extent of bile salt secretion[34,35,50].) Many details of this hypothesis remain to be clarified, e.g. some of the tenets of the earlier section (pages 145 and 146) cannot yet be fully accommodated within it.

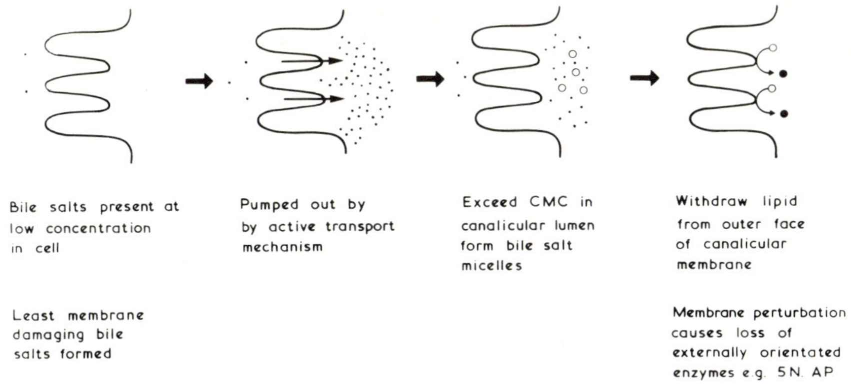

Figure 2 Hypothesis relating bile salt secretion to the presence of other components in bile

Mammalian liver secretes a bile salt pattern containing a high proportion of 'mild', i.e. less membrane-damaging, types of bile salt

In many mammalian species, conjugates of cholate form a high proportion of the total bile salts. Free bile salts of all types are rare and occur in only low concentration; most bile salts are conjugated with taurine or glycine[34,51]. Studies of the tissue-damaging effects of bile salts at both physiological and pharmacological levels correlate the damage better with the presence of dihydroxy rather than trihydroxy bile salts[30,40,52,53]. Many mammalian species possess hydroxylation and conjugation mechanisms to deal with the less hydroxylated bile acids, which are metabolites of the colonic bacteria. These bacterial metabolites are returned to the liver in the enterohepatic circulation[34,39].

Thus it may be that damage in the hepatobiliary system is reduced by the overall secretion bias towards cholate conjugates, which are adequate for an intestinal role on dietary lipids but are less damaging to membranes than other types of bile salt.

The bile salts are present in extremely low concentration inside the hepatocyte and are pumped out by an active transport mechanism

Bile salt synthesis takes place in hepatocytes and these cells are also able to take up bile salts present in low concentration, from the bloodstream[34]. Only low concentrations of bile salts have been found in liver tissue and in subcellular fractions[34,42]. Thus the intracellular membranes and the inside face of the plasma membrane would never normally be exposed to high, membrane-damaging concentrations of bile salts, and thus their structure and functions remain relatively unaffected. The characteristics of bile salt output are those of an active transport mechanism, i.e. saturation kinetics and competitive inhibition[34]. There is no evidence from electron microscopy for packaging and package secretion[54,55].

The bile salts are pumped out of the hepatocyte and rise to high concentrations only outside the cell

The concentrations in the bile canaliculus could then reach micellar levels. The only membrane to which these high concentrations would be exposed is the outer face of the plasma membrane of the bile canaliculus.

As the bile salt concentration approaches or exceeds micellar levels, materials are removed from the outer face of the plasma membrane but do not result in sufficient cell damage to cause leakage of intracellular materials

The small bile salt micelles may then incorporate phospholipids and cholesterol from the outer leaflet of the plasma membrane. By analogy with human erythrocytes, if the outer leaflet of the bilayer of the bile canaliculus membrane contained a high proportion of phosphatidylcholine and sphingomyelin (see previous section) these lipids would predominate in bile as they did in extracts from erythrocytes[33,38]. The phospholipid composition of many mammalian biles is known; it is predominantly phosphatidylcholine with small amounts of sphingomyelin and other lipids[34,35].

Another suggestion[50] for the origin of biliary phospholipid is that the lipid is drawn from the plasma membrane during the outward diffusion of bile salts micelles from inside the cell. This is less attractive since, (i) it would require a higher intracellular bile salt concentration, (ii) bile salts would be exposed to a wide variety of phospholipids and (iii) the characteristics of the externalization of bile salts would be those of large aggregates not simple molecules.

The action of bile salts on the outer leaflet of the plasma membrane may not only remove lipid but may also remove a proportion of the proteins of the plasma membrane, probably those projecting into the outer leaflet of the lipid bilayer but not penetrating through it. This may explain the presence of the plasma membrane ecto-enzymes in bile and also the relative absence of intracellular enzymes, since the plasma membrane probably would not be breached by removal of proteins embedded in only one side[33,44,45]. Evans *et al.* have also suggested that extracellular bile salts may cause the removal of plasma membrane ecto-enzymes[13].

The influence of bile salts upon the enzyme levels in the bile can be seen from studies in which the bile salt load on the liver was increased; 5′-nucleotidase and alkaline phosphatase output in bile were increased[3,11,56,57], but there was no parallel increase in intracellular enzymes[3]. Dehydrocholate, a non-micelle-forming synthetic bile salt, and secretin, which both increase bile volume without increasing phospholipid output, also had little effect upon the output of the plasma membrane enzyme; any effect in these cases was also accompanied by a parallel output of intracellular enzymes[3].

The individual patterns of the phospholipid and enzyme profiles in the bile may be accounted for partly by the composition of the plasma membrane and partly by the bile salt pattern being secreted

It is too early yet to say which is the most important; further experimentation with membranes of different composition, and with mixtures of bile salts, should be instructive.

Concentration of the forming bile takes place without further membrane damage

Bile salt micelles probably remove lipids from the outer surface of the canalicular membrane and possibly those of duct cells as they drain down the biliary tree. The replete mixed micelles may grow by condensation with one another without causing further membrane damage. This will probably occur in the lower parts of the biliary tree and also in the gallbladder. Although the composition of canalicular bile is not known, gallbladder bile, with its replete mixed micelles, can exist at a concentration of up to 7% bile salts without further damage. If however, bile-salt-only micelles are introduced into the gallbladder[52] they will cause extensive damage.

The health of the membrane of the bile canaliculus is maintained by continuous biosynthetic repair mechanisms

The plasma membrane components' continuous loss into the bile must be replaced. Without this replacement, the canalicular membrane would soon become irreversibly damaged, as did the membranes of intact cells when presented with levels of glycocholate which were too high[33,44]. The phospholipid biosynthesis capacity of the liver is very high and phospholipid output in bile is increased by an extra bile salt load[34,58,59]. The phospholipid metabolism which supplies the biliary outflow is becoming apparent[34,60], but there remains the problem of how newly synthesized phospholipid can arrive at the outer face of the leaflet. Biosynthetic replacement of the proteins lost in bile is also required if the membrane is to stay intact; this also poses the problem of how a continuous supply of protein is brought to the outer face of the plasma membrane. These problems await investigation.

Acknowledgement

We thank the MRC for financial support.

References

1. Holdsworth, G. and Coleman, R. (1975). Enzyme profiles of mammalian bile. *Biochim. Biophys. Acta,* **389,** 47
2. Slater, T. F. and Griffiths, D. B. (1965). Studies on the enzymic composition of rat bile following the administration of various toxic substances. In: W. Taylor (ed.). *The Biliary System,* pp. 431–438. (London: Blackwell)
2a. Cooper, B., Eakin, M. N. and Slater, T. F. (1976). The effects of various anaesthetic techniques on the flow rate, constituents and enzymic composition of rat bile. *Biochem. Pharmacol.,* **25,** 1711
3. Bode, J. Ch., Zelder, O. and Neuberger, H. O. (1973). Effect of taurocholate, dehydrocholate and secretin on biliary output of alkaline phosphatase and GOT. *Helv. Med. Acta,* **37,** 143
4. DePierre, J. W. and Karnovsky, M. L. (1973). Plasma membranes of mammalian cells. A review of methods for their characterization and isolation. *J. Cell Biol.,* **56,** 275
5. Chiondussi, L., Green, S. F. and Sherlock, S. (1962). Serum alkaline phosphatase fractions in hepatobiliary and bone diseases. *Clin. Sci.,* **22,** 425
6. Pineda, E. P., Goldberg, J. A., Banks, B. M. and Rutenburg, A. M. (1960). Serum leucine aminopeptidase in pancreatic and hepatobiliary diseases. *Gastroenterology,* **38,** 698
7. Kaplan, M. M. and Righetti, A. (1970). Induction of rat liver alkaline phosphatase: the mechanism of the serum elevation in bile duct obstruction. *J. Clin. Invest.,* **49,** 508
8. Hill, P. G. and Sammons, H. G. (1969). Further studies on human alkaline phosphatase and 5′-nucleotidase. *Enzymol. Biol. Clin.,* **10,** 401
9. Price, C. P., Hill, P. G. and Sammons, H. G. (1972). The nature of the alkaline phosphatases of bile. *J. Clin. Pathol.,* **25,** 149
10. Hargreaves, T. (1968). *The Liver and Bile Metabolism.* (Amsterdam: North-Holland)
11. Anderson, K. E. and Javitt, N. B. (1974). Bile formation. In: F. F. Becker (ed.). *The Liver. Normal and Abnormal Functions,* Part A pp. 371–400. (New York: Marcel Dekker)
12. Pope, C. E. and Cooperband, S. R. (1966). Protein characterization of serum and bile alkaline phosphatase. *Gastroenterology,* **50,** 631
13. Evans, W. H., Kremmer, T. and Culvenor, J. G. (1976). Role of membranes in bile formation. Comparison of bile and a liver bile–canalicular plasma-membrane subfraction. *Biochem. J.,* **154,** 589
14. Holdsworth, G. and Coleman, R. (1975). Presence of plasma membrane enzymes in bile and their partial purification by affinity chromatography with Concanavalin A. *Biochem. Soc. Trans.,* **3,** 746
15. Holdsworth, G. and Coleman. R. (19—). Carbohydrate affinity chromatography of biliary proteins. (In preparation)
16. Kaplan, M. M. (1972). Alkaline phosphatase. *Gastroenterology,* **62,** 452
17. Evans, W. H. and Gurd, J. W. (1973). Properties of a 5′-nucleotidase purified from mouse liver plasma membranes. *Biochem. J.,* **133,** 189
18. Gurd, J. W., Evans, W. H. and Perkins, H. R. (1973). Immunological characterization of proteins from mouse liver plasma membranes. *Biochem. J.,* **135,** 827
19. Song, C. S., Tandler, P. and Bodansky, O. (1967). Solubilization of 5′-nucleotidase from human liver and reassociation of its activity into membranous vesicles. *Biochem. Med.,* **1,** 100

20. Evans, W. H., Hood, D. O. and Gurd, J. W. (1973). Purification and properties of a mouse-liver plasma-membrane glycoprotein hydrolysing nucleotide pyrophosphate and phosphodiester bonds. *Biochem. J.*, **135**, 819
21. Widnell, C. C. and Unkeless, J. C. (1968). Partial purification of a lipoprotein with 5′-nucleotidase activity from membranes of rat liver cells. *Proc. Natl. Acad. Sci. USA*, **61**, 1050
22. Nakamura, S. (1976). Effect of deoxycholate on 5′-nucleotidase. *Biochim. Biophys. Acta*, **426**, 339
23. Evans, W. H. (1974). Nucleotide pyrophosphatase, a sialoglycoprotein located on the hepatocyte surface. *Nature* (*London*), **250**, 391
24. Sternes, W. L. and Behal, F. L. (1974). A human liver aminopeptidase. The amino acid and carbohydrate content and some physical properties of a sialic acid containing glycoprotein. *Biochemistry*, **13**, 3221
25. Prospero, T. D., Burge, M. L. E., Norris, K. A., Hinton, R. H. and Reid, E. (1973). Alkaline ribonuclease and phosphodiesterase activity in rat liver plasma membranes. *Biochem. J.*, **132**, 449
26. Gurd, J. W. and Evans, W. H. (1974). Distribution of liver plasma membrane 5′-nucleotidase as indicated by its reaction with anti-plasma-membrane serum. *Arch. Biochem. Biophys.*, **164**, 305
27. Bischoff, E., Liersch, M., Keppler, D. and Dekker, K. (1970). Fate of intravenously administered UDP-glucose. *Hoppe-Seyler's Z. physiol. Chem.*, **351**, 729
28. DePierre, J. W. and Karnovsky, M. L. (1974). Ectoenzymes of granulocytes: 5′-nucleotidase. *Science*, **183**, 1096
29. Emmelot, P. and Visser, A. (1971). Studies on plasma membranes XIII CA^{2+}-activated aminopeptidase(s) in the globular units locally coating rat liver plasma membranes. *Biochim. Biophys. Acta*, **241**, 273
30. Trams, E. G. and Lauter, C. J. (1974). On the sidedness of plasma membrane enzymes. *Biochim. Biophys. Acta*, **345**, 180
31. Misra, D. N., Gill, T. J. and Estes, L. W. (1974). Lymphocyte plasma membranes. Cytochemical localization of 5′-nucleotidase. *Biochim. Biophys. Acta*, **352**, 455
32. Misra, D. N., Ladoulis, C. T., Estes, L. W. and Gill, T. J. (1975). Biochemical and enzymatic characterization of thymic and splenic lymphocyte plasma membranes from inbred rats. *Biochemistry*, **14**, 3014
33. Holdsworth, G. and Coleman, R. (1976). Plasma membrane components can be removed from isolated lymphocytes by the bile salts glycocholate and taurocholate without cell lysis. *Biochem. J.*, **158**, 493
34. Heaton, K. W. (1972). *Bile Salts in Health and Disease*. (London: Churchill Livingstone)
35. Coleman, R. (1973). Phospholipids and the hepatoportal system. In: G. B. Ansell, R. M. C. Dawson and J. N. Hawthorne (eds.). *Form and Function of Phospholipids*, pp. 345–375. (Amsterdam: Elsevier)
36. Coleman, R. (1974). Solubilization of membrane components. *Biochem. Soc. Trans.*, **2**, 813
37. Helenius, A. and Simons, K. (1975). Solubilization of membranes by detergents. *Biochim. Biophys. Acta*, **415**, 29
38. Coleman, R., Holdsworth, G. and Vyvoda, O. S. (1977) Detergents as tools in the dissection of membranes. In: E. Reid (ed.). *Methodological Developments in Biochemistry, Vol. 6* (Chichester: Horwood)
39. Vyvoda, O. S., Coleman, R. and Holdsworth, G. (1977). Effects of different bile salts upon the composition and morphology of a liver plasma membrane preparation rich in bile canaliculi: deoxycholate is more membrane damaging than cholate and its conjugates. *Biochim. Biophys. Acta*, **465**, 68
40. Coleman, R. and Holdsworth, G. (1975). Effect of detergents on erythrocyte membranes: different patterns of solubilization of the membrane proteins by dihydroxy and trihydroxy bile salts. *Biochem. Soc. Trans.*, **3**, 747

41. Coleman, R., Holdsworth, G. and Finean, J. B. (1975). Detergent extraction of erythrocyte ghosts. Comparison of residues after cholate and Triton X-100 treatments. *Biochem. Biophys. Acta,* **436,** 38
42. Okishi, T. and Nair, P. P. (1966). Studies on bile acids. Some observations on the intracellular localization of major bile acids. *Biochemistry,* **5,** 3662
43. Greim, H., Trulzsch, D., Roboz, J., Dressler, K., Czygan, P., Hutterer, P., Schaffner, F. and Popper, H. (1972). Mechanism of cholestasis. 5. Bile acids in normal rat livers and those after bile duct ligation. *Gastroenterology,* **63,** 837
44. Coleman, R. and Holdsworth, G. (1976). The release of membrane components prior to haemolysis during extraction of intact erythrocytes with bile salts. *Biochim. Biophys. Acta,* **426,** 776
45. Coleman, R., Holdsworth, G. and Vyvoda, O. S. (1976). Glycocholate can remove lipid and protein components from the outer leaflet of the plasma membrane without causing cell lysis. *Biochem. Soc. Trans.,* **4,** 244
46. Zwaal, R. F. A., Roelofsen, B. and Colley, C. M. (1973). Localization of red cell membrane constituents. *Biochim. Biophys. Acta,* **300,** 159
47. Renooij, W., VanGolde, M. G., Zwaal, R. F. A. and Van Deener, L. L. M. (1976). Topological asymmetry of phospholipid metabolism in rat erythrocyte membranes. Evidence for flip-flop of lecithin. *Eur. J. Biochem.,* **61,** 53
48. Billington, D., Coleman, R. and Lusak, Y. (1977). Topographical dissection of sheep erythrocyte membrane lipids by taurocholate and glycocholate. *Biochim. Biophys. Acta,* **466,** 526
49. Pearlstein, E. and Seaver, J. (1976). Non-lytic non-ionic detergent extraction of plasma membrane constituents from normal and transformed fibroblasts. *Biochim. Biophys. Acta,* **426,** 589
50. Small, D. M. (1970). The formation of gallstones. *Adv. Intern. Med.,* **16,** 243
51. Haslewood, G. A. D. (1967). *Bile Salts.* (London: Methuen)
52. Kappas, A. and Palmer, R. M. (1963). Selected aspects of steroid pharmacology. *Pharmacol. Rev.,* **15,** 123
53. Greim, H., Schwarz, L. Czygan, P. and Popper, H. (1975). Mechanism of necrosis induced by hepatotoxic bile acids. In: D. Keppler (ed.). *Pathogenesis and Mechanisms of Liver Cell Necrosis,* pp. 199–208. (Lancaster: MTP)
54. Biava, C. G. (1964). Studies on cholestasis: a re-evaluation of the fine structure of normal human bile canaliculi. *Lab. Invest.,* **13,** 840
55. Heath. T. and Wissig, S. L. (1965). Fine structure of the surface of mouse hepatic cells. *Am. J. Anat.,* **119,** 97
56. Javitt, N. (1965). 5′-Nucleotidase activity of rat bile. *Clin. Res.,* **13,** 355
57. Eakins, M. (1971). (Personal communication)
58. Swell, L., Bell, C. C. and Entenmann, C. (1968). Bile acids and lipid metabolism. III. Influence of bile acids on phospholipids in liver and bile. *Biochim. Biophys. Acta,* **164,** 278
59. Nillson, S. and Schersten, T. (1969). Importance of bile acids for phospholipid secretion into human hepatic bile. *Gastroenterology,* **57,** 525
60. Yousef, I. M., Bloxan, D. L., Phillips, M. J. and Fisher, M. M. (1975). Liver cell plasma membrane lipids and the origin of biliary lipid. *Can. J. Biochem.,* **53,** 989

DISCUSSION

Dr N. B. Javitt, New York (USA): Several years ago, we (see reference [56] in text) demonstrated that infusion of the naturally occurring bile acid conjugates in the rat was followed by an increase in the excretion rate of the amount of 5′-nucleotidase in bile which is consistent with the described enzyme solubilization with bile acids. However, the critical question is whether the bile acids are prepackaged in the cell together with these enzymes, or if solubilization occurs following the transport into bile.

Dr S. Erlinger, Clichy (France): Since the bile acid concentration in duct bile often far exceeds the critical micellar concentration, is it possible that membrane alterations occur in biliary epithelial cells and that some of the enzymes referred to might originate from the biliary cells rather than the canaliculi?

Dr Coleman, Birmingham (England): Yes.

Dr R. Holzbach, Cleveland (USA): How can the presented hypothesis regarding bile salt–mixed lipid micelle formation be reconciled with observations[1] that in labelling experiments newly synthesized biliary phosphatidylcholine and cholesterol are synthesized in hepatocytic microsomes and thereafter appear rapidly in canaliculi in micellar form? These microsomal lipids would have to rapidly migrate after synthesis to be inserted stereospecifically into the *outer* lining of the canalicular membrane, there to be cleared out by detergent action of bile salt micelles and then to be rapidly replaced from within the cell presumably by specific out-migration of newly synthesized phosphatidylcholine from inside the cell, through the *inner* part of the plasma membrane. Such a complex and rapid process is difficult to accept despite its attractiveness in explaining other phenomena.

Dr Coleman, Birmingham (England): I fully agree with Dr Holzbach's diagnosis. There is a specific phosphatidylcholine transfer protein in liver cells which may very well be the vehicle for such rapid and specific movements.

Dr R. O. Soloway, Philadelphia (USA): Our data[2] showed a hyperbolic relationship between cholesterol *or* phospholipid and bile salt excretion. There is a point at which bile salt excretion continues to increase with only a slight increase in phospholipid and no increase in cholesterol excretion. Your theory might explain this by a limit to the replacement of lecithin in the membrane from which it is extracted by bile salts. In your diagram of the red cell membrane, most of the lecithin was located on the outer side of the membrane. If this is also true for the hepatocyte, it would support your theory. From your diagram, lecithin accounted for only about 40% of the phospholipid in the outer membrane. Since lecithin accounts for more than 90% of the phospholipid in bile, either lecithin is selectively removed by bile salts, or lecithin aggregates in microdomains which are dissected out by the luminal bile salts. Also, why do bile salts not harm the interior of the hepatocyte as during passage from plasma to bile? Is this because bile salts are protein-bound in the cell?

Dr Coleman, Birmingham (England): Too much loss balanced by too little replacement might cause cell death. We have some preliminary evidence that phosphatidylcholine is more readily dissected from the membranes of intact cells than is sphingomyelin. Intracellular damage may be low because the bile salts are inside the cell in very low concentrations (i.e. $\sim 10^{-6}$ M) and may also be protein-bound.

References

1. Gregory, D. H., Vlahcevic, Z. R., Schatzki, P. and Swell, L. (1975). Mechanism of secretion of biliary lipids. I. Role of bile canalicular and microsomal membranes in the synthesis and transport of biliary lecithin and cholesterol. *J. Clin. Invest.*, **55,** 105
2. Wagner, C. I., Trotman, B. W. and Soloway, R. O. (1976). Kinetic-analysis of biliary lipid excretion in man and dog. *J. Clin. Invest.*, **57,** 473

10
Insulin and Glucagon Receptors in Isolated Cells and Cell Membranes of Liver

P. FREYCHET

Interactions between polypeptide hormones and their specific receptors have been extensively studied over the past six years. It has been generally recognized that the first step in the action of a polypeptide hormone is binding of the hormone to a specific recognition site, usually referred to as the 'receptor', on the plasma membrane of the target cell. Since the receptor has thus far not been isolated as a 'pure' physicochemical entity, it has been defined in most instances by the properties exhibited in binding a given hormone, i.e. specificity, affinity, saturability and reversibility. Relating hormone binding and hormone effect(s) has often proven difficult, especially in the case of insulin[1]. As a result, the term receptor has been generally employed to designate only that component of the cell membrane which selectively recognizes a hormone in a specific binding reaction. At least two sequential aspects of hormone action evolve from such a definition of receptor: (1) hormone binds to receptor to form a hormone–receptor complex; (2) subsequent events occur which eventually result in cell biological response(s).

The liver has been widely used as a source of receptors for insulin and glucagon in direct studies of hormone binding to membranes and cells prepared from this organ[1–4]. This paper will summarize and discuss some of the following aspects of hormone–receptor interactions, mainly insulin, in the liver: (1) methodology; (2) properties of hormone binding; (3) alterations of hormone binding.

METHODOLOGY

Preparation and use of liver membranes and cells as a source of receptors

Purified liver plasma membranes, prepared by the method of Neville (step 15 in reference[5]), were used in the studies reported here. Plasma membranes (PM) prepared by this method retain insulin-binding and glucagon-binding abilities and other properties such as basal and glucagon-stimulated adenylate cyclase activity after storage for several months at −70 °C. This PM preparation appears to be 80–90% pure, based on purification ratios of contaminating organelle marker[7]. The specific activities (amount of hormone bound, or level of enzyme activity, per mg of protein) are increased 40- to 60-fold for insulin binding[2,8,9], 17-fold for glucagon-stimulated adenylate cyclase[10], and 20-fold for 5′-nucleotidase[7,9,10] over the values found in the homogenate.

At least three reasons favour the use of a purified PM preparation in studies of hormone binding: (1) The higher concentration of specific binding sites, which makes it possible to use low concentrations of membrane protein, thereby decreasing non-specific binding and hormone degradation[6,11]; (2) The presence of binding sites on intracellular organelles[9,12,13] which contaminate less purified preparations. They contribute a source of non-specific adsorption and hormone degradation[11]; (3) These sites may be of different affinity or concerned with different functions with respect to the hormone[7]. Thus, the marked difference in insulin-binding capacity of purified liver PM from genetically obese ob/ob mice and their thin littermates (see below) was not observed in fractions of smooth or rough endoplasmic reticulum[9]. Therefore, quantitation of membrane receptors in comparative studies requires the use of the most purified and best characterized PM preparation. Glucagon-stimulated adenylate cyclase, on the other hand, has been found at a relatively higher level of activity in the partially purified (step 11 in reference[5]) PM than in the fully purified (step 15 in reference[5]) PM fraction of liver[10]. This may reflect a relatively higher enrichment of the partially purified fraction in blood sinusoidal fronts[14].

Isolated rat liver cells, obtained by enzymatic digestion of the liver, were also used in the studies reported here. They were prepared following the method of Seglen[15] with certain modifications[16]. The proportion of parenchymal cells, i.e. hepatocytes, in the purified cell suspension exceeds 90–95%. The viability of freshly isolated cell suspensions, estimated by the trypan blue exclusion method, ranges between 80–90%. The periphery of intact cells appears highly refractive under the light microscope. Their ATP content is maintained at about 20 nmol/10^6 cells, i.e. about 5 mmol/ℓ cell water[18] for at least 2 hours at 30 °C. Their K^+ content (130–150 mEq/ℓ cell water) is not significantly affected after 1 hour of incubation at 37 °C[18]. Freshly isolated hepatocytes retain, in most instances, membrane selectivity towards various substrates, preserved ultrastructure of membranes and organelles, and metabolic capabilities[16]. They possess specific transport systems for neutral amino acids[18] and respond to glucagon by increasing cyclic AMP production[17], amino acid transport[19], and gluconeogenesis[16]. They also possess specific binding sites for insulin[16,20,21] and glucagon[20–22].

Isolated hepatocytes have a major potential advantage over membrane preparations since they may permit measurement of both hormone binding and hormone effect(s). This is of particular interest in the case of insulin because of the lack of a consistent (or reproducible) direct effect of this hormone on membrane-bound enzymes[6]. Furthermore, whole cells are presumably more representative of the hepatocyte total surface than plasma membrane fractions which may be relatively enriched in a specialized area of the cell surface[14]. Another advantage of isolated cells in hormone–receptor binding studies is the possibility of measuring and expressing the binding directly per cell. These potential advantages may, however, be counterbalanced by the fact that liver cells usually exhibit relatively high hormone-degrading activity and may themselves be altered by the collagenase during the isolation procedure. The latter problem can be solved by the careful selection of the collagenase used, and, possibly, by maintaining hepatocytes for a few hours or days in primary culture.

Preparation and use of radiolabelled hormones

Evidence has accumulated to indicate that monoiodoinsulin retains *in vitro* both the receptor-binding affinity and the biological potency of native insulin[6,23,24]. The preparation, isolation and characterization of monoiodoinsulin have been described elsewhere[6,23,25]. The specific radioactivity of purified mono[^{125}I]insulin (380 μCi/μg or 2300 Ci/mmol) permits the use of low concentrations (0.01–0.1 nM, i.e. 0.06–0.6 ng/ml or about 1.5–15 μU/ml) in studies of [^{125}I]insulin binding to receptor preparations. Mono[^{125}I]glucagon (650 μCi/μg) prepared in a manner similar[26] to that of mono[^{125}I]insulin exhibits, in contrast to monoiodoinsulin, a higher biological potency than native glucagon in stimulating adenylate cyclase in liver membranes[1]. This higher activity is, however, only partly accounted for by a higher binding affinity of the monoiodohormone for the glucagon receptor in liver membranes[1,27].

SOME PROPERTIES OF INSULIN AND GLUCAGON BINDING IN LIVER CELLS AND LIVER MEMBRANES

Specificity

The biological specificity of insulin and glucagon binding in isolated hepatocytes and liver PM is demonstrated by the data shown in Figure 1 and Table 1. Thus, proinsulin and proinsulin-like molecules compete with the binding of labelled insulin to liver cells (Figure 1, left) and membranes (Table 1) in direct proportion to their biological potency *in vitro* (Table 1)[8,25,28]. Insulin A and B chains are ineffective. Deshistidine glucagon is about 10% as potent as native glucagon in competing with the binding of [^{125}I]glucagon, whereas secretin and vasoactive intestinal polypeptide (VIP) do not affect the binding of labelled glucagon (Figure 1, right)[29,30]. Two conclusions can be drawn from these observations: (1) The binding of insulin and glucagon involves sites that possess the specificity of biologically important receptors to the hormone; (2) These specific hormone-binding systems are interesting tools for the study of hormone structure–activity relationships and for biospecific competitive assays ('radioreceptorassay') of endogenous forms of the hormone[1].

The residual binding of labelled hormone at high concentration of unlabelled hormone (Figure 1) is considered to be 'non-specific' and is usually subtrac-

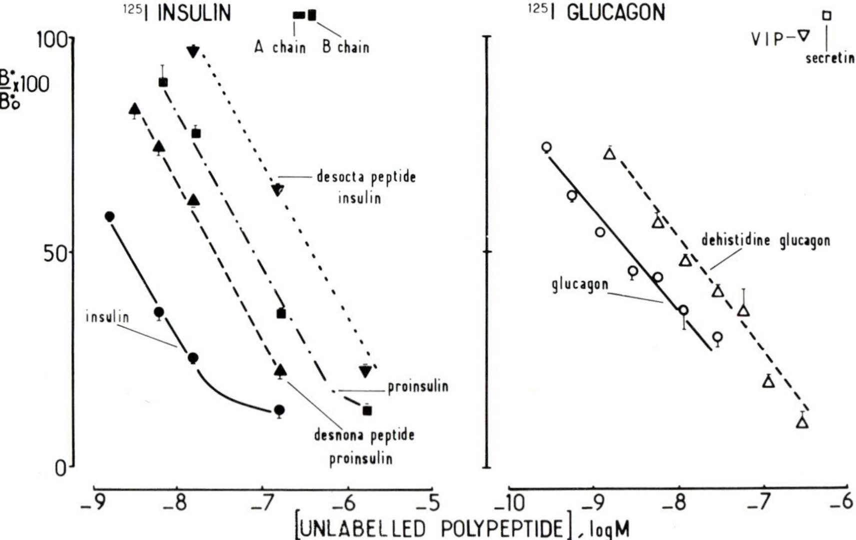

Figure 1 Specificity of binding of [^{125}I]insulin and [^{125}I]glucagon to isolated rat hepatocytes. Binding is expressed as percent of initial binding (Bo) of labelled hormone (i.e. the binding of [^{125}I]hormone in the absence of unlabelled polypeptide), and is plotted against the concentration of unlabelled polypeptide. See reference[20] for experimental conditions (from reference[20], with permission)

Table 1 Specificity of insulin binding in rat liver plasma membranes. The binding affinity (measured in rat liver plasma membranes, see references[8,25] and[28]) and biological potency *in vitro* (measured in isolated rat adipocytes, see references[8,23] and[25]) were determined for insulins from various animal species, proinsulins and proinsulin-like molecules. The data are expressed as percent of porcine insulin (100%)

Insulin or analogue	*Binding affinity in liver membranes (% of native insulin)*	*Biological potency in fat cells (% of native insulin)*
Human, bovine	100	100
Guinea-pig	1.5	2.8
Chicken	180	190
Fish (bonito)	53	50
Proinsulin (porcine)	3.3	2.1
Desdipeptide proinsulin	14	19
Desnonapeptide proinsulin	14.3	20
Diarginine insulin	88.9	73
Proinsulin (rat)	5.5	n.d.

ted from the total to give specific binding. This non-specific component of hormone binding accounts for only 2–5% of total binding in purified liver PM but may represent a higher proportion (e.g. 10–20%) of total binding when whole cells (or less purified fractions of cells) are used[1].

Kinetic and quantitative aspects

The specific binding of [^{125}I]insulin and [^{125}I]glucagon is rapid and temperature-dependent both in hepatocytes and in liver PM[1,20,31]. The binding is more stable at 30 °C than at 37 °C. This is due, at least in part, to accelerated degradation of hormone[11,31] and receptor[31] at higher temperature. [^{125}I]Insulin– and [^{125}I]glucagon–receptor complexes in whole cells can be dissociated by addition of homologous unlabelled hormone in excess (Figure 2). Three points deserve mention: (1) Dissociation is not complete. Thus, 65 to 75% of the cell-bound [^{125}I]hormone is dissociated after 2 hours at 30 °C but dissociation does not appear to significantly proceed afterwards (Figure 2). The nature of this residual bound radioactivity (which is observed even after subtraction of the non-specific component) is unknown. It may represent some intracellular penetration or translocation of hormone (or hormone–receptor complex), or some irreversible binding of hormone since it has also been observed in membranes[25,31]. (2) In liver membranes (and in the absence of EDTA) the rate of dissociation of [^{125}I]glucagon is markedly reduced[32] when compared to that observed in whole cells; guanyl nucleotides markedly enhance and accelerate the dissociation of membrane-bound glucagon[33]. (3) In many of the insulin-binding systems investigated, including intact cells and cell membranes, the rate of dissociation of labelled insulin is accelerated as the occupancy

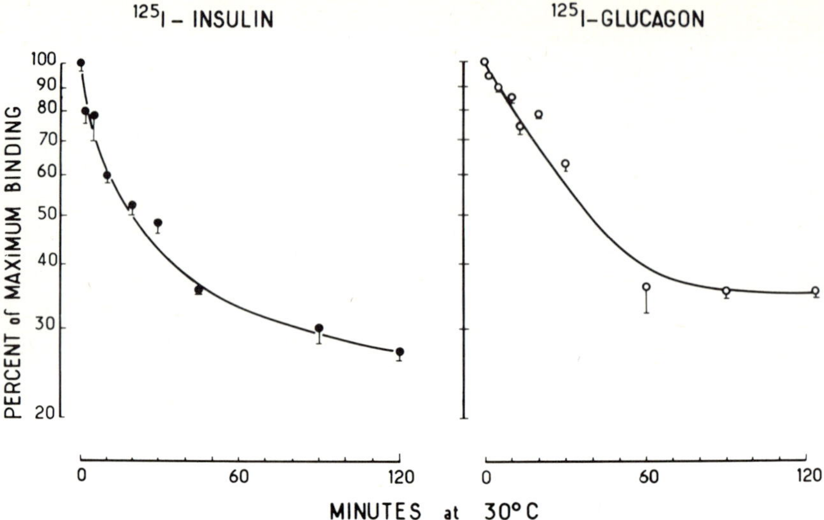

Figure 2 Dissociation of [^{125}I]insulin and [^{125}I]glucagon from isolated hepatocytes at 30 °C. Maximum binding refers to the binding of labelled hormone measured at steady state immediately before the addition (time 0 on figure) of an excess of unlabelled homologous hormone. See reference [20] for other experimental conditions (from reference [20], with permission)

of receptors by the hormone increases; this observation has been interpreted as reflecting insulin-induced negative cooperativity between the receptor sites[34, 35].

In addition to the interaction of insulin and glucagon with their receptors, two processes occur during the exposure of the hormone to liver cells or cell membranes, i.e. degradation[11, 31] (or inactivation[36]) of hormone and 'degradation' of receptor sites[31]. In liver PM, the insulin receptor sites and the insulin-degrading sites appear to differ by several criteria[11]. These observations do not exclude, however, the possibility that the binding of hormone to its receptor may favour or even initiate a specific hormone-degrading process in the intact cell[37] or in the whole organ[38]. On the other hand, insulin itself may, under certain conditions, decrease the concentration of its own receptors on target cells (see below).

The specific binding of insulin to isolated rat hepatocytes (Figure 3) and liver PM (Figure 4, right) is saturable. At apparent steady-state of hormone binding, the major part of the binding occurs over a range of insulin concentrations in the incubation medium (0.35–10 nM, i.e. 2–60 ng/ml or 50–1500 μU/ml) similar to the hormone concentrations in hepatic portal blood under conditions of basal and stimulated insulin secretion[39]. When the data are subjected to Scatchard analysis, the resulting plot is not linear but curvilinear upward (Figure 3, inset). This apparent heterogeneity of binding has been observed in many insulin-binding systems (including both intact cell and cell membrane preparations) and in a number of binding systems for some other

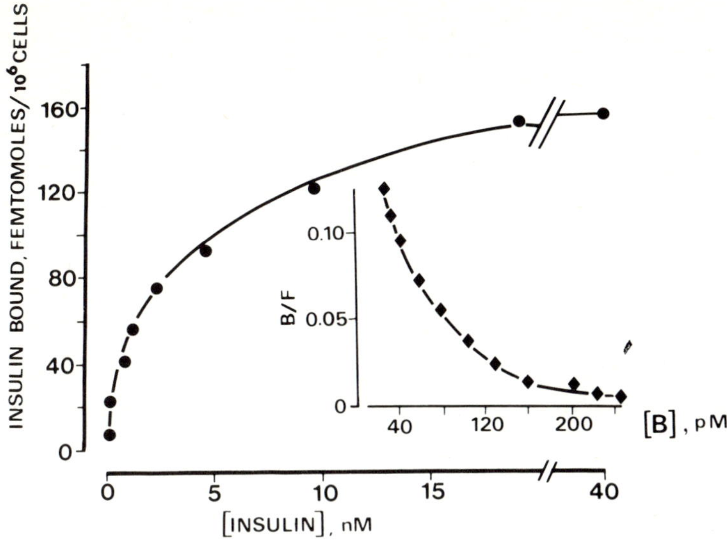

Figure 3 Saturability of insulin binding to isolated rat hepatocytes. The amount of insulin specifically bound at steady state is plotted against the concentration of insulin in the incubation medium. See reference[16] for experimental conditions. Inset: Scatchard plot of the data (from reference[16], modified)

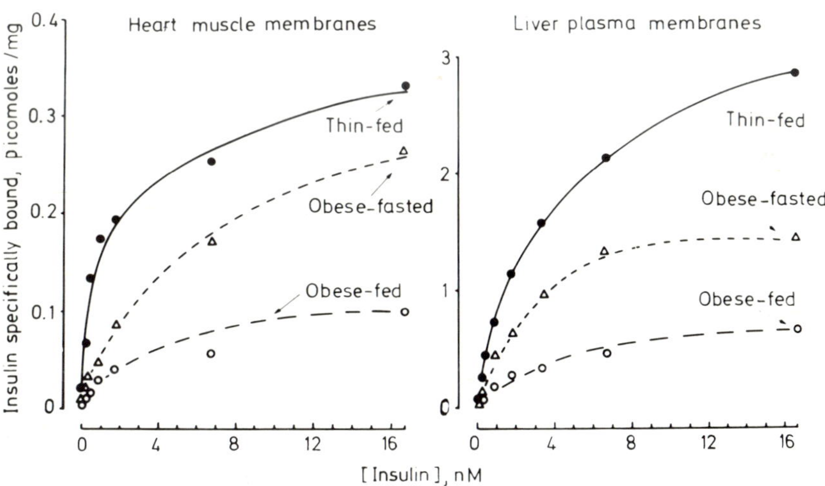

Figure 4 Insulin binding to heart muscle membranes and liver plasma membranes from ob/ob mice fed *ad libitum* (obese-fed) or fasted for 40 hours (obese-fasted), and from their thin littermates (thin-fed). The amount of insulin specifically bound at steady state is plotted against the concentration of insulin in the incubation medium. See refeence[41] for experimental conditions (from reference[41], with permission)

polypeptide hormones including glucagon (see reference[1]). Along with the kinetic properties of dissociation discussed above, these observations are compatible with insulin-induced negative cooperativity between the binding sites of a single receptor population[34, 35]. An alternative, or coexisting, possibility is the presence of several orders of sites of varying affinity[31]. Both models represent possible mechanism(s) whereby the target cell (particularly the liver cell) may acutely adjust its sensitivity to varying concentrations of insulin *in vivo*.

ALTERATIONS OF HORMONE BINDING

Changes in the number (and/or affinity) of the binding sites is another mechanism by which the hormone–receptor interaction can be altered and the sensitivity of the cell to hormone may be modulated.

Insulin binding in animal models of obesity

The binding of insulin is markedly impaired in the genetically obese hyperglycaemic (ob/ob) mouse (see reference[1]). Plasma membranes prepared from livers of ob/ob mice bind only 15–20% as much insulin as do membranes of non-obese mice (Figures 4 and 5). This binding defect is not restricted to the liver; it has also been observed in fat cell membranes[40], heart muscle membranes[41] (Figure 4) and in thymic lymphocytes[42] of the ob/ob mouse. The main characteristics of the defect can be summarized as follows: (1) Insulin binding is involved predominantly[9, 43]. Thus, glucagon binding is decreased by only about 20–30% in liver membranes of the obese mouse, in contrast to the 70–80% decrease in insulin binding (Figure 5); (2) the defect is more apparent in the fully purified PM than in less purified fractions[9]; (3) a decrease in receptor concentration rather than an alteration in receptor affinity mainly accounts for the decreased binding[9,40–43]. Not only the receptor concentration but also the number of binding sites per cell appears to be decreased, as indicated in studies with isolated hepatocytes[9] and thymocytes[42] of the ob/ob mouse; (4) the decreased binding is not due to a direct occupation of receptors by the endogenous insulin, nor can it be attributed to differences in membrane purification or insulin-degrading activity[9,41,45,46]; (5) insulin binding is reversed towards normal when ob/ob mice are fasted (Figure 4) or treated with streptozotocin (Figure 5). The higher degree of increase in insulin binding is observed in mice in which streptozotocin has been the most effective in reducing the hyperinsulinaemia[43]; in contrast to insulin binding, glucagon binding is not affected (Figure 5). Furthermore, fasting has opposite effects on insulin and glucagon binding in liver membranes of ob/ob mice: after a prolonged (12 days) fast, glucagon binding is decreased by 30–40% whereas insulin binding

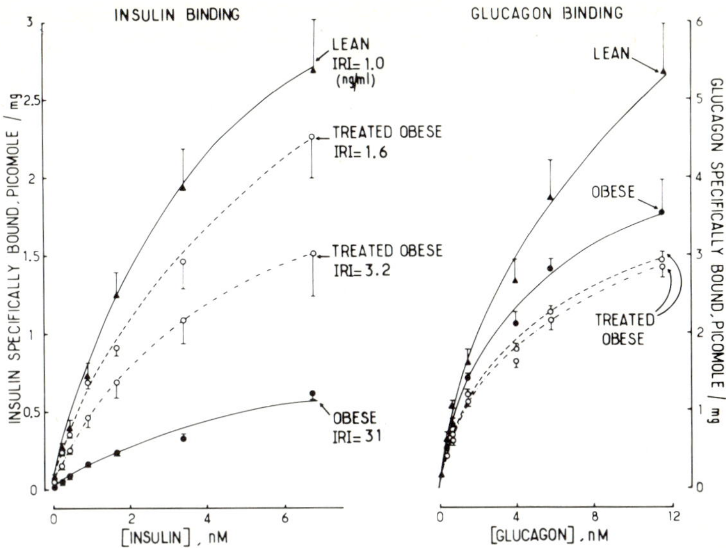

Figure 5 Effect of streptozotocin on the binding of insulin and glucagon to liver plasma membranes from ob/ob mice. Binding data are expressed as in Figure 4. 'Treated obese' refers to streptozotocin-treated animals, which are compared to untreated obese ('obese') and untreated lean ('lean') mice. IRI is the plasma immunoreactive insulin (in ng/ml) assayed in each group. See reference[43] for other experimental conditions (from reference[43], modified)

is enhanced 3- to 4-fold[43]. Fasting is accompanied by a decrease in glucagon binding in non-obese[21,22] and obese[21] rats (see below).

A similar defect in insulin binding has been observed in liver membranes from other genetically obese and hyperglycaemic mice: db/db mouse at the hyperinsulinaemic stage[44], the New Zealand obese mouse[47] and the KK mouse[48].

The insulin-binding defect is not restricted to genetic obesities. Insulin binding is also impaired in liver membranes of goldthioglucose obese mice[44]. In this hypothalamic obesity, the appearance of the binding defect closely parallels the development of hyperinsulinaemia and insulin resistance[46]. Insulin binding is also impaired in liver membranes[49] and hepatocytes of glucocorticoid-treated rats[50]. Conversely, severely diabetic Chinese hamsters[51] and streptozotocin-treated rats[52] that are hypoinsulinaemic, have increased insulin-binding capacity.

All of these data indicate an inverse relationship between the concentration of insulin *in vivo* and the number of insulin receptors on target cells. *In vitro* studies have demonstrated more directly a time-dependent loss of insulin receptors in human lymphocytes cultured in the presence of high concentrations of insulin[53]. Not every hyperinsulinaemic state, however, has been found to be accompanied by a decrease in insulin binding[21]. Other factors besides insulin are most probably involved in the regulation of receptors, which may also be diversely affected by varying physiological or pathological conditions, and differently modulated from tissue to tissue or species to species.

Glucagon binding during fasting

The binding of glucagon is decreased in liver membranes and isolated hepatocytes of fasting animals[1,21,22,43]. This decrease is observed both in obese mice[43] and rats[21] and in non-obese rats[21,22] that have been fasted for varying periods of time (24 hours–12 days). In isolated hepatocytes from 48-hour fasted rats, the decrease in glucagon binding is accompanied by a rightward shift in the dose-response curve of glucagon-stimulated cyclic AMP accumulation[22], i.e. a decrease in the cell sensitivity to hormone; the latter, however, is still capable of exerting maximal stimulation at appropriate concentrations. These observations are compatible with the 'spare' receptor concept, i.e. the cell is equipped with a great number of receptors, all of which are fully functional but only part of which are occupied when the hormone exerts its maximal biological effect(s) (see reference[1]). As a result, any decrease in receptor number (or affinity) will decrease the cell sensitivity to the hormone, as observed with glucagon in hepatocytes from fasting rats[22]. The possibility exists that, in fasting animals, endogenous hyperglucagonaemia may result in a decrease in glucagon receptor number thorugh a 'down' regulatory process similar to that hypothesized in hyperinsulinaemic states[1].

In a similar fashion and in keeping with the spare receptor theory, the cell sensitivity to insulin (but not the hormone maximal effect) will be altered in states (e.g. the animal models of obesity discussed above) where the number of insulin receptors is decreased. In the presence of spare receptors, a decrease in hormone action at maximally stimulating level implies that the defect is distal to the receptor. It should indeed be kept in mind (see page 157) that steps that are subsequent to hormone binding may be rate-limiting in the biological response. Thus, besides (or even aside from) a binding defect, intracellular metabolic alteration(s) appear to be, in many instances, major cause(s) of insulin resistance[1,43].

Antibodies directed against the insulin receptor

Recent investigations have pointed to an entirely different mechanism of decreased binding, namely a profound alteration in insulin–receptor interaction due to antibodies that compete with insulin for binding to the insulin receptor and are present in the serum of patients with a syndrome of extremely severe insulin resistance[54]. Rat liver PM appear to be, in most instances, an appropriate material for the detection of such antibodies directed against the insulin receptor[54].

SUMMARY AND CONCLUSIONS

The binding of insulin and glucagon to hepatocytes and liver PM involves sites that possess the specificity of biologically important receptors to the hormone. The binding is highly specific, saturable and reversible; the latter property is, for glucagon, more apparent in whole cells than in cell membranes. Dissociation kinetics and steady-state data of insulin binding suggest insulin-induced site–site interaction of the negative type. Insulin binding is altered (mainly due to a reduction in the number of sites) in several genetic and acquired obesities. Hyperinsulinaemia appears to be a common denominator in these insulin-resistant states, in which the insulin-binding defect can be reversed when the hyperinsulinaemia is corrected. Other factors, however, are probably also involved in receptor regulation. This binding defect results in decreased cell sensitivity to the hormone, but intracellular defect(s) appear(s) to be, in many instances, major cause(s) of hormone resistance. Antibodies directed against the insulin receptor may also be responsible for extreme insulin resistance.

References

1. Freychet, P. (1976). Interactions of polypeptide hormones with cell membrane specific receptors: studies with insulin and glucagon. The Minkowski Award Lecture 1975. *Diabetologia,* **12,** 83
2. Kahn, C. R. (1975). Membrane receptors for polypeptide hormones. In: E. D. Korn (ed.). *Methods in Membrane Biology, Vol. 3,* pp. 81–146. (New York and London: Plenum Press)
3. Roth, J. (1973). Peptide hormone binding to receptors: a review of direct studies *in vitro. Metabolism,* **22,** 1059
4. Rodbell, M. (1973). The problem of identifying the glucagon receptor. *Fed. Proc.,* **32,** 1854
5. Neville, D. M. Jr. (1968). Isolation of an organ-specific protein antigen from cell-surface membrane of rat liver. *Biochim. Biophys. Acta,* **154,** 540
6. Freychet, P. (1977). Insulin receptors. In: M. Blecher (ed.). *Methods in Receptor Research,* (New York: Marcel Dekker) (In press)
7. Neville, D. M., Jr. and Kahn, C. R. (1974). Isolation of plasma membranes for cell surface membrane receptor studies. In: A. I. Laskin and J. A. Last (eds.). *Subcellular Particles, Structures and Organelles,* pp. 57–88
8. Freychet, P., Roth, J. and Neville, D. M., Jr. (1971). Insulin receptors in the liver: specific binding of [^{125}I]insulin to the plasma membrane and its relation to insulin bioactivity. *Proc. Natl. Acad. Sci. USA,* **68,** 1833
9. Kahn, C. R., Neville, D. M., Jr. (1973). Insulin–receptor interaction in the obese-hyperglycemic mouse. A model of insulin resistance. *J. Biol. Chem.,* **248,** 244
10. Pohl, S. L., Birnbaumer, L. and Rodbell, M. (1971). The glucagon-sensitive adenyl cyclase system in plasma membranes of rat liver. I. Properties. *J. Biol. Chem.,* **246,** 1849

11. Freychet, P., Kahn, R., Roth, J. and Neville, D. M., Jr. (1972). Insulin interactions with liver plasma membranes. Independence of binding of the hormone and its degradation. *J. Biol. Chem.*, **247,** 3953
12. Bergeron, J. J. M., Evans, W. H. and Geschwind, I. I. (1973). Insulin binding to rat liver Golgi fractions. *J. Cell Biol.*, **59,** 771
13. Horvat, A., Li, E. and Katsoyannis, P. G. (1975). Cellular binding sites for insulin in rat liver. *Biochim. Biophys. Acta,* **382,** 609
14. Wisher, M. H. and Evans, W. H. (1975). Functional polarity of the rat hepatocyte surface membrane. Isolation and characterization of plasma membrane subfractions from blood-sinusoidal, bile-canalicular and contiguous surfaces of the hepatocyte. *Biochem. J.*, **146,** 375
15. Seglen, P. O. (1973). Preparation of rat liver cells. III. Enzymatic requirements for tissue dispersion. *Exp. Cell Res.*, **82,** 391
16. Le Cam, A., Guillouzo, A. and Freychet, P. (1976). Ultrastructural and biochemical studies of isolated adult rat hepatocytes. *Exp. Cell Res.*, **98,** 382
17. Rosselin, G., Freychet, P., Fouchereau, M., Rançon, F. and Broer, Y. (1974). Interactions of insulin and glucagon with isolated rat liver cells. II. Dynamic changes in the cyclic AMP induced by hormones. *Horm. Metab. Res.*, **5** (Suppl. Series), 78
18. Le Cam, A. and Freychet, P. (1977). Neutral amino acid transport: characterization of the A and L systems in isolated rat hepatocytes. *J. Biol. Chem.*, **252,** 148
19. Le Cam, A. and Freychet, P. (1976). Glucagon stimulates the A system for neutral amino acid transport in isolated hepatocytes of adult rat. *Biochem. Biophys. Res. Commun.*, **72,** 893
20. Freychet, P., Rosselin, G., Rançon, F., Fouchereau, M., and Broer, Y. (1974). Interactions of insulin and glucagon with isolated rat liver cells. I. Binding of the hormone to specific receptors. *Horm. Metab. Res.*, **5** (Suppl. Series), 72
21. Broer, Y., Freychet, P. and Rosselin, G. (1976). Insulin and glucagon–receptor interactions in the genetically obese Zucker rat: studies of hormone binding and glucagon-stimulated cyclic AMP levels in isolated hepatocytes. *Endocrinology* (In press)
22. Fouchereau-Peron, M., Rançon, F., Freychet, P. and Rosselin, G. (1976). Effect of feeding and fasting on the early steps of glucagon action in isolated rat liver cells. *Endocrinology*, **98,** 755
23. Freychet, P., Roth, J. and Neville, D. M., Jr. (1971). Monoiodoinsulin: demonstration of its biological activity and binding to fat cells and liver membranes. *Biochem. Biophys. Res. Commun.*, **43,** 400
24. Sodoyez, J. C., Sodoyez-Goffaux, F., Goff, M. M., Zimmerman, A. E. and Arquilla, E. R. (1975). [^{125}I]- Or carrier-free [^{125}I]monoiodoinsulin. Preparation, physical, immunological and biological properties and susceptibility to 'insulinase' degradation. *J. Biol. Chem.*, **250,** 4268
25. Freychet, P. (1974). The interactions of proinsulin with insulin receptors on the plasma membrane of the liver. *J. Clin. Invest.*, **54,** 1020
26. Nottey, J. J. and Rosselin, G. (1971). Monoiodoglucagon: Préparation, isolement, identification, contrôle radioimmunologique. *C.R. Acad. Sci.* (*Paris*), **273,** 2118
27. Desbuquois, B. (1975). Iodoglucagon. Preparation and characterization. *Eur. J. Biochem.*, **53,** 569
28. Freychet, P., Brandenburg, D. and Wollmer, A. (1974). Receptor-binding assay of chemically modified insulins: comparison with *in vitro* and *in vivo* bioassays. *Diabetologia*, **10,** 1
29. Rodbell, M., Birnbaumer, L., Pohl, S. L. and Sundby, F. (1971). The reaction of glucagon with its receptor: evidence for discrete regions of activity and binding in the glucagon molecule. *Proc. Natl. Acad. Sci. USA*, **68,** 909
30. Bataille, D., Freychet, P. and Rosselin, G. (1974). Interactions of glucagon, gut glucagon, vasoactive intestinal polypeptide and secretin with liver and fat cell plasma membranes: binding to specific sites and stimulation of adenylate cyclase. *Endocrinology*, **9,** 713

31. Kahn, C. R., Freychet, P., Roth, J. and Neville, D. M., Jr. (1974). Quantitative aspects of the insulin–receptor interaction in liver plasma membranes. *J. Biol. Chem.*, **249,** 2249
32. Rodbell, M., Krans, H. M. J., Pohl, S. L. and Birnbaumer, L. (1971). The glucagon-sensitive adenyl cyclase system in plasma membranes of rat liver. III. Binding of glucagon: method of assay and specificity. *J. Biol. Chem.*, **246,** 1861
33. Rodbell, M., Krans, H. M. J., Pohl, S. L. and Birnbaumer, L. (1971). The glucagon-sensitive adenyl cyclase system in plasma membranes of rat liver. IV. Effects of guanyl nucleotides on binding of [^{125}I]glucagon. *J. Biol. Chem.*, **246,** 1872
34. De Meyts, P., Roth, J., Neville, D. M., Jr., Gavin, J. R., III, and Lesniak, M. A. (1973). Insulin interactions with its receptors: experimental evidence for negative cooperativity. *Biochim. Biophys. Res. Commun.*, **55**, 154
35. De Meyts, P., Bianco, A. R. and Roth, J. (1976). Site–site interactions among insulin receptors. Characterization of the negative cooperativity. *J. Biol. Chem.*, **251,** 1877
36. Pohl, S. L., Krans, H. M. J., Birnbaumer, L. and Rodbell, M. (1972). Inactivation of glucagon by plasma membranes of rat liver. *J. Biol. Chem.*, **247,** 2295
37. Terris, S. and Steiner, D. F. (1975). Binding and degradation of [^{125}I]insulin by rat hepatocytes. *J. Biol. Chem.*, **250,** 8389
38. Terris, S. and Steiner, D. F. (1976). Retention and degradation of [^{125}I]insulin by perfused livers from diabetic rats. *J. Clin. Invest.*, **57,** 885
39. Blackard, W. G. and Nelson, N. C. (1970). Portal and peripheral vein immunoreactive insulin concentrations before and after glucose infusion. *Diabetes,* **19,** 302
40. Freychet, P., Laudat, M. H., Laudat, P., Rosselin, G., Kahn, C. R., Gorden, P. and Roth, J. (1972). Impairment of insulin binding to the fat cell plasma membrane in the obese hyperglycemic mouse. *FEBS Lett.*, **25,** 339
41. Forgue, M. E. and Freychet, P. (1975). Insulin receptors in the heart muscle: demonstration of specific binding sites and impairment of insulin binding in the plasma membrane of the obese hyperglycemic mouse. *Diabetes,* **24,** 715
42. Soll, A. H., Goldfine, I. D., Roth, J., Kahn, C. R. and Neville, D. M., Jr. (1974). Thymic lymphocytes in obese (ob/ob) mice: a mirror of the insulin receptor defect in liver and fat. *J. Biol. Chem.*, **249**, 4127
43. Le Marchand, Y., Loten, E. G., Assimacopoulos-Jeannet, F., Forque, M. E., Freychet, P. and Jeanrenaud, B. (1976). Effect of fasting and streptozotocin in the obese hyperglycemic (ob/ob) mouse: apparent lack of a direct relationship between insulin binding and insulin effects. *Diabetes* (In press)
44. Soll, A. H., Kahn, C. R., Neville, D. M., Jr. and Roth, J. (1975). Insulin receptor deficiency in genetic and acquired obesity. *J. Clin. Invest.*, **56,** 769
45. Soll, A. H., Kahn, C. R. and Neville, D. M., Jr. (1975). Insulin binding to liver plasma membranes in the obese hyperglycemic (ob/ob) mouse. *J. Biol. Chem.*, **250,** 4702
46. Le Marchand, Y., Freychet, P. and Jeanrenaud, B. (1977). Longitudinal study on the establishment of insulin resistance in hypothalamic obese mice. (In preparation)
47. Bacter, D. and Lazarus, N. R. (1975). The control of insulin receptors in the New Zealand obese mouse. *Diabetologia,* **11**, 261
48. Kern, P., Picard, J., Caron, M. and Veissière, D. (1975). Decreased binding of insulin to liver plasma membrane receptors in hereditary diabetic mice. *Biochim. Biophys. Acta*, **389,** 281
49. Kahn, C. R., Goldfine, I. D., Neville, D. M., Jr., Roth, J., Garrison, M. M. and Bates, R. (1973). Insulin receptor defect: a major factor in the insulin resistance of glucocorticoid excess. In *55th Annual Meeting of the Endocrine Society*, p. 168
50. Olefsky, J. M., Johnson, J., Liu, F., Jen, P. and Reaven, G. M. (1975). The effects of acute and chronic dexamethasone administration in insulin binding to isolated rat hepatocytes and adipocytes. *Metabolism,* **24,** 517
51. Hepp, K. D., Langley, J., Von Fucke, H. J., Renner, R. and Kemmler, W. (1975). Increased

insulin binding capacity of liver membranes from diabetic Chinese hamsters. *Nature (London)*, **258,** 154

52. Davidson, M. B. and Kaplan, S. A. (1975). Effect of insulin therapy and caloric restriction in the increased insulin binding by hepatic plasma membranes from diabetic rats. *Diabetes,* **24** (Suppl. 2), 394 (abstract)
53. Gavin, J. R., III, Roth, J., Neville, D. M., Jr., De Meyts, P. and Buell, D. N. (1974). Insulin-dependent regulation of insulin receptor concentration: a direct demonstration in cell culture. *Proc. Natl. Acad. Sci. USA,* **71,** 84
54. Flier, J. S., Kahn, C. R., Roth, J. and Bar, R. S. (1975). Antibodies that impair insulin receptor binding in an unusual diabetic syndrome with severe insulin resistance. *Science,* **190,** 63

11
HBsAg: a Target Antigen on the Liver Cell?

F. GUDAT and L. BIANCHI

It is well established that — especially in non-cytolytic viral infections — the expression of viral antigens at the cell surface may be the most important event for the recognition and subsequent elimination of infected cells by the immune response[1]. The tolerance of the hepatitis B (HB) virus by liver cells in immunosuppressed patients and healthy carriers has clearly shown that the HB virus infection is primarily not cytopathic and therefore supposedly determined by viral antigen expression and the specific immune response[2]. Among the known hepatitis B (HB) associated antigens the surface antigen (HBsAg) is a prime candidate for a target antigen on infected cells for two reasons:

(1) A specific humoral and cellular immune insufficiency for HBsAg has been repeatedly shown in persistent HB infections[3].
(2) HBsAg has been localized by immunofluorescence at the liver cell membrane in tissue sections[4–7] and on single liver cells in suspension[8]

In the present study the membrane expression of HBsAg in chronic HB was related to the formation of virus core (HBcAg) in the liver as well as to the appearance of DANE particles and of specific antibodies in the blood.

MATERIAL AND METHODS

HBAg seropositive patients, 93 in number, were studied for the presence of HBsAg and HBcAg in the liver by immunofluorescence and for DANE particles in blood by immune electron microscopy as described elsewhere[9]. For the identification of spontaneous agglutinates of HBAg particles the pellet of 0.3 ml serum (60 min, 20 000 *g*) was screened using electron microscopy by the negative staining method as described before[9]. Anti-HBsAg was determined by radioimmunoassay (Aus-AB of Abbot Laboratories) and anti-HBcAg by an indirect immunofluorescent test on HBcAg positive human liver sections.

RESULTS AND CONCLUSIONS

Figure 1 shows a typical example of the honeycomb-like pattern of membrane-associated fluorescence specific for HBsAg which was noted facultatively in

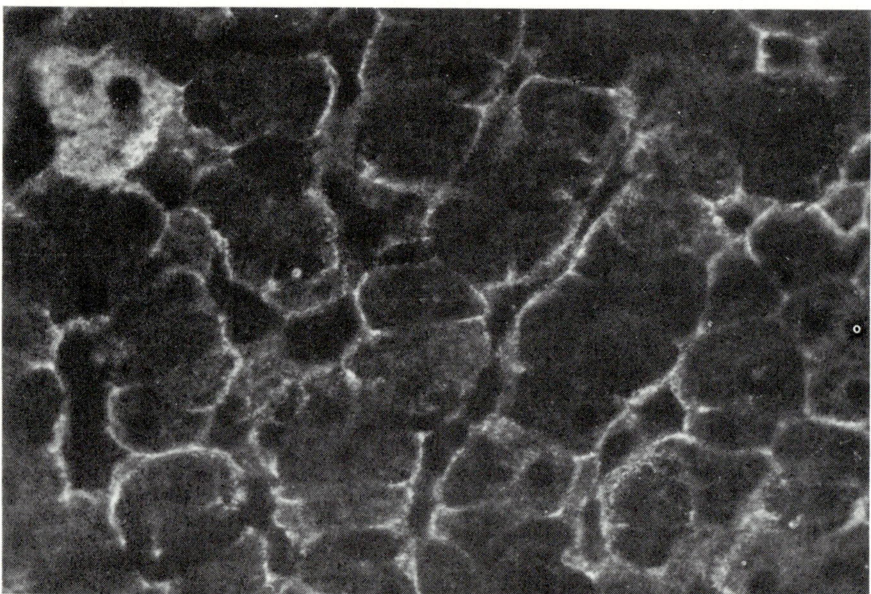

Figure 1 Focal cytoplasmic and diffuse honeycomb-like membrane-associated fluorescence specific for HBsAg in a case of chronic aggressive hepatitis (× 560)

addition to the more common intracytoplasmic expression. Figure 2 demonstrates the abolition of the specific fluorescence after absorption of the specific antiserum with HBsAg in an example of focal membrane staining for HBsAg.

The incidence of diffuse or focal membrane-associated HBsAg is shown in Figure 3 in which the patients have been grouped not only according to the histological reaction but essentially according to the expression of HBcAg in the liver. Core formation was regularly associated with the appearance of DANE particles.

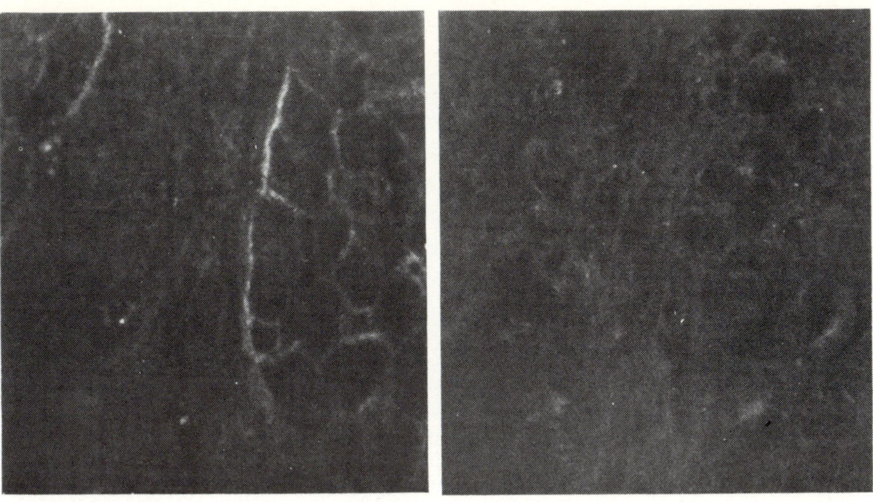

Figure 2 Focal membrane-associated HBsAg (left) and absorption control with HBsAg (right) (× 320)

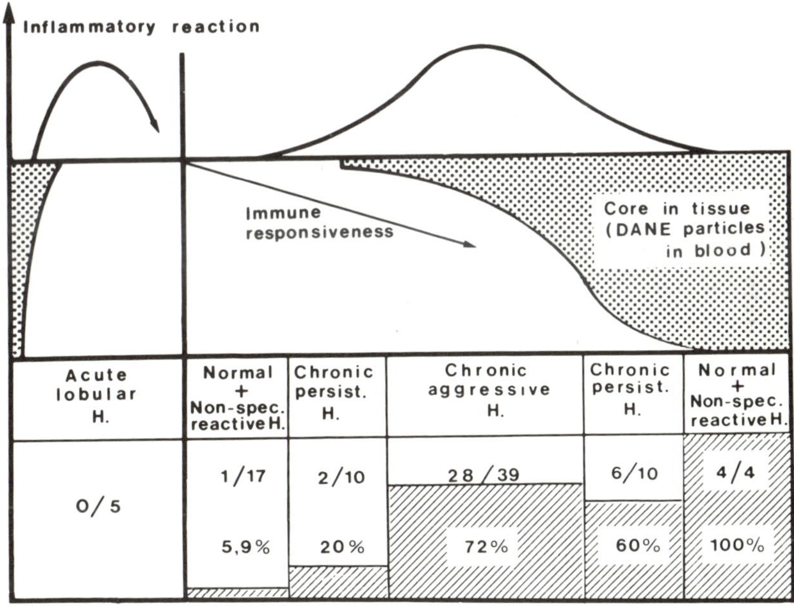

Figure 3 Number of patients with membrane-associated HBsAg in relation to histology, core and DANE particle formation

The most extensive core formation detectable in 60–100% of liver cell nuclei was found in chronic persistent or non-specific reactive hepatitis under effective immunosuppression or, in some patients, spontaneously (HBc-predominance type[2]). The same histological reaction, however, may be associated with the absence or extremely rare occurrence of HBcAg and DANE particles (HBs-predominance type[2]). Spotty expression of HBcAg, up to approximately 60% of liver cell nuclei, is mainly associated with chronic aggressive hepatitis or borderline to chronic persistent hepatitis (focal HBc + s type[2,9]).

We have assumed[2,9] that the core expression rather quantitatively reflects an elimination insufficiency of the immune system among a scale graded between complete immune competence and incompetence (indicated by arrow in Figure 3). This assumption is based on the observation that the HBc-predominance type with generalized core formation and persistence of infection is regularly found under strong immunosuppressive therapy, whereas the normergic response of acute lobular hepatitis is associated with elimination of HBAg expressing cells and termination of the infection. In other words, the lower the immune responsiveness the more extensive will be the tolerance and expression of viral antigens *in* and *on* liver cells.

Against this background we found a striking increase of membrane staining for HBsAg within the histological groups as core and DANE particle formation became more prominent: 5.9% of patients with non-specific reactive hepatitis and 20% of patients with chronic persistent hepatitis, both associated with the HBs-predominance type; 72% of patients with both chronic aggressive hepatitis and the focal HBc+s type; 60% of patients with chronic persistent hepatitis and all four patients with non-specific reactive hepatitis (three on immunosuppression because of transplantation, one on anti-tumour chemotherapy), both groups exhibiting the HBc-predominance type. These findings are compatible with the interpretation that the formation of complete virus particles is tolerated by an incompetent immune state and is associated with HBsAg at the level of the liver cell membrane.

It is not yet known whether the immunofluorescent membrane staining reflects:

(1) an intracellular submembranous accumulation, e.g. as part of a presecretory process,
(2) an integrated component of the cell membrane as known for other enveloped viruses,
(3) or an intercellular accumulation of secreted HBsAg which is more or less associated with the cell surface.

The mentioned single cell studies[8] and our own preliminary experiments showing HBsAg positive spots on the liver cell circumference (Figure 4) in chronic forms of hepatitis would support the view that HBsAg is, indeed, a target antigen on the liver cell surface, the presence and persistence of which would be largely determined by the current immune response.

The antibody determinations are also in line with this interpretation (Table 1). By radioimmunoassay, anti-HBsAg was not detectable in most cases of chronic hepatitis irrespective of membrane staining for HBsAg and irrespective

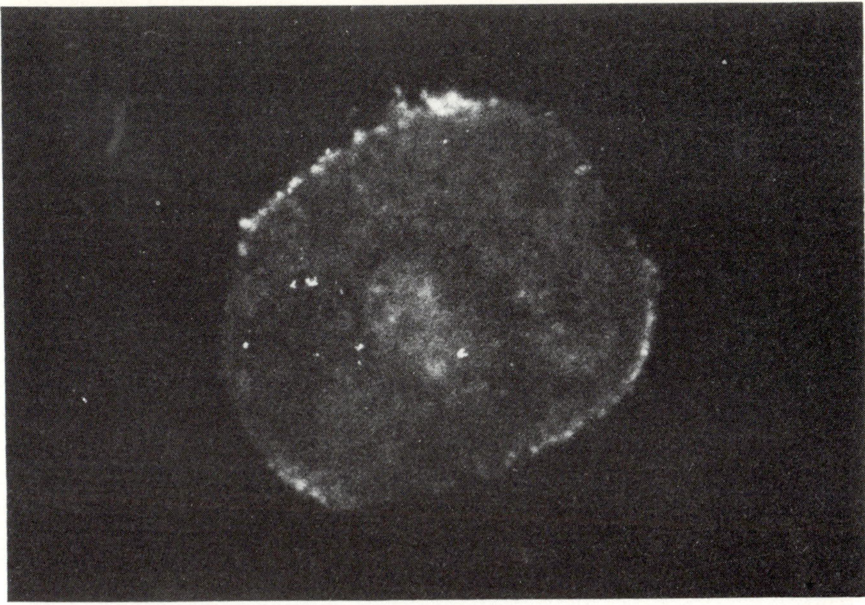

Figure 4 Liver cell with spotty demonstration of HBsAg on the cell surface obtained after incubation of liver cell suspension with anti-HBsAg (× 2100)

Table 1 Humoral immune response in chronic hepatitis in relation to membrane-associated HBsAg

Membrane-associated HBsAg	*Patient number*	*Anti-HBs positive (RIA)*	*Anti-HBc positive (IF)*
Positive	46	4.3%	96%
Negative	47	5.0%	100%

of detectable core formation. The apparently paradoxical finding that — except for immunosuppressed patients — all developed antibodies against HBcAg speaks against any role of this system in the elimination of core-forming cells.

A new aspect for the elimination of core-forming cells arises from the electron microscopic study of spontaneously occurring HBAg complexes in these sera (Figure 5). Similar to the frequency of membranous HBsAg fluorescence we could identify HBAg complexes mainly in the sera of patients with core and DANE particle formation, i.e. in chronic aggressive hepatitis (73% of all patients) and in chronic persistent hepatitis with the HBc predominance type of core expression (64% of all patients). The low incidence among patients with non-specific reactive hepatitis is probably due to the im-

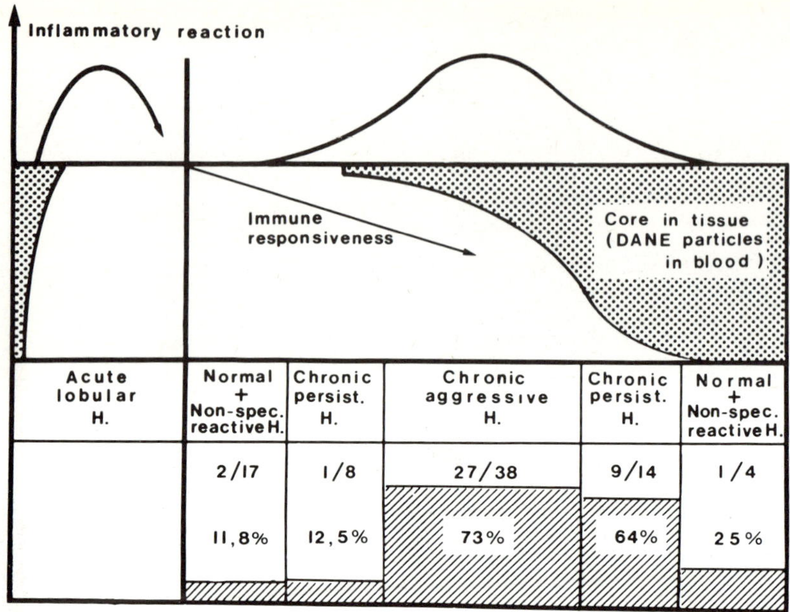

Figure 5 Number of patients with spontaneous HBAg complexes in blood in relation to histology, core and DANE particle formation

munosuppression and the relative antigen excess in this group (one of four patients). The interesting point is that these complexes are preferentially or exclusively composed of tubules and/or DANE particles (Figures 6 and 7).

On the basis of these findings we offer the working hypothesis that — in addition to the classical HBsAg target system — there are distinct antigens on the surface of tubules and DANE particles which are under separate control of the immune system. Their relationship to the antigens described by Moodie *et al.*[10] and the e-antigen[11] remains to be established. This or these DANE particle-associated antigen(s) are possibly additional HB antigens coexisting with the classical HBsAg on the cell surface during certain core-dependent phases of the virus replication and release. Accordingly, they may likewise play a role as target antigens for the elimination of virus core replicating cells.

Acknowledgements

This study has been supported by a grant from the Schweizerische Nationalfond (Nr. 6.164–0.75). The radioimmunoassay for anti-HBsAg was kindly performed by Dr M. Gasser, Institute for Microbiology, Basel.

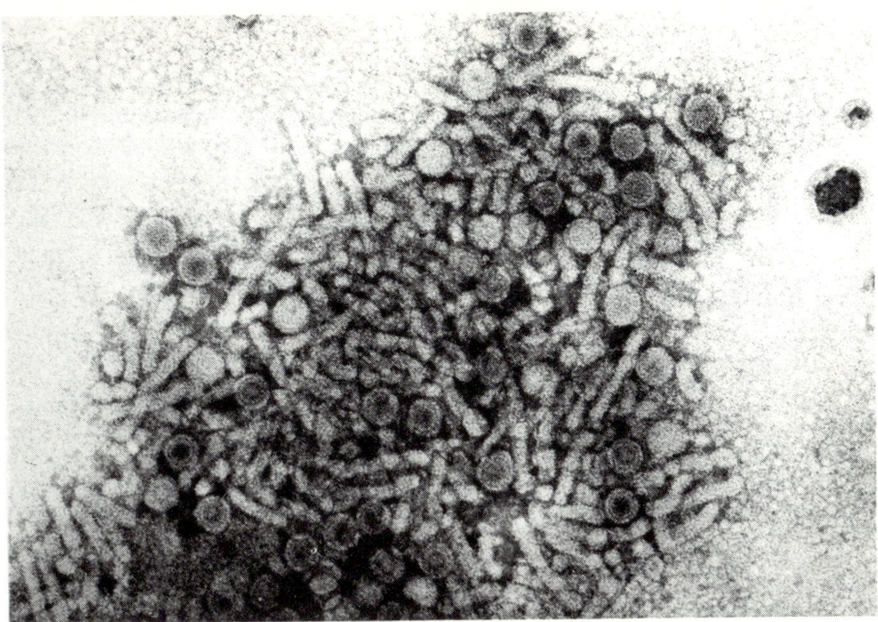

Figure 6 Spontaneous HBAg complex, composed preferentially of tubules and DANE particles. Obtained from the blood of a patient with chronic aggressive hepatitis (× 100 000)

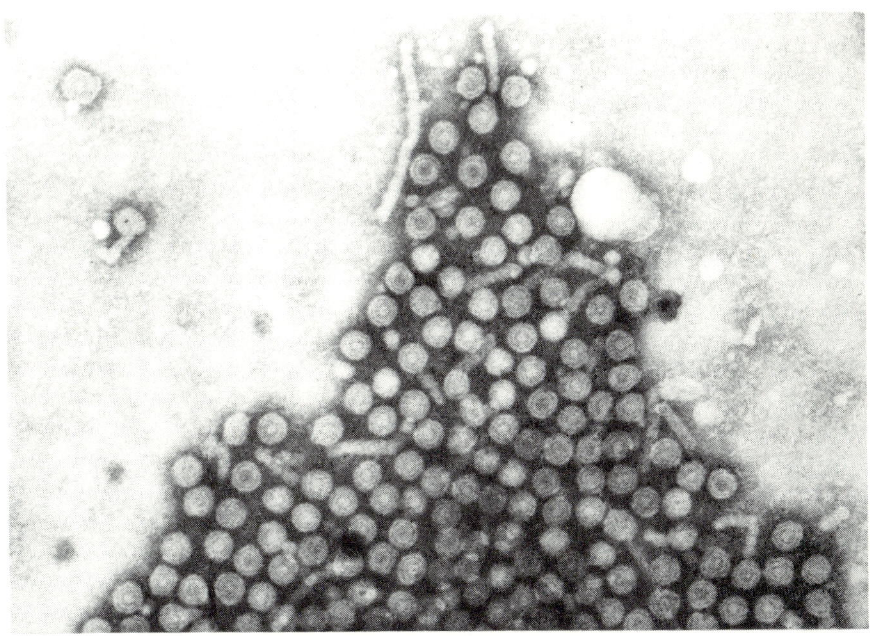

Figure 7 Spontaneous HBAg complex, consisting of DANE particles. Obtained from blood of a patient with chronic aggressive hepatitis (× 120 000)

References

1. Burns, W. H. (1975). Viral antigens. In: A. L. Notkins (ed.). *Viral Immunology and Immunopathology*, pp. 43–56. (New York, San Francisco, London: Academic Press)
2. Gudat, F., Bianchi, L., Sonnabend, W., Thiel, G., Aenishaenslin, W. and Stalder, G. A. (1975). Pattern of core and surface expression in liver tissue reflects state of specific immune response in hepatitis B. *Lab. Invest.*, **32,** 1
3. Edgington, T. S. and Chisari, F. V. (1975). Immunological aspects of hepatitis B virus infection. *Am. J. Med. Sci.*, **270,** 213
4. Edgington, T. S. and Ritt, D. J. (1971). Intrahepatic expression of serum hepatitis virus-associated antigens. *J. Exp. Med.*, **134,** 871
5. Nowoslawski, A., Krawczynski, K., Brzosko, W. J. and Madalinski, K. (1972). Tissue localization of Australia antigen immune complexes in acute and chronic hepatitis and liver cirrhosis. *Am. J. Pathol.*, **68,** 31
6. Roos, C. M., Feltkamp-Vroom, T. M. and Helder, A. W. (1976). The localization of hepatitis B antigen and immunoglobulin G in liver tissue: an immunofluorescence, light and electron microscopic study. *J. Pathol.*, **118,** 1
7. Ray, M. B., Desmet, V. J., Bradburne, A. F., Desmyter, J., Fevery, J. and De Groote, J. (1976). Differential distribution of hepatitis B surface antigen and hepatitis B core antigen in the liver of hepatitis B patients. *Gastroenterology*, **71,** 462
8. Alberti, A., Realdi, G., Themolada, F. and Cadrobbi, P. (1975). HBAg on liver-cell surface in viral hepatitis. *Lancet*, **i,** 346
9. Gudat, F., Bianchi, L., Stalder, G. A. and Schmid, M. (1976). Klassifizierung und Infektionsität der chronischen Hepatitis B, definiert durch Dane-Partikel im Blut und Virus-Komponenten in der Leber. *Schw. Med. Wochenschr.*, **106,** 812
10. Moodie, J., Stannard, L. M. and Kipps, A. (1974). Dane complexes in hepatitis B antigen. *J. Gen. Virol.*, **24,** 375
11. Magnius, L. O. and Espmark, A. (1972). A new antigen complex co-occurring with Australia antigen. *Acta Pathol. Microbiol. Scand., Section B.*, **80,** 335

12
Cell Surface Carbohydrates and Cell Transformation: a General Change Signifying Tumorigenicity

P. EMMELOT, W. P. van BEEK and L. A. SMETS

The cell surface seems to be intimately involved in the establishment of functional contact relations between cells leading to recognition, adhesion and regulation of proliferation. Since these properties may be changed in neoplasia (loss of/escape from immunological, positional and growth controls), cell surface alterations could be instrumental in neoplastic behaviour[1–7].

In the carbohydrate domain, a number of changes that affect glycoprotein and glycolipid components of the surface of transformed cells have been noted[1–7]. These changes may or do include:

(1) decreased glycosyl extension (shorter carbohydrate chains) in gangliosides and neutral glycolipids,
(2) increased agglutination by Concanavalin A,
(3) changes in glycoprotein : glycosyltransferases,
(4) reduction or loss of large, external, transformation-sensitive (LETS) glycoprotein; ~250 000 daltons,
(5) increase in early-eluting, fucose-labelled and sialic acid-containing glycopeptides.

Although these five types of changes are not the only ones, they are the ones most extensively studied. Yet the pertinent research has been confined mainly to cells of the fibroblastic type transformed by oncogenic viruses and cultured *in vitro*. Accordingly the problem as to the general occurrence of these cell surface changes arises:

Do they occur irrespective of cell type, oncogenic determinant and mode of cell growth?

Do they relate to the expression of transformation or of tumorigenicity which are not necessarily equivalent?

In the present state of our knowledge, heterogeneity rather than uniformity or specificity of change seems to prevail in respect of most of the five afore-

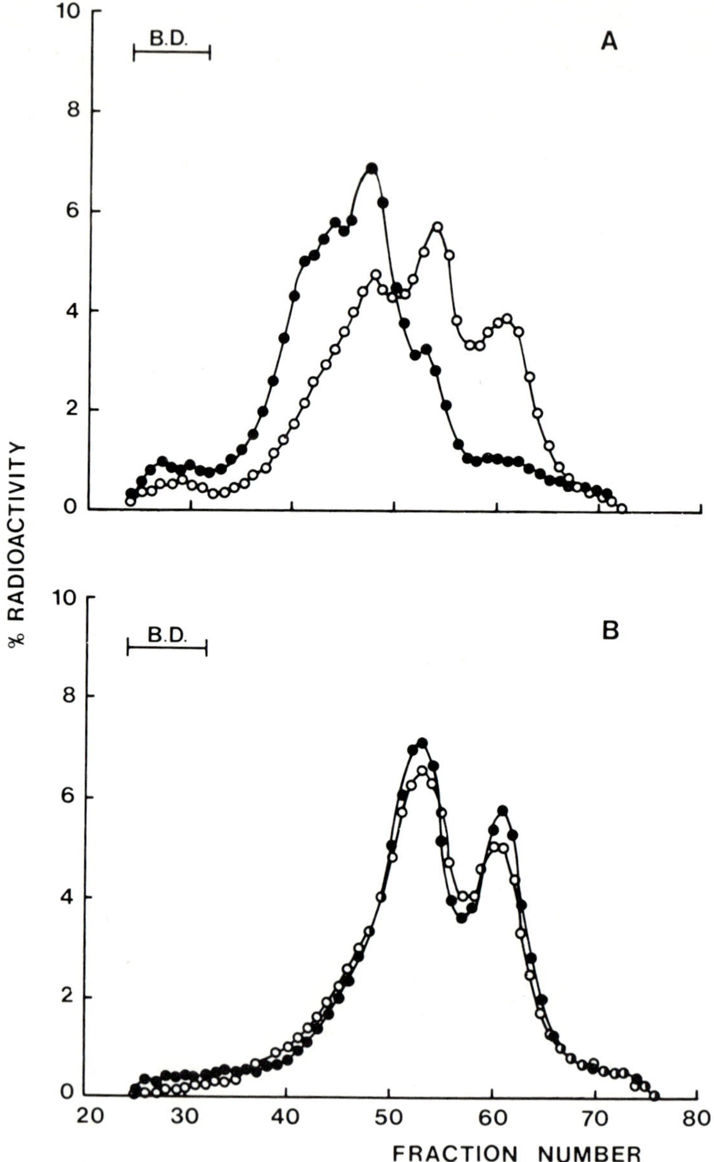

Figure 1A, B Elution profile of surface glycopeptides from spontaneously transformed 3T3-F cells (black circles) compared with that from normal 3T3 cells (open circles). A before, B after neuraminidase pretreatment

mentioned properties. Only in respect of the fifth change can a reasonably firm conclusion be drawn as to the consistency of the change and its significance as a marker of tumorigenicity.

Change number 5 is monitored as follows. Normal cells are labelled with [^{14}C]fucose and neoplastic ones with [^{3}H]fucose. Trypsinization and exhaustive pronase digestion of the combined trypsinates of the pair of cells to be compared, followed by gel filtration over Sephadex G-50 fine–Biogel P10 (1 : 2), shows an increase in early-eluting glycopeptides from neoplastic as compared with normal cells, which is frequently accompanied by a decrease in late-eluting glycopeptides (Figure 1). This change in elution profile was discovered by Buck, Glick and Warren[8–12] using various virus-transformed fibroblastic cell lines, and was later confirmed and extended in our laboratory (Tables 1, 2 and 3)[13–15].

Table 1 Fibroblastic system

Transformed in vitro	*Grown* in vitro
Virus transformed	Mouse, hamster, chicken fibroblasts RSV*, MSV*; polyoma*† (Normal profile for ts mutant at non-permissive temperature)*
Spontaneously transformed	3T3 (3T3f)†

* Buck, Glick and Warren[8,9]
† van Beek, Smets and Emmelot[13]

Transformation *in vitro* is known to result in a spectrum of alterations such as loss of the contact inhibitions of movement and growth, decreased serum requirement, expression of T antigen, and anchorage independence of growth. However, it has also been shown[16] that these phenotypic traits may not be coordinately expressed, and that the anchorage independence of growth is actually the only one of the aforementioned changes that correlates with the tumorigenicity of virus-transformed cells (in immunologically neutral mice). Can a similar distinction be made in the case of the early-eluting glycopeptides?

First of all a non-selected population of transformed cells may not only contain a heterogeneity of phenotypes but in fact may consist of only a few tumorigenic cells. If so, any biochemical change that might be related to tumorigenicity as distinct from cell transformation, would be missed by studying the non-selected population *in vitro*, but would be picked up after selection of the cells *in vivo*.

Thus, non-selected dimethylnitrosamine-transformed hamster cells[17] and SV40-transformed 3T3 mouse cells[18] did *not* show the increase in early-eluting glycopeptides, but, after these cells had been selected by transplantation and the tumours passaged and labelled *in vitro*, the increase was very distinct (Table 2, Figures 2 and 3). As a further control the SV40-3T3 cells were nonselectively grown *in vivo* in a diffusion chamber; these cells showed a normal glycopeptide profile.

Furthermore different SV40-3T3 clones were selected *in vitro*. These clones varied widely and progressively exhibiting transformed phenotypic characters such as abnormal morphology, nuclear overlap, saturation density, serum requirement, DNA synthesis under stationary conditions and Concanavalin A agglutinability. However, all these transformed cell populations showed normal glycopeptide profiles and the conclusion was therefore drawn[18] that the increase in early-eluting glycopeptides does not relate to transformation, as judged by most of the altered cellular properties *in vitro* that one commonly associates with the transformed state, but instead relates to tumorigenicity.

Table 2 Fibroblastic system

Transformed in vitro	*Grown* in vivo	*Labelled* in vitro
Py- and dimethylnitrosamine-transformed hamster fibroblasts*		
SV40-transformed 3T3†		

* Glick, Rabinowitz and Sachs[17]
† Smets, van Beek and van Rooy[18]

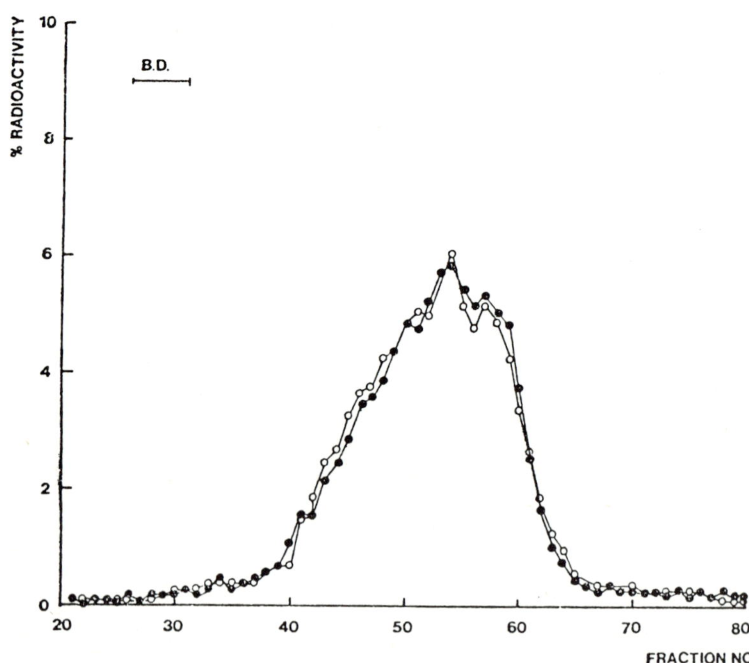

Figure 2 Elution profile of surface glycopeptides from SV40-transformed 3T3 cells (black circles) compared with that from 3T3 cells (open circles)

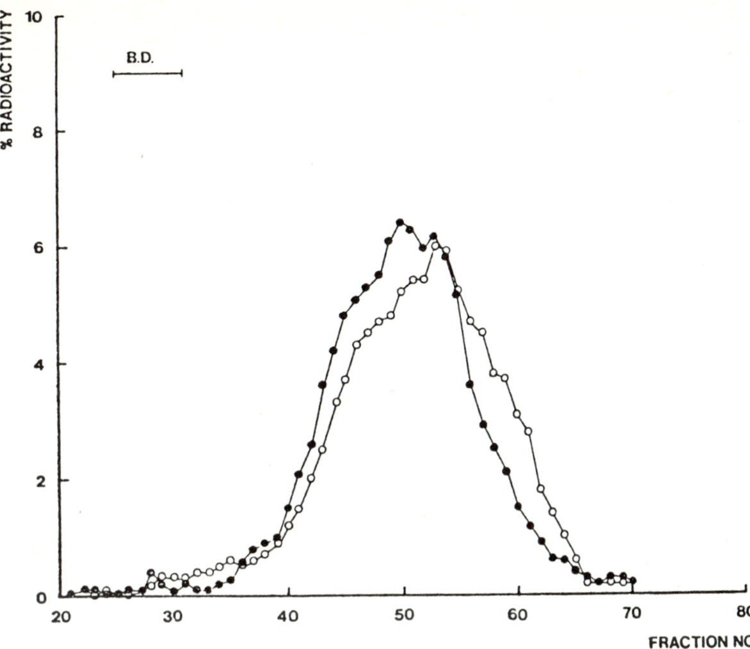

Figure 3 Elution profile of surface glycopeptides from SV40-transformed 3T3 cells selected by transplantation (black circles) as compared with that from 3T3 cells (open circles)

We tried to show that the increase in early-eluting glycopeptides is exhibited by as many tumours and tumour types as possible, independent of conditions. It was deemed more important to demonstrate the universality of the change at this stage, than to study more precisely the chemical composition of the material.

After the fibroblast system the mouse lymphoid system was studied (Table 3). First, malignant lymphoblasts (spontaneously arisen *in vivo*, but explanted

Table 3 Mouse lymphoid system

Spontaneous formation	*Tumour*	*Grown/ labelled*
In vivo	Malignant lymphoblasts (MBVIA)	*in vitro* or *in vivo*
	versus non-malignant lymphoblasts (MBIII)	*in vitro*
In vivo	GRSL lymphoma	*in vivo*
	versus thymocytes (GR)	*in vivo*, *in vitro*

van Beek, Smets and Emmelot[13], and unpublished

and maintained in tissue culture for many years) compared to their non-malignant homologues grown and labelled *in vitro*, showed the increase in early glycopeptides irrespective of whether the neoplastic cells had been grown and labelled *in vitro* or *in vivo* (as ascites cells, Figure 4A). Thus, the growth and labelling condition did not determine the change.

Secondly, the increase in early glycopeptides was also found for thymus-

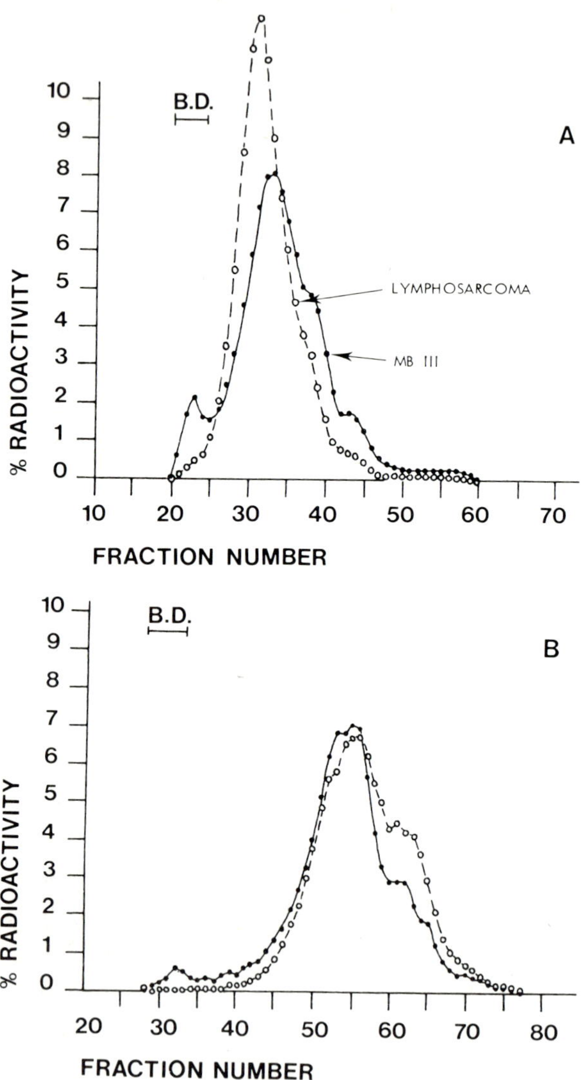

Figure 4A, B Elution profile of surface glycopeptides from mouse lymphosarcoma cells, MBVIA, grown in (C57Black × DBA)F_1 females as ascites cells and labelled *in vivo* (open circles) as compared with that from non-malignant M13III lymphoblasts (black circles) grown and labelled *in vitro*. A before, B after neuraminidase pretreatment

derived leukaemia cells spontaneously arisen recently and maintained as ascites cells in GR mice and labelled *in vivo*, in comparison with the corresponding material from GR thymocytes (Figure 5).

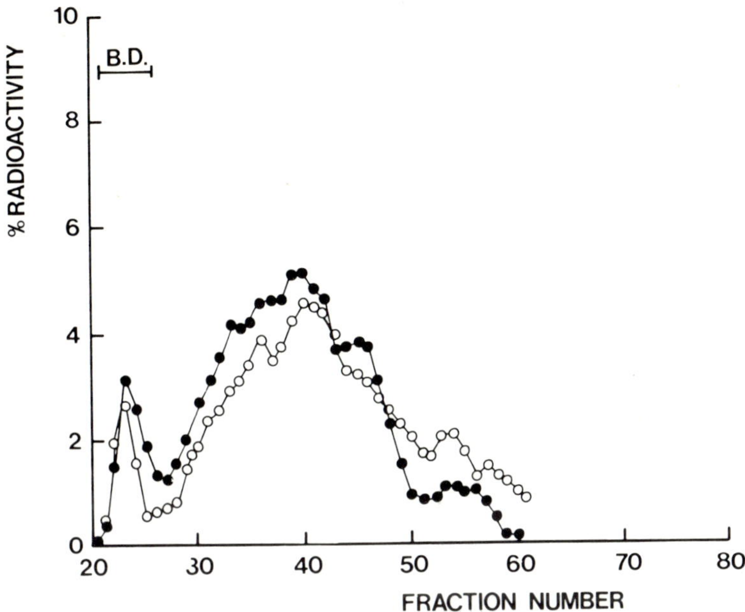

Figure 5 Elution profile of surface glycopeptides from mouse leukaemia cells, GRSL cells, thymus derived, maintained as ascites cells and labelled *in vivo* (black circles) as compared with that from normal thymocytes from GR/A mice (open circles)

The rat normal liver cell/hepatoma system was chosen as an example of an epithelial system, and was extensively studied in three types of experiments using four different chemically-induced tumours. The tumours were grown *in vitro* and *in vivo* — under the latter condition as ascites as well as solid hepatomas (Table 4 — two Novikoff hepatoma strains, one maintained *in vitro*

Table 4 Rat hepatoma/liver

Grown	*Tumours/controls*		
In vitro	Novikoff[1]; H-35; MH_1C_1		versus rat liver cells *in vitro*
In vivo	Novikoff[2] ascites (AT) Novikoff[2] solid (ST_2) Novikoff[1] solid (ST_1) Controls: regenerating and late-embryonic liver	Labelled *in vivo*	versus RLC *in vitro*
In vivo	ST_1; ST_2; H35	Labelled *in vivo*	late-embryonic liver labelled *in vivo*

van Beek, Smets and Emmelot, 1976 and unpublished.

and the other *in vivo*, were used next to the Reuber H-35 hepatoma, and MH_1C_1 hepatoma cells).

First, for the *in vitro* grown hepatoma cells, rat liver cells (RLC) cultured *in vitro* served as controls[13]. Secondly, *in vivo* grown hepatoma cells were also labelled *in vivo* and were then compared with controls as follows[15]: *In vivo* [^{3}H]fucose labelled hepatomas compared with [^{14}C]labelled RLC *in vitro* exhibited the increase in early-eluting glycopeptides (Figure 6).

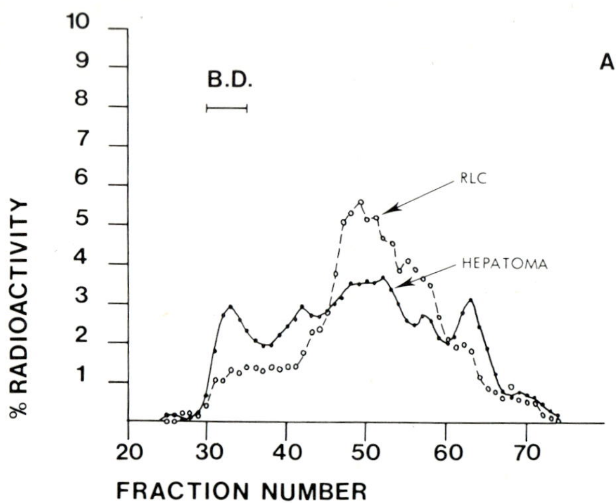

Figure 6 Elution profile of surface glycopeptides from Novikoff rat hepatoma cells, strain N_1S_1-67, grown as ascites cells in Sprague–Dawley females and labelled *in vivo* (black circles) as compared with that from rat liver cells (RLC) cultured and labelled *in vitro* (open circles). After neuraminidase pretreatment the two profiles coincided (not illustrated)

Next, *in vivo* [^{3}H]fucose labelled regenerating liver or late embryonic liver had similar profiles to [^{14}C]labelled RLC *in vitro* (Figures 7 and 8). This dual comparison (indirect via RLC *in vitro*) allowed a comparison of *in vivo* growing hepatoma with *in vivo* growing liver.

Thirdly, notwithstanding the costs involved in using [^{14}C]fucose for *in vivo* labelling, we finally compared *in vivo* [^{3}H]fucose labelled hepatomas with *in vivo* [^{14}C]labelled late-embryonic liver.

In all cases (Table 4) the early-eluting glycopeptides were increased in the tumour material.

The above results indicate that the constant difference in glycopeptide profile, i.e. the increase in higher molecular weight fucose-labelled material, is:

(1) related to tumorigenicity but not related to
  (i) the more common traits of transformation,

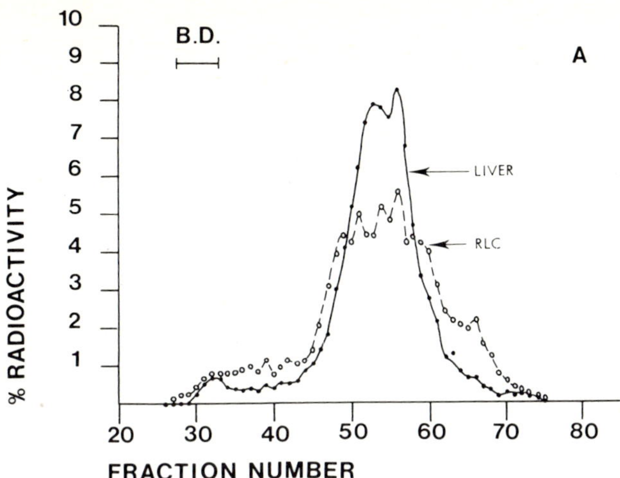

Figure 7 Elution profile of surface glycopeptides from late-embryonic rat liver (pregnant Sprague–Dawley females were injected with tritiated L-fucose 48 and 24 hours before delivery — 6–14 hours after birth livers of the young were collected) (black circles) as compared with that from RLC cultured and labelled *in vitro*

(ii) proliferation *per se*,

(iii) late-embryonic expression;

(2) independent of species, mode of growth and labelling conditions of the cells (*in vitro*/*in vivo*; solid tumour/ascites cells);

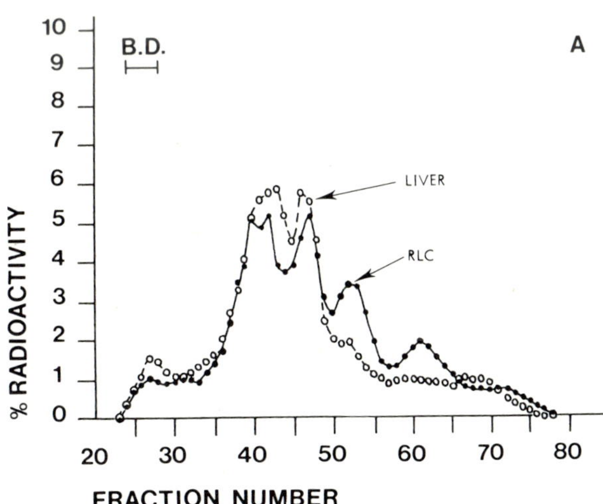

Figure 8 Elution profile of surface glycopeptides from regenerating liver cells (two-thirds hepatectomized Sprague–Dawley rats received tritiated L-fucose at 24 and 48 hours after the operation and were sacrificed 24 hours later) (open circles) as compared with that from RLC cultured and labelled *in vitro*

(3) independent of oncogenic determinant — viral, chemical, spontaneous;
(4) independent of tissue type — fibroblastic, lymphoid, epithelioid.

Mouse mammary tumour compared with lactating breast has recently been found in this laboratory to be another example of an epithelial type of tissue that exhibits the glycopeptide change (for melanoma compare[19, 20]).

This raises the question of the nature of the increase in the early-eluting glycopeptides. In the elution profile a vertical shift indicates a change in amount of material, and a horizontal one a change in size or molecular weight. A profile is obtained by plotting the percentage radioactivity present in each fraction, the radioactivity of the total eluate amounting to 100. Thus the profile illustrates relative amounts of fucose-labelled glycopeptides of various molecular weights, and since the column was not overloaded in our experiments, the profile is — and was checked to be — independent of the amount of material brought on the column. It is, therefore, a useful parameter for comparative analysis.

If the glycopeptide material is pretreated with neuraminidase (*Vibrio cholerae*, multispecific), the elution profile (which exhibits four peaks, if the first is disregarded) is shifted to the right, to the lower molecular weight region (containing only two peaks). By this pretreatment normal and tumour profiles do become identical (Figures 1B, 4B, 10B and 12B). This behaviour is shown by most cultured cells which generally are loosely associated, and by free cells *in vivo* such as ascites tumour and leukaemic cells. However, exceptions do exist to the rule that neuraminidase pretreatment abolishes the difference between normal and tumour glycopeptide profiles. Exceptions appear to be tumour strain or type specific (the H-35 hepatoma and Burkitt lymphoma are cases in point), but may also depend on the mode of tissue growth. The

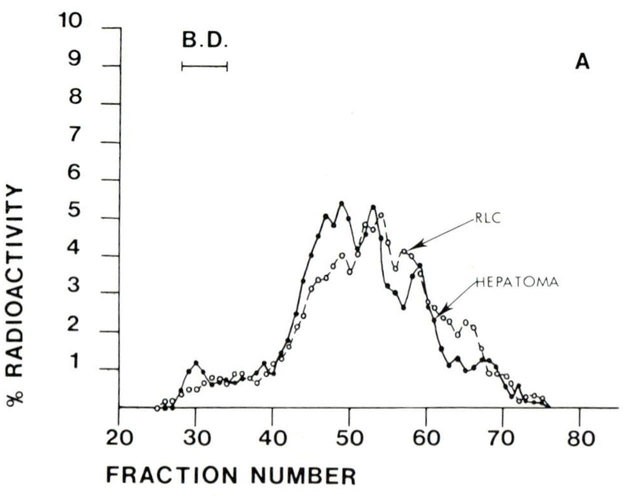

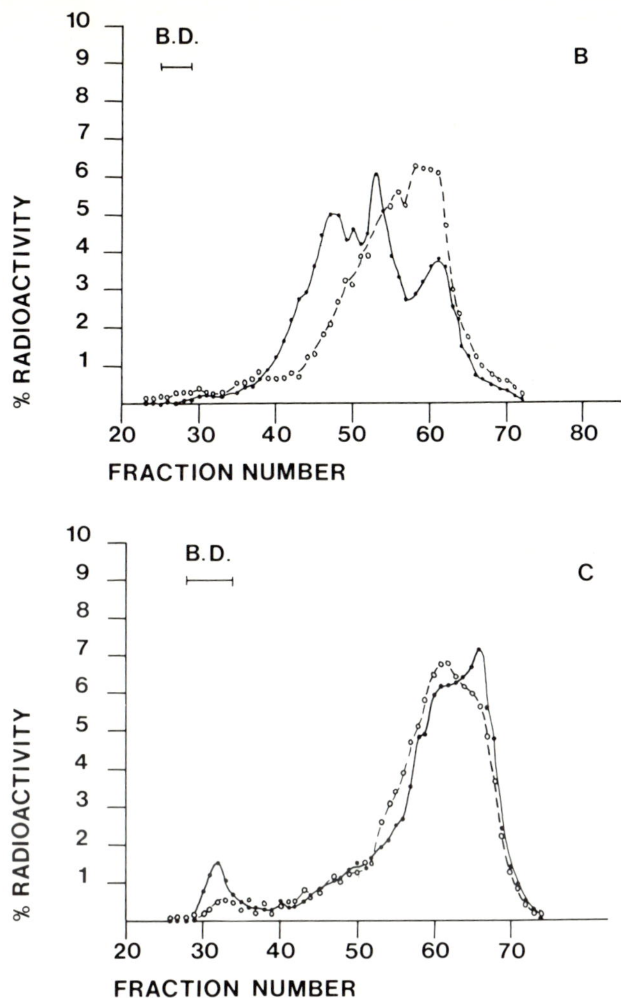

Figure 9A, B, C Elution profile of surface glycopeptides from Novikoff rat hepatoma cells, grown as solid tumour following subcutaneous injection of ascites cells (Figure 6) and labelled *in vivo* (black circles) as comlared with that of RLC cultured and labelled *in vitro* (open circles). A before, B after neuraminidase pretreatment; C after mild chemical hydrolysis as mentioned in the text

latter is illustrated by the solid hepatoma (ST-2) obtained by subcutaneous injection of the intraperitoneally maintained ascites hepatoma. Whereas neuraminidase treatment led to the coincidence between *ascites* hepatoma and normal liver cell profiles (Figure 6), this was not so in the case of the *solid* hepatoma compared to normal liver cells (Figure 9A, B).

Some 30% of the sialic acid in isolated liver cell and hepatoma plasma membranes is insensitive to neuraminidase[21] and this might explain the lack of effect of neuraminidase in the last-mentioned solid hepatoma. We therefore used a very mild acid hydrolysis to remove sialic acid (90 min in 0.01 N HCl at 80 °C). This procedure was specific for sialic acid, since, firstly, sialyl bonds of the glycosyl bonds are the most sensitive to acid, followed by fucosyl bonds, and, secondly, 10% of the fucose label at the most was lost by the chemical hydrolysis. Applying the mild hydrolysis to the solid hepatoma–liver system indeed led to the coincidence of profiles (Figure 9C).

Coincidence of elution profiles after neuraminidase treatment or very mild acid hydrolysis specific for sialic acid indicates that the original difference in elution profiles resulted from an increased sialic acid density (concentration) in fucose-labelled surface glycoprotein of tumour cells[10,11,22,23,24].

The sialic acid is apparently bound in two forms: (a) neuraminidase sensitive, and (b) neuraminidase insensitive, but sensitive to mild chemical hydrolysis. The latter has been found in the case of Burkitt lymphoma and solid hepatomas, and it might be a property of solid tumours in general.

There is a third category of change, in which neither neuraminidase nor mild acid hydrolysis brings the elution profiles to coincidence. This has been found for rat hepatoma H-35 (independent of whether cultured *in vitro* or grown as solid tumour *in vivo*), and chronic myelocytic leukaemia in man.

The human material studied consists of biopsy specimens (Table 5) and a large number of established tumour cell strains.

Table 5 Human biopsies

Leukaemias:		
	ALL, CLL, AML, CML* (20/20)	
	(profile*: specific disease characteristic)	
	unknown diagnosis: 5/5	
	uncertain diagnosis: 2/4	
Lymphosarcomas (2/2)		
	controls:	a–Peripheral lymphocytes
		b–Peripheral lymphocytes, PHA-stimulated: lymphoblasts
		c–Infectious mononucleosis (12 patients)
		d–Leukaemoid reaction by virus infection (2)
		e–Remissions (3)

* van Beek, Smets and Emmelot[14], and unpublished

Some 20 patients, both adults and children, suffering from the four main leukaemias — acute and chronic, lymphocytic and myelocytic[14] — were studied, and without exception showed the increase in early-eluting glycopeptides, which in all cases studied except chronic myelocytic leukaemia (CML) could be abolished by neuraminidase (Figure 10). Five patients in which the diagnosis of leukaemia at the time of our analysis was incomplete had

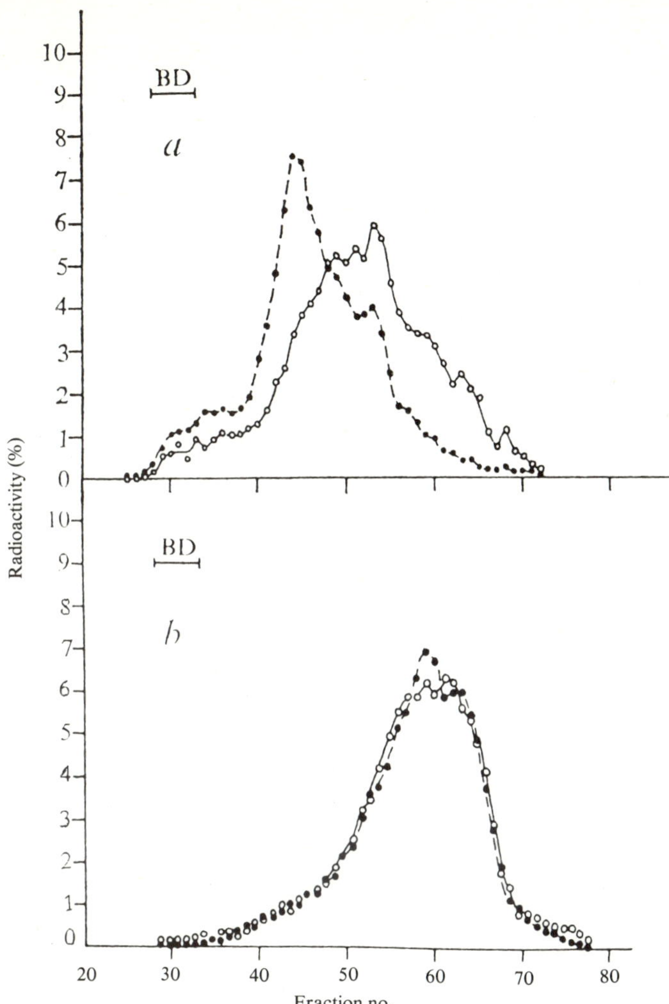

Figure 10 Elution profile of surface glycopeptides from human acute myelocytic leukaemia cells (black circles) compared with that from normal peripheral lymphocytes (open circles). *a* before, *b* after neuraminidase pretreatment. Fresh cells were labelled for 2 days under tissue culture conditions. (From reference [14], by courtesy of Macmillan)

malignant profiles and turned out to be leukaemic. In addition, of five patients with diagnosis of leukaemia uncertain, two were ranked (on account of their glycopeptide profiles) as malignant and this was in the course of the disease confirmed by the treating haematologists. Patients who had gone into remission by therapy showed a normal profile. In all these cases peripheral lymphocytes, either resting or stimulated into proliferation by phytohaemagglutinin served as controls, their profiles being similar in the ascending limb[14]. Similarly, 12 patients with the non-malignant lymphoproliferative disorder infectious mononucleosis[14] and a number of patients with leukaemoid reaction (leukocytosis) from virus infections had a normal profile. Thus, as shown here and earlier, the change in early glycopeptide is not due to proliferation *per se*.

Finally 39 different human tumour cell strains were studied. Increase of early-eluting glycopeptides was shown by (a) 20 out of 20 Burkitt lymphoma strains. The profile of Burkitt lymphoma was characteristic for this disease (Figure 11) and was not due to the expression of EB virus antigen; as mentioned previously, it is neuraminidase resistant, but sensitive to mild acid hydrolysis; (b) four out of five non-Burkitt, non-Hodgkin lymphoma strains

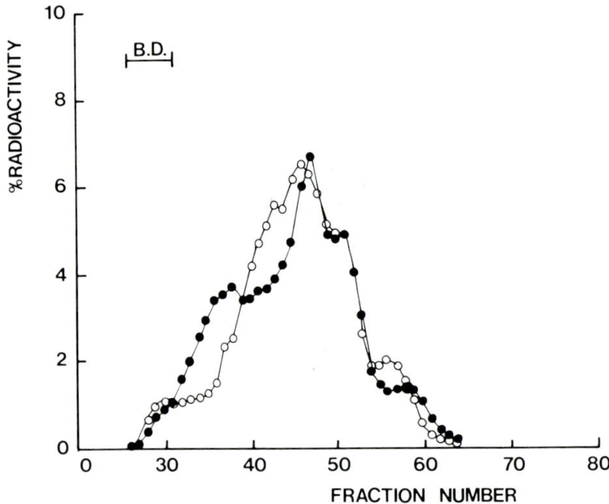

Figure 11 Elution profile of surface glycopeptides from Burkitt lymphoma cells (Bjab/HRSK; EBNA positive cultured *in vitro*) (black circles) as compared with that from normal peripheral lymphocytes (open circles)

(Figure 12); (c) six out of seven leukaemia strains. All cases (a–c) were run against normal lymphocytes; (d) six out of seven gliomas compared with low-passage glial cells. Altogether 36 of the 39 human tumour cell lines showed an increase in early-eluting glycopeptides. The three negative specimens were either not the cell types assumed or these cells have lost their malignant potency by long-term culturing or at least consist of too few tumorigenic cells.

CONCLUDING REMARKS

The phenomenon — increase of early-eluting glycopeptides — appears to be a general marker of malignancy and thus may serve diagnostic purposes.

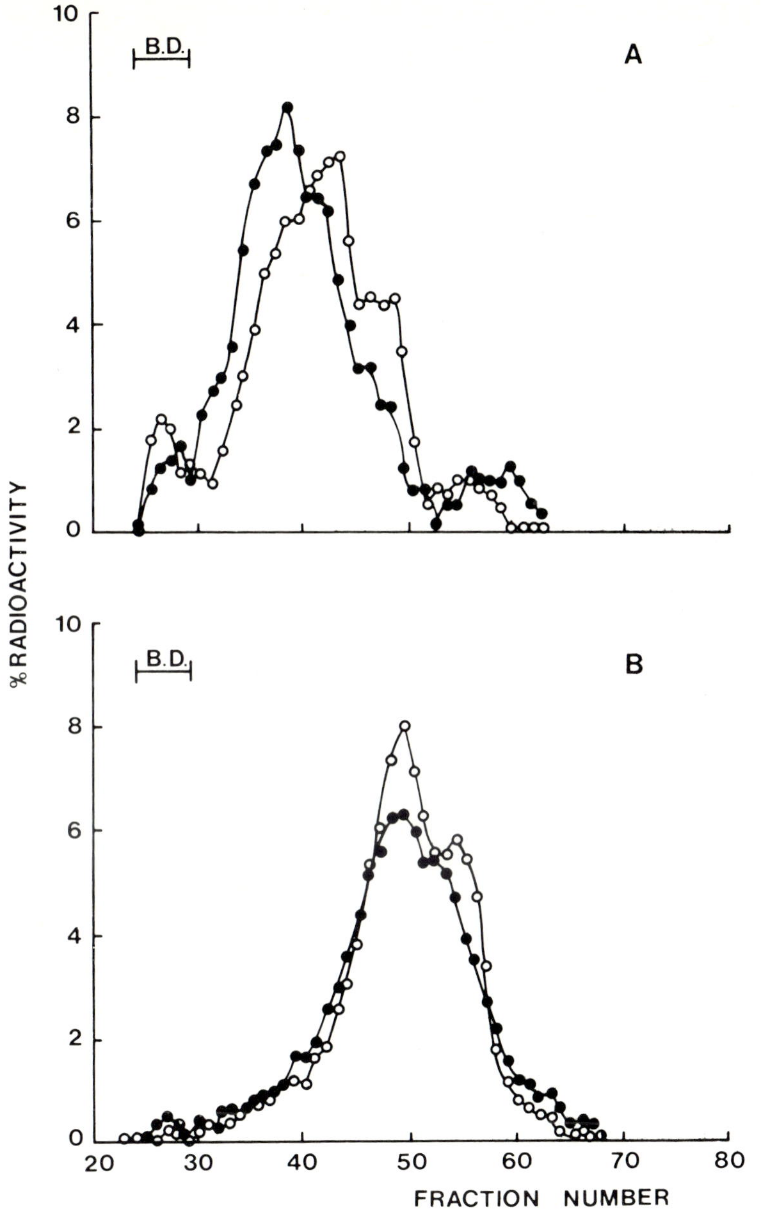

Figure 12A, B Elution profile of surface glycopeptides from non-Burkitt, non-Hodgkin lymphoma cells (line K562, cultured *in vitro*) (black circles) as compared with that from normal peripheral lymphocytes (open circles). A before, B after neuraminidase pretreatment

Moreover, being an inseparable attribute of tumour cells, one ought to assume that the particular glycoprotein components — which are also coordinately expressed on intracellular membranes[25] — are in some way instrumental in neoplastic cell behaviour. Mitotic normal fibroblasts also show the increase in early-eluting glycopeptides as compared with normal interphase cells[26] and the mitotic surface expression appears to be retained on tumour cells[1,7]. This may signify the lack of the tumour cell (surface) to adapt by steric (structure-modifying) and/or synthetic processes to a condition that allows for normal functional contact relation between cells.

References

1. Emmelot, P. (1973). Biochemical properties of normal and neoplastic cell surfaces: a review. *Eur. J. Cancer,* **9,** 317
2. Pardee, A. B. (1975). The cell surface and fibroblast proliferation: some current research trends. *Biochim. Biophys. Acta* (*Reviews on Cancer*), **417,** 153
3. Nicolson, G. L. (1976). Transmembrane control of the receptors on normal and tumor cells. II. Surface changes associated with transformation and malignancy. *Biochim. Biophys. Acta* (*Reviews on Cancer*), **458,** 1
4. Brady, R. O. and Fishman, P. H. (1975). Membranes of transformed mammalian cells. In: C. F. Fox (ed.). *Biochemistry of Cell Walls and Membranes*, pp. 61–96. (London: Butterworths)
5. Hynes, R. O. (1976). Cell surface proteins and malignant transformation. *Biochim. Biophys. Acta* (*Reviews on Cancer*), **458,** 73
6. Hakomori, S. I. (1975). Structures and organization of cell surface glycolipids. Dependency on cell growth and malignant transformation. *Biochim. Biophys. Acta* (*Reviews on Cancer*), **417,** 55
7. Emmelot, P. (1974). Cell surface changes and neoplastic phenotype(s). In: W. Davis and C. Maltoni (eds.). *Characterization of Human Tumours. Proceedings Fifth International Symposium on the Biological Characterization of Human Tumours*, pp. 3–9. (Amsterdam: Excerpta Medica)
8. Buck, C. A., Glick, M. C. and Warren, L. (1970). A comparative study of glycoproteins from the surface of control and Rous sarcoma virus-transformed hamster cells. *Biochemistry,* **9,** 4567
9. Buck, C. A., Glick, M. C. and Warren, L. (1971). Glycopeptides from the surface of control and virus-transformed cells. *Science,* **172,** 169
10. Warren, L., Fuhrer, J. P., Buck, C. A. and Walborg, E. F., Jr. (1974). Membrane glycoproteins in normal and virus-transformed cells. In: J. Schultz and R. E. Bock (ed.). *Membrane Transformation in Neoplasia*, Miami Winter Symposia, Vol. 8, pp. 1–20. (New York: Academic Press)
11. Glick, M. C. (1974). Chemical components of surface membranes related to biological properties. In: E. J. C. Lee and E. E. Smith (eds.). *Biology and Chemistry of Eukaryotic Cell Surfaces*, Miami Winter Symposia, Vol. 7, pp. 213–237. (New York: Academic Press)
12. Warren, L., Critchley, D. and Macpherson, I. (1972). Surface glycoproteins and glycolipids of chicken embryo cells transformed by a temperature-sensitive mutant of Rous sarcoma virus. *Nature* (*London*), **235,** 275
13. van Beek, W. P., Smets, L. A. and Emmelot, P. (1973). Increased sialic acid density in surface glycoproteins of transformed and malignant cells: a general phenomenon? *Cancer Res.,* **33,** 2913

14. van Beek, W. P., Smets, L. A. and Emmelot, P. (1975). Changed surface glycoprotein as a marker of malignancy in human leukaemic cells. *Nature* (*London*), **235**, 457
15. Smets, L. A., van Beek, W. P., Collard, J. G., Temmink, H., van Gils, B. and Emmelot, P. (1975). Comparative evaluation of plasma membrane alterations associated with neoplasia. In: *Cellular Membranes and Tumor Cell Behavior.* The University of Texas System Cancer Center M.D. Anderson Hospital and Tumor Institute, 28th Annual Symposium on Fundamental Cancer Research, pp. 269–287 (Baltimore, Maryland: Williams and Wilkins)
16. Shin, S. I., Freedman, V. H., Risser, R. and Pollack, P. (1975). Tumorigenicity of virus-transformed cells in nude mice is correlated specifically with anchorage independent growth *in vitro*. *Proc. Natl. Acad. Sci. USA*, **72**, 4435
17. Glick, M. C., Rabinowitz, Z. and Sachs, L. (1973). Surface membrane glycopeptides correlated with tumorigenesis. *Biochemistry*, **12**, 4869
18. Smets, L. A., van Beek, W. P. and van Rooy, H. (1976). Surface glycoproteins and Concanavalin A-mediated agglutinability of clonal variants and tumor cells obtained from SV40-virus transformed mouse 3T3 cells. *Int. J. Cancer*, **18**, 462
19. Warren, L., Zeidman, I. and Buck, C. A. (1975). The surface glycoproteins of a mouse melanoma growing in culture and as a solid tumor *in vivo*. *Cancer Res.*, **35**, 2186
20. Bhavanandan, V. P. and Davidson, E. A. (1976). Characteristics of a mucin-type sialoglycopeptide produced by B16 mouse melanoma cells. *Biochem. Biophys. Res. Commun.*, **70**, 139
21. Emmelot, P. and Bos, C. J. (1972). Studies on plasma membranes. XVII. On the chemical composition of plasma membranes prepared from rat and mouse liver and hepatomas. *J. Membr. Biol.*, **9**, 83
22. Warren, L., Fuhrer, J. P. and Buck, C. A. (1972). Surface glycoproteins of normal and transformed cells: a difference determined by sialic acid and a growth-dependent sialyl transferase. *Proc. Natl. Acad. Sci. USA*, **69**, 1838
23. Ogata, S.-I., Muramatsu, T. and Kobata, A. (1976). New structural characteristic of the large glycopeptides from transformed cells. *Nature* (*London*), **259**, 580
24. Critchley, D. R., Wyke, J. A. and Hynes, R. O. (1976). Cell surface and metabolic labeling of the proteins of normal and transformed chicken cells. *Biochim. Biophys. Acta*, **436**, 335
25. Buck, C. A., Fuhrer, J. P., Soslan, G. and Warren, L. (1974). Membrane glycopeptides from subcellular fractions of control and virus-transformed cells. *J. Biol. Chem.*, **249**, 1541
26. Glick, M. C. and Buck, C. A. (1973). Glycoproteins from the surface of metaphase cells. *Biochemistry*, **12**, 85

13
An Influence of the Cell Surface on Growth Properties of Normal and Neoplastic Cells

M. M. BURGER

Since Virchow's days the cell has perhaps been too much considered as a microcosm by itself. Many a cell function is seen by the cell biologist in the narrow context of the individual cell rather than in the context of its neighbouring cells and the extracellular matrix. Thus phagocytosis, migration, secretion and division are all functions that are treated in the textbook as individual functions of an individual cell. Whether a cell will initiate such a process and at what time it will do so depend to a large degree on influences by neighbouring cells and a cell's environment in general.

Kalckar created the term 'cell sociology'[1], implying thereby that cell communities exist which communicate with each other, thereby modifying their mutual behaviour and possibly also their function. Morphogenesis and organogenesis are thought to be processes where cell–cell interaction may play an important rôle. Three possible mechanisms of cell–cell recognition may illustrate some types of cell–cell interactions in general.

CELL INTERACTION MECHANISMS

Early in embryonal life poorly differentiated cells may have to recognize each other while moving around at random or establishing temporary contacts during morphogenesis. Such recognition can be mediated by macromolecules located at the cell surface as shown in the upper part of Figure 1[2,3]. Later on epithelial as well as mesenchymal cells become embedded in a matrix of specific proteoglycan and collagen macromolecules[4] which will certainly provide a certain cohesion to the conglomerate of cells and might be responsible — together with the cellular cytoskeleton — for the tissue architecture of an organ (see extracellular linkages, i.e. upper bridge in the last two adhering cells on the right in Figure 1). Finally a third means of cell–cell interaction, besides that of surface recognition and matrix-mediated cohesion, is of course cell–cell communication via the exchange of intracellular material, as for instance via the gap junctions (lower portion of Figure 1).

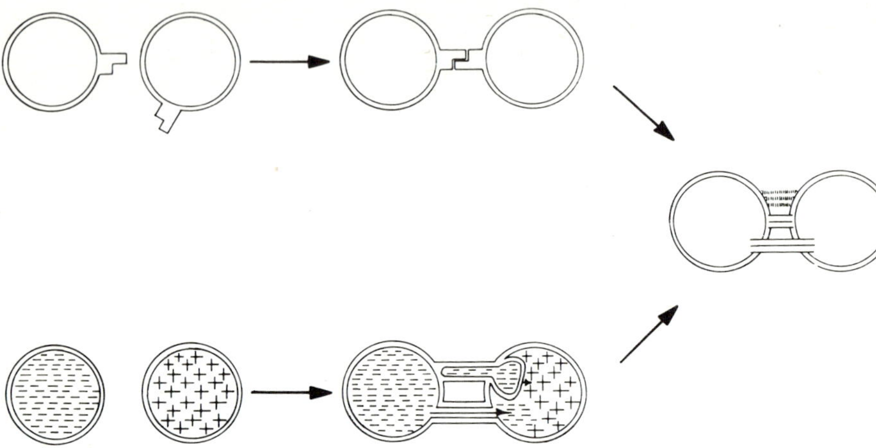

Figure 1 Two possible recognition mechanisms. In the upper portion of this figure a mechanism is depicted where recognition occurs directly via surface macromolecules. In the lower portion an alternative mechanism is considered where information from within the cell is passed on to the neighbouring cell, either via vesicular exchange (upper part) or via gap or other junctions (lower part). Such connections are suggested to have a short lifespan and can be considered as a mutual probing seen often at the ruffling edge of a moving cell. In both cases such cell interactions will on the one hand either lead to secondary stable linkages (intercytoplasmic, intermembranous or extracellular linkages, i.e. lower, middle and upper bridge in the last two adhering cells) or on the other hand to the disengagement of the two partner cells (not shown). (From Burger *et al.*[2])

All of these interactions are either based on plasma membrane contact, interactions between the extracellular matrix and the plasma membrane or are mediated via plasma membrane macromolecules or cellular structures that are anchored or connected with the plasma membrane.

MEMBRANE ALTERATIONS

It is expected therefore that alterations in the cell surface should also alter phenomena where cell sociology plays a rôle. Pathologists have long ago suspected that embryopathies may be due to an infection by a membrane-budding virus of the organ that is at the same time in an active phase of morphogenesis.

If the cell surface is important in organogenesis we should expect interesting results correlating some embryopathies with virus-induced cell-surface defects. On the other hand pathologists have suspected for a long time that neoplasia is connected with cell-surface defects, primarily since metastasizing was in part considered to be due to a loss of cohesion among the tumour cells[5]. This correlation is still the basis for a working hypothesis since relatively little experimental work has been carried out and since whatever solid work has been done so far is insufficient for sweeping conclusions that could embrace all neoplastic growth. Recently a mouse melanoma line, which was selected for its increased metastatic properties, displayed lower adhesion than its parent cell line which was less metastasizing and showed better adhesion[6]. Not only should many more cell lines be tested, but, since so many adhesion assays exist

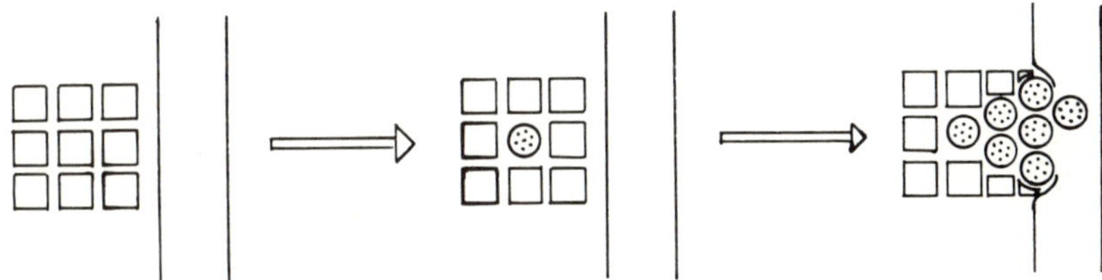

Figure 2 At the left nine untransformed cells are shown close to a capillary and in a state of growth control. In the middle one of these cells is transformed by whatever mechanism. In the case of a liver cell, surface alterations can already be detected in minimal deviation hepatomas[7]. The morphological alteration can be almost undetectable and the schematic illustration of a round cell merely serves the purpose of indicating that the cell has been altered, particularly its surface. Having lost growth control such cells begin to grow invasively. In the third part on right the tumour cells break through the vascular wall and if the cells have a potential for metastasizing, a property that does not have to be identical to the first alteration connected with a loss of growth control, then they may break loose from the primary site and establish a new tumour cell colony somewhere else in the body

now, they and their direct relevance for the metastatic properties will have to be carefully and critically assessed. Surface changes in neoplastic cells may, however, not only be relevant for the phenomenon of metastasis, they may have as much importance for growth control and invasive growth, so typical for earlier stages of neoplasia already.

Figure 2 schematically illustrates that surface changes — with and without morphological changes — do occur already at the point of transition to the cell that escapes growth control which is prior to and independent of its eventual development into a highly metastasizing or non-metastasizing tumour. Surface changes typical but not exclusively typical for neoplastic cells have also been found in cells that have been abortively transformed by tumour viruses (and will never become transformed), in precancerous cells as well as in minimal deviation hepatomas[7].

PRINCIPLES OF GROWTH CONTROL

Growth control at a biochemical level is presently studied extensively under *in vitro* conditions, in epithelial or fibroblastic cells and cell lines. Growth studies on cell cultures are not immediately applicable to the *in vivo* growth control problem for several reasons. But, if conclusions about the *in vivo* situation are not indiscriminately drawn, cell-culture studies at least allow us to formulate questions and working hypotheses which can be applied, and hopefully later tested with *in vivo* systems.

Termination of growth in a culture dish can be observed at a certain density of a particular cell line. Fibroblasts, as well as epithelial cells (hepatocytes) will in general reach a lower density level than transformed derivative cells[7]. This difference in contact inhibition of growth is observed provided optimal and identical amounts of nutrients and serum are used, and depends on pH as well as other parameters. It has been well documented in the last ten years that serum factors and an ever-increasing list of other factors (epithelial growth factor, insulin, somatomedin, fibroblast growth factor, etc.), some of which might contribute to the serum activity, can provoke untransformed as well as transformed cells to grow to a higher density. However, termination of growth at a certain cell density is not simply due to a total exhaustion of serum growth factors or nutrients but to the high cell density itself which may create a microenvironment within and around the cultured cells diminishing the ease with which individual cells can take up serum factors and nutrients. In sparse cultures individual free cells would thus have better access to these 'nutrients'. This is, however, only one possible mechanism explaining density inhibition of growth. Moreover, cells show an increased tendency to stop growing if they get into crowded contact, a phenomenon that could be due to a shutdown in serum factor and nutrient uptake which in itself could be caused by membrane-mediated phenomena. Similarly the higher ultimate cell density reached by transformed cells may be caused by an alteration of the cell surface that leads, for instance, to an increased serum or nutrient uptake. But other consequences of the altered surface are not ruled out, some of which could influence growth by other means. Anchorage dependence of growth — typical for untransformed cells — is unexplained so far, and besides the cytoskeleton the cell surface may contribute to the mechanics of anchorage. Consequently the plasma membrane may, through its glycocalixportion (adhesion) or the cytoskeleton, influence a growth control distinctly different for transformed and untransformed cells. Treatments of cultured cells will be reviewed in the following which altered growth control, both decrease and increase. In some cases the primary site of action is probably the cell surface, in others it is only suspected that the surface is the target.

STIMULATION OF GROWTH

Resting populations of fibroblasts can be activated to enter the DNA synthesizing S-phase of the cell cycle by many agents, some most probably acting inside the cell like colchicine[8], some probably at the cell surface like insulin[9] and some where the primary site of action is not clear at all, like serum factors. Some of these stimulating agents can bring the cell up to mitosis (colchicine), most others lead to completion of at least one full round of a cell cycle including cell division.

The necessary duration of interaction between the stimulating agent and the target cell has not always been sufficiently studied. In some cases the interaction has to be continuous and in others a finite incubation, less than the duration of a cell cycle, promotes the completion of the cycle. Treatment of chick embryo fibroblasts and mouse 3T3 fibroblasts with proteolytic enzymes led to a full round of growth after relatively short incubations. Primary chick embryo fibroblasts require several hours (6–12 hours) while some established mouse fibroblasts need some minutes and others require several hours. Moreover, the number of cells that could be triggered with a given dose of pronase for instance varied among the cell lines. Some could not be triggered at all. This apparent inertia to proteases may be due simply to a variable degree of serum sensitivity of different cell lines. Thus cells that required higher doses of serum to grow in the absence of proteases similarly required also more serum after being triggered with protease[10]. The serum can simply be considered as a nutrient of which some cell lines require much, others little.

Although enzymes like proteases have an effect on the cell surface no hard evidence rules out that they are taken up and exert an additional effect inside the cell. Some preliminary experiments with a bead-bound protease indicated a primary site of action at the cell surface but immediate uptake of the enzyme at the contact site between the bead and the cell surface has not been excluded.

Other agents were similarly triggers, that is, they can be incubated with the resting cells for times less than the duration of the cell cycle (about 3 hours). Another enzyme, neuraminidase, and a purified bacterial lipopolysaccharide, which has previously been shown to activate resting lymphocytes as a mitogen seem to belong to this class[11].

INTENSIFIED GROWTH CONTROL

According to Goto *et al.*[12] the addition of sulphated polydextran shuts down growth of untransformed cells at cell densities lower than confluency[12]. This is not a toxic phenomenon since removal of the sulphated polydextran permits further growth and thus reverses the inhibition.

Similarly we recently demonstrated that the addition of a mannose-binding plant protein, Concanavalin A (Con A), inhibited growth of untransformed fibroblasts. This lectin had to be modified with succinyl groups since the unmodified lectin killed the cells. According to an earlier prediction[13] these untransformed cells were only susceptible to the effect of the succinylated lectin at the mitotic and the early G_1 phase of the cell cycle, when the cell surface proceeds through a cell cycle alteration reflected in an increased agglutinability for that lectin.

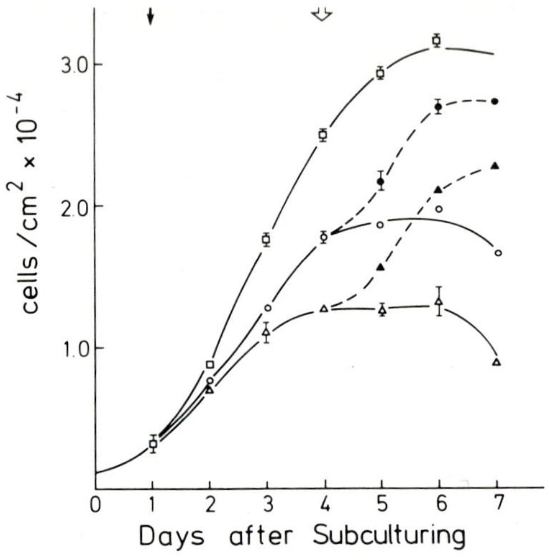

Figure 3 Characteristics of succinyl-Con A-induced growth inhibition of 3T3 mouse fibroblasts. 3T3 cells were subcultured into 3.5 cm tissue culture dishes containing Dulbecco's modified Eagle's medium (DME) + 10% calf serum; the medium was changed after 1 and 4 days of growth to DME + 5% calf serum ± succinyl-Con A as indicated. Cell numbers were determined using a Coulter counter after removal with a buffered EDTA (0.54 M)-trypsin (0.05%) solution; each point represents duplicate samples counted twice. □, control; ○, 250 μg/ml^{-1} succinyl-Con A; △, 500 μg ml^{-1} succinyl-Con A; ● and ▲, succinyl-Con A removed on day 4 and replaced by fresh medium containing 5% calf serum. (From Mannino and Burger[14])

The shutdown (Figure 3) could be overcome by the addition of the hapten α-methylmannose or by fresh serum, indicating that the cells were not damaged or killed[14]. They were arrested at the G_1 stage, a point in the cell cycle where untransformed cells are seen to halt growth under circumstances of regular density inhibition.

Inhibition of growth prior to confluency does not necessarily mean that the phenomenon is increased growth control, even if the inhibition is reversible and non-toxic. To consider this observation as a growth control phenomenon it has to be density dependent. Even in sparse cultures the cells are staying or sticking together in groups after addition of the modified lectin, while under normal sparse growth conditions they are homogeneously dispersed. This was corroborated in quantitative terms. Regardless of the density at which cells were seeded, they always reached a certain density with a certain amount of lectin. If they were seeded above the density level reached with a given dose of lectin, the cells came off until they reached the same density as was achieved with that particular dose of lectin, but starting with lower cell densities[14]. Possibly succinylated Con A increases the sensitivity of treated cells to density-dependent inhibition of growth or it mimics the density-inhibited state in cells that are not yet dense.

The purpose of succinylation of the lectin is to prevent the formation of the tetramer which is toxic. The succinylated form is a non-toxic dimer. Since succinylation introduces some eleven anionic groups per monomer while abolishing eleven basic amino groups, and particularly since the sulphated polydextran, that inhibits growth, is also heavily anionic, one might assume that any strong polyanion would inhibit growth. Preliminary experiments with succinylated bovine serum albumin indicated that this macromolecule, about the size of succinylated Con A but even more charged with succinyl groups, does not inhibit growth. As to possible mechanisms for this succinyl-Con A effect we only had the opportunity to study two. Preliminary experiments (Otten, Mannino and Burger), do not indicate that a generalized increase in cAMP seems to be the cause. Furthermore the extensive surface cover with such a charged macromolecule does not reduce a single cell's capacity to move around on the culture plate (Riddle, Mannino and Burger, preliminary observation). Possibly the cells display an increased stickiness to each other which could either be directly due to the lectin cover or which could also be induced by the lectin. The transformed cell may miss some surface macromolecules or coat which, if replaced, can bring about some growth control properties that are more pronounced in untransformed cells still possessing these surface macromolecules.

Although Trowbridge and Hilborn[15] found no inhibitory effect on the growth of transformed cells, we conclude that receptor saturating doses of the modified lectin inhibit also transformed cells in a reversible fashion (Ballmer, Mannino and Burger, unpublished data). The surface effective dose of lectin can only be achieved if serum glycoproteins and nutrient glucose, which bind and remove lectins from its potential surface site, are either kept low or if the lectin dose is adjusted to fairly high levels.

Whether this is an effect upon the cell surface is not yet established. Such evidence is much better for another type of experiment; based on the notion that some of the surface alterations in neoplastic and transformed cells could be due to increased proteolytic activity in or around such cells we investigated the possibility of inhibiting growth of transformed cells more than that of untransformed cells with inhibitors of proteolytic enzymes. Some low molecular inhibitors like tosyl-lysyl-chloromethylketone (TLCK) and the phenylalanine analogue (TPCK) turned out to be, together with the protein ovomucoid, good growth inhibitors[16]. But the small molecular inhibitors were toxic, for protein biosynthesis, for instance[17]. The protease inhibitor ovomucoid was thought to be a more likely candidate for a specific effect at the cell surface. We coupled it, therefore, covalently to polyacrylamide and not to Sepharose beads in order to reduce leakage of the protein from the bead during the long-term incubation in the culture dishes. The reversible inhibition of growth of transformed cells that was observed was even more convincing than the one achieved with the molecule in solution[18]. This effect was not simply a mechanical effect by heads or an unspecific effect by protein-covered beads since haemoglobin-substituted beads displayed no inhibitory activity. These experiments were not followed up since we tried to gain direct evidence that ovomucoid acted as a protease inhibitor and was cleaved thereby, as was known from pure biochemical model systems. Nevertheless this preliminary experiment[18] arrives again at the same conclusion as the others, namely that surface manipulations can alter growth and growth control in a reversible fashion.

SUMMARY

Very little is known about the macromolecular control of cell growth in culture. In particular, density-dependent inhibition of growth in culture can still not be explained on a solid and ubiquitous biochemical basis. Loss or reduction of this control in transformed cells is a phenotypic expression of some carcinogenic alteration inside the cell but it may be based on a surface membrane defect which leads, among others, to the aberrant growth control and social behaviour of the tumour cell. Manipulations at the cell surface, which alter behaviour, need first to be defined in molecular terms, but they may eventually contribute to the elucidation of the molecular basis of growth control and its aberration in tumour cells.

Acknowledgement

This work was supported by grant No. 3.1330-73 from the Swiss National Foundation for Scientific Research.

References

1. Kalckar, H. M. (1965). Galactose metabolism and cell 'sociology'. *Science,* **150,** 305
2. Burger, M. M., Turner, R. S., Kuhns, W. J. and Weinbaum, G. (1975). A possible model for cell–cell recognition via surface macromolecules. *Phil. Trans. R. Soc. Lond. B,* **271,** 379

3. Sing, V., Yeh, Y. and Ballou, C. E. (1976). Isolation of a *Hansenula wingei* mutant with an altered sexual agglutinin. In: R. A. Bradshaw *et al.* (eds.). *Surface Membrane Receptors*, pp. 87–97. (New York: Plenum Press)
4. Toole, B. P., Okayama, M., Orkin, R. W., Yoshimura, M., Muto. M. and Kaji, A. (1977). Developmental roles of hyaluronate and chondroitin sulfate-proteoglycans. In: M. M. Burger and J. Lash (eds.). *Cell and Tissue Interactions* (New York: Raven Press)
5. Coman, D. R. (1944). Decreased mutual adhesiveness, a property of cells from squamous cell carcinomas. *Cancer Res.*, **4,** 625
6. Fidler, I. J. (1973). Selection of successive tumor lines for metastasis. *Nature New Biology*, **242,** 148
7. Borek, C., Grob, M. and Burger, M. M. (1973). Surface alterations in transformed epithelial and fibroblastic cells in culture: a disturbance of membrane degradation versus biosynthesis? *Exp. Cell Res.*, **77,** 207
8. Vasiliev, J. M., Gelfand, I. M. and Guelstein, V. I. (1971). Initiation of DNA synthesis in cell cultures by Colcemid. *Proc. Natl. Acad. Sci. USA*, **68,** 977
9. Temin, H. M. (1966). Studies on carcinogenesis by avian sarcoma viruses VI. Differential multiplication of uninfected and of converted cells in response to insulin. *J. Cell Physiol.*, **69,** 377
10. Noonan, K. D. (1976). Role of serum in protease-induced stimulation of 3T3 cell division past the monolayer stage. *Nature* (*London*), **259,** 573
11. Vaheri, A., Ruoslahti, E. and Hovi, T. (1974). Cell surface and growth control of chick embryo fibroblasts in culture. In: B. Clarkson and R. Baserga (eds.). *Control of Proliferation in Animal Cells*, pp. 305–312. (Cold Spring Harbor Laboratory)
12. Goto, M., Kataoka, Y., Kimura, T., Goto, K. and Sato, H. (1973). Decrease of saturation density of cells of hamster cell lines after treatment with dextran sulfate. *Exp. Cell Res.*, **82,** 367
13. Burger, M. M. (1971). The significance of surface structure changes for growth control under crowded conditions. In: G. E. W. Wolstenholme and J. Knight (eds.). *Growth Control in Cell Cultures*, pp. 45–63. (Edinburgh: Churchill Livingstone)
14. Mannino, R. J. and Burger, M. M. (1975). Growth inhibition of animal cells by succinylated Concanavalin A. *Nature* (*London*), **256,** 19
15. Trowbridge, I. S. and Hilborn, D. A. (1974). Effects of succinyl-Con A on the growth of normal and transformed cells. *Nature* (*London*), **250,** 304
16. Schnebli, H. P. and Burger, M. M. (1972). Selective inhibition of transformed cells by protease inhibitors. *Proc. Natl. Acad. Sci. USA*, **69,** 3825
17. Chou, I., Black, P. H. and Roblin, R. O. (1974). Non-selective inhibition of transformed cell growth by a protease inhibitor. *Proc. Natl. Acad. Sci. USA*, **71,** 1748
18. Talmadge, K. W., Noonan, K. D. and Burger, M. M. (1974). The transformed cell surface: an analysis of the increased lectin agglutinability and the concept of growth control by surface proteases. In: B. Clarkson and R. Baserga (eds.). *Control of Proliferation in Animal Cells*, pp. 313–326. (Cold Spring Harbor Laboratory)

DISCUSSION

Dr L. Bianchi, Basel (Switzerland): Do other glycosidases have similar effects like neuraminidase?

Dr Burger, Basel (Switzerland): In a study of glycosidases in various sera in the growth medium, α-glucosidase and β-*N*-acetylglucosaminidase of calf serum as well as β-*N*-acetylglucosaminidase, β-glucosidase, β-galactosidase and β-fucosidase of fetal calf serum (the chief glycosidases occurring in these two sera) showed no growth-promoting, growth-sustaining or growth-controlling properties since they could be inactivated specifically without any relevant effects (Boch, Betschart and Burger: *Exp. Cell Res.* (1977) in press). Additions of large doses of 'purified' glucosidase preparations were similarly inefficient, particularly for β-galactosidase and α-mannosidase. Glycosidase preparations that give rise to cell triggering, however, might do so because of protease impurities always present in the starting preparation from which the glycosidase was prepared. To rule out such impurities in the final preparation is difficult since activity against common protease substrates may be absent, although activities against specific cell surface proteins may still be present, a dilemma not even solved for most of the so-called pure neuraminidase preparations.

14
Fucoprotein Metabolism in the Plasma Membrane of Liver and Morris Hepatomas

Ch. BAUER, P. VISCHER, H. P. MORRIS and W. REUTTER

It is well-established that L-fucose and *N*-acetylneuraminic acid are constituents of membrane glycoproteins and glycolipids. Investigations have centred on the metabolism of *N*-acetylneuraminic acid, because considerable evidence suggests that this sugar is associated with a variety of alterations observed in malignant tissue and cells. However, only limited information is available concerning the significance and function of fucose-containing lipids and proteins. Gesner and Ginsburg[1] have already shown a decade ago that the specificity of the so-called homing of lymphocytes is partly dependent on the presence of membrane-bound fucose, and a few years later Buck *et al.*[2] purified a fucopeptide which is enriched in BHK cells upon transformation by Rous sarcoma virus. Quite recently Watanabe *et al.*[3] described the accumulation of a fucosylceramide in highly malignant, metastatic human colon carcinoma, and Bryant *et al.*[4] demonstrated that human lung tumour cells synthesize elevated amounts of fucoproteins. These observations stimulated comparative investigations of the metabolism of L-fucose in Morris hepatomas of different growth rate and degree of differentiation.

INCORPORATION OF [^{14}C]FUCOSE INTO LIVER AND HEPATOMA PLASMA MEMBRANES

Fucose is probably one of the most suitable precursors available at present for studying synthesis and secretion of glycoproteins, because only negligible amounts are converted to other sugars[5]. Moreover, within 2 hours of the injection of the labelled compound, 90% of the protein-bound radioactivity can be detected in the plasma membrane. Rats were given a single injection of 1 mCi [^{14}C]fucose/kg body weight via the tail vein. Ninety minutes later plasma membranes were isolated from both liver[6] and hepatoma 7777[7] and finally subjected to SDS-polyacrylamide electrophoresis. The results are shown in an autoradiogram (Figure 1). The content of fucose-labelled proteins is substantially higher in hepatoma plasma membranes than in host liver and certain fucoproteins are only visible in the tumour preparation.

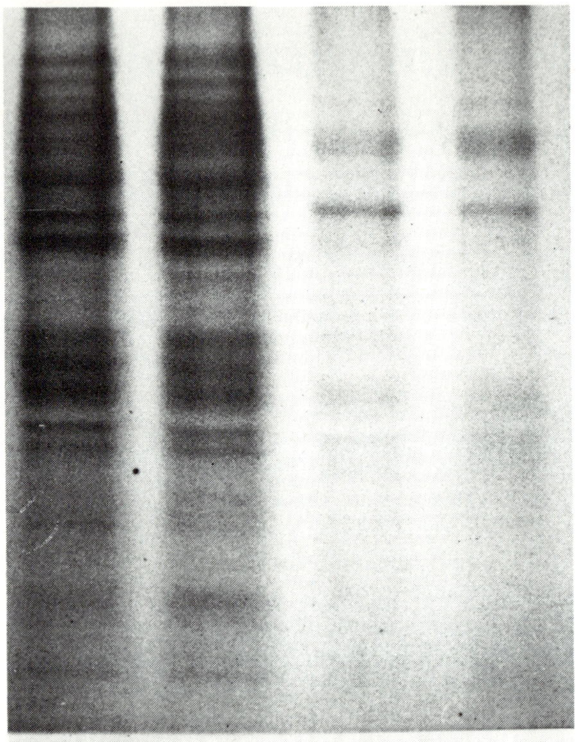

Figure 1 Autoradiogram of host liver and hepatoma plasma membranes labelled with [^{14}C]fucose

CONTENT OF PROTEIN-BOUND CARBOHYDRATES

In order to exclude the possibility that the autoradiogram may only reflect artefacts, the sugar content of liver and hepatoma plasma membranes[8] was determined (Table 1). The data showed only small differences in the concen-

Table 1 Protein-bound carbohydrate content of liver and hepatoma plasma membranes

Carbohydrate	*Host liver*	*Hepatoma 7777*
	nmol/mg protein	
Fucose	6.1 ± 1.7	26.1 ± 3.7
N-Acetylneuraminic acid	50.1 ± 8.7	60.7 ± 8.7
Galactose	81.4 ± 8.4	100.3 ± 9.1
Mannose	35.2 ± 2.5	44.6 ± 4.2
N-Acetylglucosamine	95.8 ± 8.8	127.4 ± 32.9
N-Acetylgalactosamine	49.8 ± 6.6	78.4 ± 12.9

Prior to hydrolysis internal standards for the individual sugars were added. Fucose was determined according to Dische and Shettles[9]; *N*-acetylneuraminic acid by the resorcinol method of Jourdian *et al.*[10]; mannose and galactose enzymatically; *N*-acetylglucosamine and *N*-acetylgalactosamine were separated by using an automatic amino acid analyser[11,12].

tration of *N*-acetylneuraminic acid or mannose, while the content of fucose was increased 4-fold in hepatoma membranes. Similarly, the specific radioactivity of protein-bound fucose was increased to a similar extent (data are not shown).

CONCENTRATION OF GDP-FUCOSE AND ACTIVITY OF FUCOSYLTRANSFERASE

These results raised the question why plasma membranes of the Morris hepatoma show a higher content of protein-bound fucose and fucoprotein bands. The observed alterations may be due either to a higher rate of synthesis or to a higher rate of degradation (Figure 2). Cytoplasmic enzymes convert fucose to

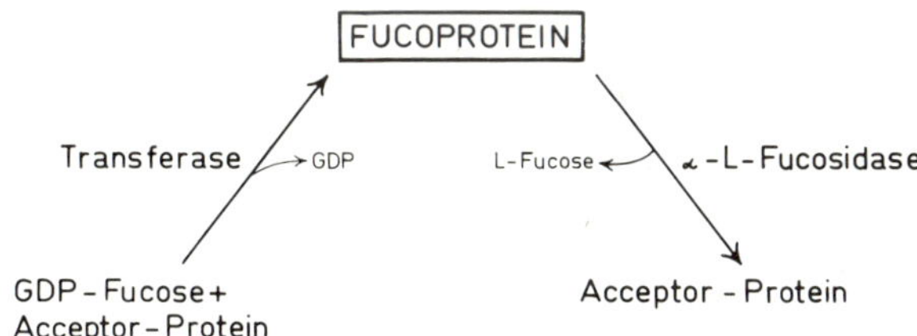

Figure 2 Hepatic fucoprotein metabolism

GDP-fucose, the substrate for fucosyltransferases which attach the fucose molecule to nascent glycoproteins within the Golgi complex. Figure 3 shows the concentration of GDP-fucose in normal rat liver of Buffalo or ACI strain (controls), different hepatoma lines and the corresponding host livers. In control liver 6 to 10 nmol/g wet weight were found, a value which is much lower than those found for all other sugar nucleotides (Table 2). In rapidly growing tumours a 2-fold higher level of GDP-fucose was found with changes in the pool size of a sugar donor affecting the rate of glycosylation. This

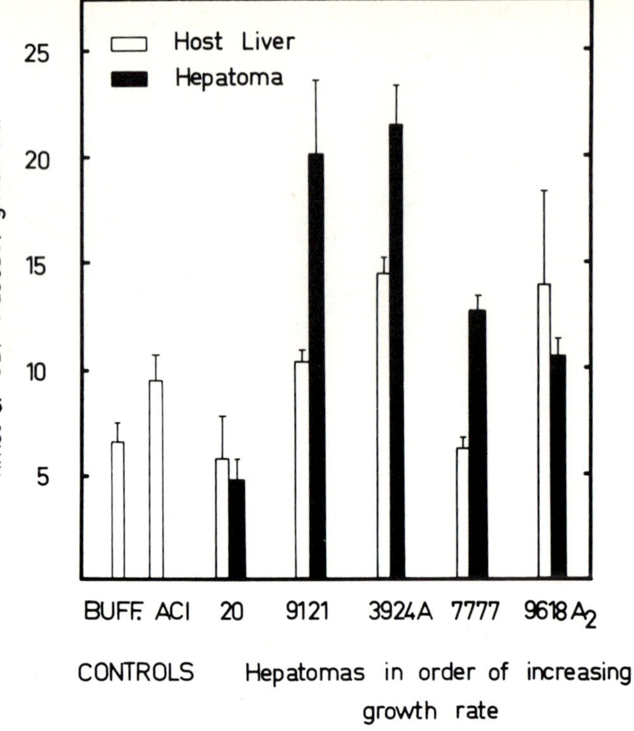

Figure 3 Concentration of GDP-fucose in liver and different hepatoma lines. For the determination of the sugar nucleotide liver samples were obtained *in situ* by the freeze clamp technique[13]. GDP-fucose was extracted from the frozen tissue by 65% (v/v) ethanol and separated from other sugars by chromatography

Table 2 Concentration of sugar nucleotides in liver and hepatoma

Sugar nucleotide	*Liver* (*nmol/g wet weight*)	*Hepatoma 7777* (*nmol/g wet weight*)
GDP-Fucose	6.5 ± 0.9	12.8 ± 0.7
CMP-*N*-Acetylneuraminic acid	42 ± 10	37 ± 5
UDP-Galactose	90 ± 10	86 ± 23
UDP-*N*-Acetylglucosamine	410 ± 19	201 ± 9
UDP-*N*-Acetylgalactosamine	182 ± 6	97 ± 9
GDP-Mannose	21 ± 6	19 ± 3
UDP-Glucose	320 ± 49	130 ± 25

UDP-Glucose and UDP galactose were determined according to Keppler *et al.*[14], CMP-*N*-acetylneuraminic acid by the isotope dilution technique of Harms *et al.*[15]; for the other nucleotides see footnote to Table 1

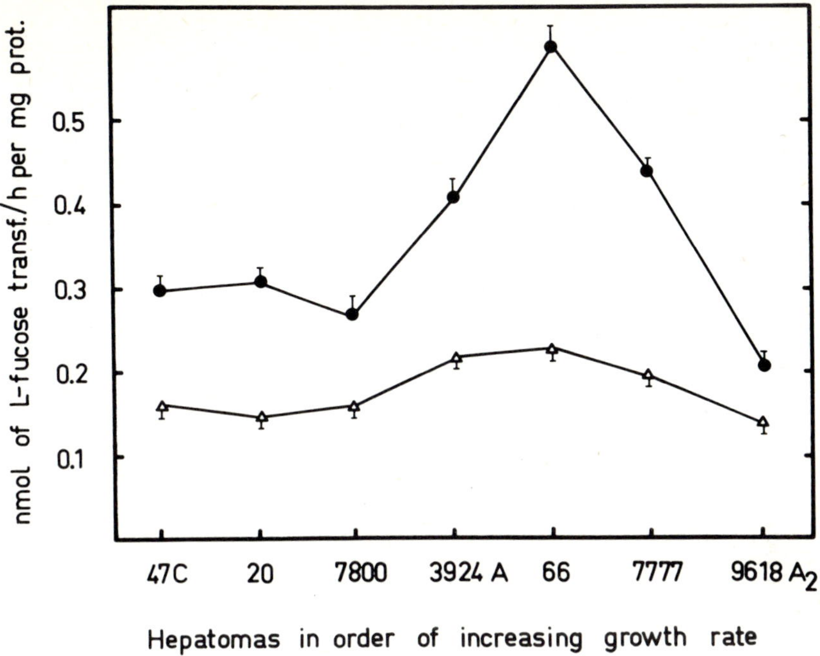

Figure 4 Activity of GDP-fucose: glycoprotein fucosyltransferase in Morris hepatomas and host liver. Enzyme activity was determined in a radiochemical assay by measuring the transfer of [^{14}C]fucose to desialized fetuin[17]. Protein-bound radioactivity was determined as described for UDP-galactose: glycoprotein galactosyltransferase[18]

increase favours the incorporation of fucose into glycoproteins, because the Michaelis constant (K_m) of the fucosyltransferase has been calculated to be 75–100 μM GDP-fucose[16,17]. This effect is amplified by a higher fucosyltransferase activity (Figure 4). The specific activity of this enzyme is raised 2- to 3-fold in all hepatoma lines investigated. However, when the activity is expressed in nmol/g wet weight instead of nmol/mg protein this increase is reduced by 20 to 30%, a result of the lowered protein concentration in Morris hepatomas. In the liver of some tumour-bearing rats a considerable increase of GDP-fucose and a small increase in fucosyltransferase activity was observed.

ACTIVITY OF α-FUCOSIDASE

As to degradation the specific activity of α-L-fucosidase was in all hepatomas with a rapid or intermediate growth rate twice as high as in control or host liver, except for the hepatoma 9618 A_2 (Figure 5). In this tumour enzyme activity was increased 7-fold. At first sight this finding may be regarded as

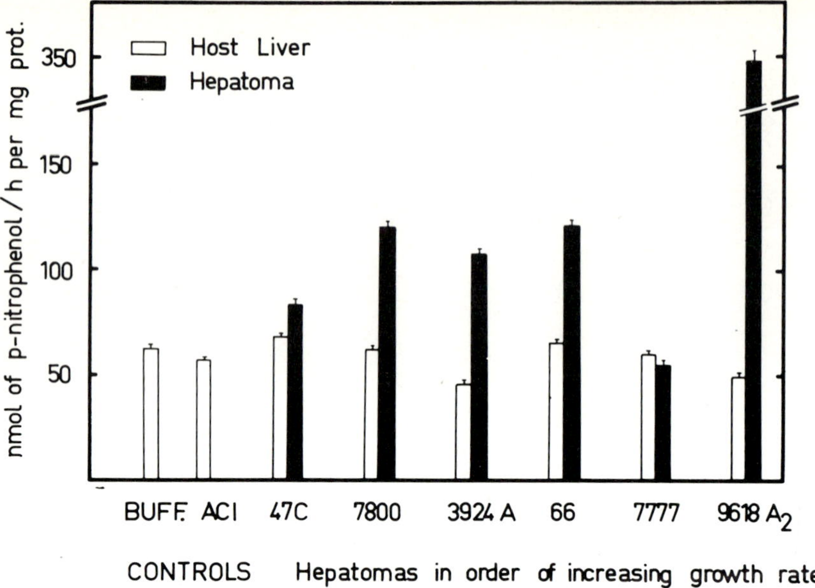

Figure 5 Activity of α-L-fucosidase in normal rat liver, different hepatoma lines and the corresponding host livers. Fucosidase was determined by measuring the liberated *p*-nitrophenol at 405 nm[19]

inconsistent with the results summarized in the first paragraph. Fucosidases are involved in the catabolism of biologically active molecules containing fucose. Unfortunately most fucosidases do not split fucosidic linkages of glycoproteins in appreciable amounts *in vitro*. Consequently fucosidase activity is usually assayed either by artificial substrates like *p*-nitrophenyl-α-L-fucopyranoside, 4-methylumbelliferyl-α-L-fucopyranoside or fucose-containing oligosaccharides, e.g. 2′-fucosyllactose or lacto-*N*-fucopentanose[20]. But most fucosidases are localized in the lysosomes whereas fucosylation takes place in the cisternae of the Golgi apparatus[5]. Therefore the physiological meaning of our finding is not clear.

ISOLATION OF A FUCOPROTEIN

Alterations of the cell surface are conventionally held responsible for the cardinal properties of the malignant cell and glycoproteins especially have an important role in mediating interactions between cells. Therefore, when the basic investigations on the fucose metabolism were completed, the next step was to isolate a fucoprotein. The plasma membranes were solubilized by low concentration of lithium diiodosalicylate (LIS). Marchesi and Andrews[21] originally used this salt, which has detergent-like properties, to extract glyco-

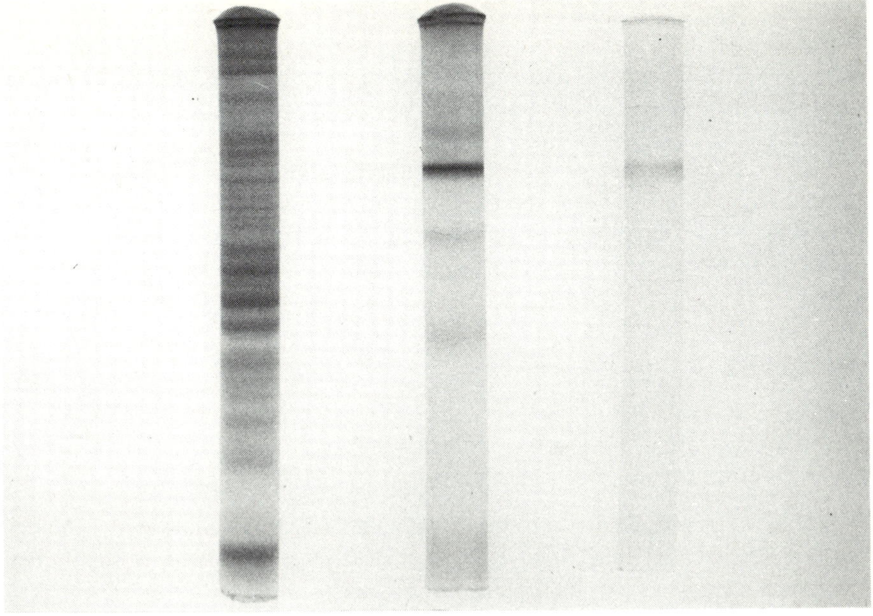

Figure 6 Electrophoretic comparison of a fucoprotein during the purification procedure from liver plasma membranes. The running gel contained 10% (w/v) acrylamide. The two gels on the left were stained with Coomassie brilliant blue, the gel on the right with periodic acid–Schiff reagent

proteins from red blood cells. After solubilization, followed by centrifugation to remove insoluble material, an equal volume of freshly prepared 50% phenol was added. The mixture was stirred thoroughly for 30 min, then centrifuged again and the upper phase, which contained soluble fucoproteins, was dialysed overnight. The dialysate was freeze-dried, dissolved in a few ml of water and applied to a Sephadex G-200 column. Protein was eluted by 50 nM Tris/HCl, pH 7.5, containing 0.02% sodium azide. Finally the fucoprotein was purified by SDS-polyacrylamide electrophoresis (Figure 6). The gel on the left side — stained with Coomassie brilliant blue — represents the protein pattern of liver plasma membranes; the gel in the middle shows that after the phenol extraction one major fucoprotein band has been enriched. However, some minor bands with predominantly high molecular weight glycoproteins are still detectable. One of these gels run after phenol extraction was stained for carbohydrates (right).

The same procedure applied to hepatoma plasma membranes showed: (a) considerable differences in the protein pattern of hepatoma and host liver membranes; (b) after phenol extraction only one fucoprotein band which showed a heavy staining for carbohydrates. After gel chromatography on Sephadex G-200 and electrophoresis the gel was cut and those slices containing the fucoprotein were eluted with 0.1% SDS in Tris-buffer for 24 hours.

Afterwards the fucoprotein was subjected to a second electrophoresis with the result that only one protein band was found. Though the fucoproteins from liver and hepatoma plasma membranes exhibited a similar migration behaviour they may differ slightly in their molecular weight, which was about 130 000.

Further chemical and biological characterizations of this fucoprotein are required. Individual fucoproteins of the plasma membrane have to be studied. However, specific alterations of fucose metabolism are a characteristic feature of Morris hepatomas.

SUMMARY

The following parameters were studied in Morris hepatomas of different growth rate and degree of differentiation.

(1) Measurement of the protein-bound carbohydrate content of plasma membranes revealed that none of the neutral sugars showed a significant difference, except for fucose which was greatly increased from 6.1 $\pm$ 1.7 nmol/mg protein in host or normal liver to 26.1 $\pm$ 3.7 nmol/g protein in hepatoma plasma membranes. Moreover, a higher number of fucose-containing glycoproteins could be demonstrated in tumour membranes.

(2) In normal liver 6 to 10 nmol GDP-fucose/g were determined, a value which is much lower than that measured for all other sugar nucleotides, whereas in rapidly growing hepatomas up to 21.5 nmol/g were found.

(3) The specific activity of GDP-fucose: glycoprotein fucosyltransferase was elevated in all hepatoma lines investigated. The increase varied between 2 to 3-fold.

(4) In all hepatomas with a short or intermediate growth rate the specific activity of α-L-fucosidase was 2 to 7-fold higher than in host liver.

(5) A fucoprotein with a molecular weight of about 130 000 has been isolated from both host liver and hepatoma plasma membranes.

Our results indicate that specific alterations of fucose metabolism are a characteristic feature of Morris hepatomas.

Acknowledgements

This work was supported by the Deutsche Forschungsgemeinschaft, Bonn–Bad Godesberg and Müller-Fahnenberg-Stiftung, Freiburg, Germany. We thank Cancer Research Inc. for permission to reproduce Figures 3 and 4 which will appear in a forthcoming paper entitled ‘Glycosyltransferases and Glycosidases in Morris Hepatomas’.

Abbreviations

The following abbreviations are used in this chapter: BHK, bay hamster kidney; SDS, sodium dodecylsulphate; GDP, guanosine diphosphate; CMP, cytidine monophosphate.

References

1. Gesner, B. M. and Ginsburg, V. (1964). Effect of glycosidases on the fate of transfused lymphocytes. *Proc. Natl. Acad. Sci. USA*, **52,** 750
2. Buck, C. A., Glick, M. C. and Warren, L. A. (1970). A comparative study of glycoproteins from the surface of control and Rous sarcoma virus-transformed hamster cells. *Biochemistry*, **9,** 4567
3. Watanabe, K., Matsubara, T. and Hakomori, S.-I. (1976). α-L-Fucopyranosylceramide, a novel glycolipid accumulated in some of the human colon tumors. *J. Biol. Chem.*, **251,** 2385
4. Bryant, M. L., Stoner, G. D. and Metzger, R. P. (1974). Protein-bound carbohydrate content of normal and tumorous human lung tissue. *Biochim. Biophys. Acta*, **343,** 226
5. Sturgess, J. M., Minaker, E., Mitranic, M. M. and Moscarello, M. A. (1973). The incorporation of L-fucose into glycoproteins in the Golgi apparatus of rat liver and in serum. *Biochim. Biophys. Acta*, **320,** 123
6. Ray, T. A. (1970). A modified method for the isolation of the plasma membrane from rat liver. *Biochem. Biophys. Acta,* **196,** 1
7. Harms, E. and Reutter, W. (1974). Half-life of *N*-acetylneuraminic acid in plasma membranes of rat liver and Morris hepatoma 7777. *Cancer Res.*, **34,** 3165
8. Vischer, P., Bauer, Ch. Grünholz, H.-J. and Reutter, W. (1976). Metabolism of L-fucose in rat liver and Morris hepatoma. Presented at the *10th International Congress of Biochemistry*, July 25–31, Hamburg, Germany
9. Dische, Z. and Shettles, L. B. (1948). A specific colour reaction of methylpentoses and a spectrophotometric micromethod for their determination. *J. Biol. Chem.*, **175,** 595
10. Jourdian, G. W., Dean, L. and Roseman (1971). A periodate–resorcinol method for the quantitative estimation of free sialic acids and their glycosides. *J. Biol. Chem.*, **246,** 430
11. Benson, J. V., Gordon, M. J. and Patterson, J. A. (1967). Accelerated chromatographic analysis of amino acids in physiological fluids containing glutamine and asparagine. *Anal. Biochem.*, **18,** 228
12. Bauer, Ch., Bachmann, W. and Reutter, W. (1972). Determination of galactosamine metabolites in the developing rat liver. *Hoppe Seyler's Z. Physiol. Chem.*, **353,** 1053
13. Wollenberger, A., Ristau, O. and Schoffa, G. (1960). Eine einfache Technik der extrem schnellen Abkühlung größerer Gewebestücke. *Pflueger's Arch. Gesamte Physiol. Menschen Tiere*, **270,** 399
14. Keppler, D., Rudigier, J. and Decker, K. (1970). Enzymatic determination of uracil nucleotides in tissues. *Anal. Biochem.*, **38,** 105
15. Harms, E., Kreisel, W., Morris, H. P. and Reutter, W. (1973). Biosynthesis of *N*-acetylneuraminic acid in Morris hepatomas. *Eur. J. Biochem.*, **32,** 254
16. Bella, A., Jr. and Kim, Y. S. (1971). Biosynthesis of intestinal glycoprotein: a study of $\alpha(1 \rightarrow 4)$fucosyltransferase in rat small intestinal mucosa. *Arch. Biochem. Biophys.*, **147,** 753
17. Jabbal, I. and Schachter, H. (1971). Pork liver guanosine diphosphate-L-fucose glycoprotein fucosyltransferase. *J. Biol. Chem.*, **246,** 5154

18. Bauer, Ch., Hassels, B. and Reutter, W. (1976). Galactose metabolism in regenerating rat liver. *Biochem. J.*, **154,** 141
19. van Hoof, F. and Hers, H. H. (1968). The abnormalities of lysosomal enzymes in mucopolysaccharidoses. *Eur. J. Biochem.*, **7,** 34
20. Alhadeff, J. A., Miller, A. L., Wenaas, H., Vedvick, T. and O'Brien, J. (1975). Human liver α-L-fucosidase. *J. Biol. Chem.*, **250,** 7106
21. Marchesi, V. T. and Andrews, E. (1971). Glycoproteins: isolation from cell membranes with lithium diiodosalicylate. *Science*, **174,** 1247

15
Contractile Protein Systems

H. E. HUXLEY

I intend to review the main facts about the contractile proteins which we have learnt from muscle, and discuss briefly the relationship, which seems to be a close one, between these proteins in muscle and analogous ones which seem to be associated with a wide variety of different kinds of motile processes in non-muscle cells. The two areas have been very well covered in recent Cold Spring Harbor publications[1,2].

The proteins in question are the principal structural and contractile proteins actin and myosin, the regulatory proteins tropomyosin and troponin and some accessory proteins of which the most important is α-actinin.

MUSCLE MOTILE SYSTEMS

Muscle actin has a molecular weight of 42 000, and the individual G-actin monomers, which contain a single polypeptide chain, are polymerized into filaments having helical symmetry. The monomers are globular in form, and about 50 Å in diameter, and the actin helix which they form can most easily be thought of as consisting of two strings of beads, twisted round each other so as to cross over at about 360–370 Å intervals, in which the individual beads repeat at 55 Å intervals with a half-period stagger in the position of the beads in the two chains (Figure 1). These actin-containing filaments are known as the thin filaments in muscle as distinct from the 'thick filaments' which contain myosin.

In vertebrate striated muscle, the regulatory proteins tropomyosin and troponin are also built into the thin filaments. Molecules of tropomyosin, which is a two-chain α-helical structure, form two continuous strands in the two long-pitch helical grooves of the actin structure. Troponin, which is a complex of three single-chain globular subunits — troponin I (inhibitory), chain weight

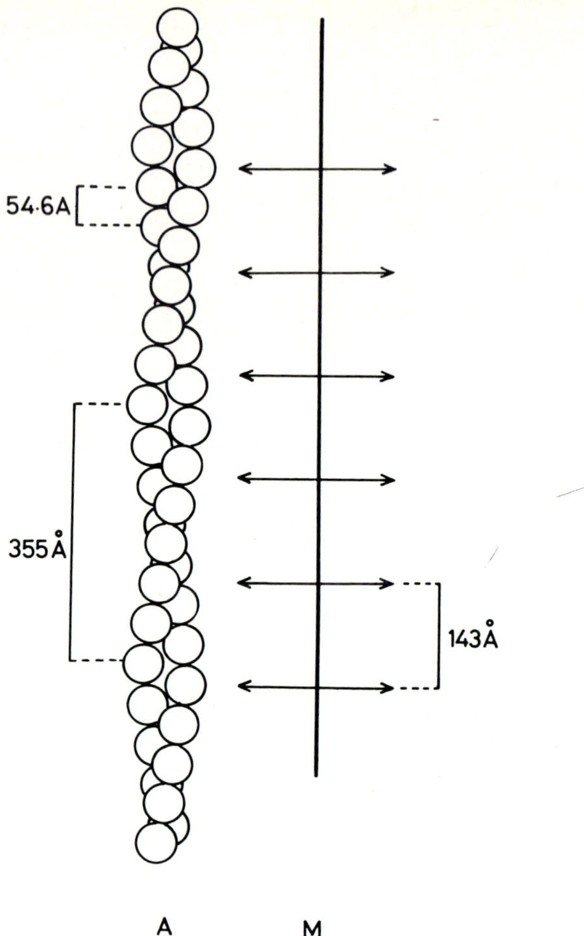

Figure 1 Diagram showing side-view of arrangement of G-actin monomers in an F-actin filament, together with representation of period of longitudinal repeat of myosin cross-bridges (orientation not indicated)

23 000; troponin C (calcium-binding), chain weight 19 000; troponin T (tropomyosin-binding), chain weight ~ 37 000 (in rabbit muscle), or ~ 44 000 (in chicken muscle) — binds to the actin–tropomyosin filaments at 385 Å intervals, there being two troponin units per interval, i.e. one per tropomyosin strand. The function of the tropomyosin–troponin system is to switch muscle contraction on and off in response to changes in the free calcium ion concentration in the sarcoplasm. The switching is achieved by controlling the ability of actin to interact with myosin, the other principal structural and contractile protein in muscle.

Myosin is a protein of unusually high molecular weight, about 480 000, consisting of two almost identical 'heavy' chains of molecular weight about 200 000 and four light chains (two or three different types) of molecular weight about 20 000. The two heavy chains are twisted round each other for about half their total length in a coiled-coil α-helical structure, which makes up the rod portion of the myosin molecule, and then, at one end, fold up separately into two globular subunits, the 'subfragment 1' (S_1) head subunits of myosin. The S_1 subunits contain the ATPase and actin-binding sites of myosin, and the light chains seem to be located there too.

In striated muscle, the myosin molecules are organized into the so-called 'thick filaments'. In these, part of the rod portions of the myosins are bonded together and form the backbone of the filaments, while the S_1 head subunits project out sideways as cross-bridges which can attach to the actin filaments. Part of the rod (the S_2 portion) subunit of myosin may function as a sort of flexible hinge to allow the myosin heads to project outwards from the backbone by varying amounts under different conditions. The myosin filaments have a very characteristic bipolar structure, which arises because the myosin

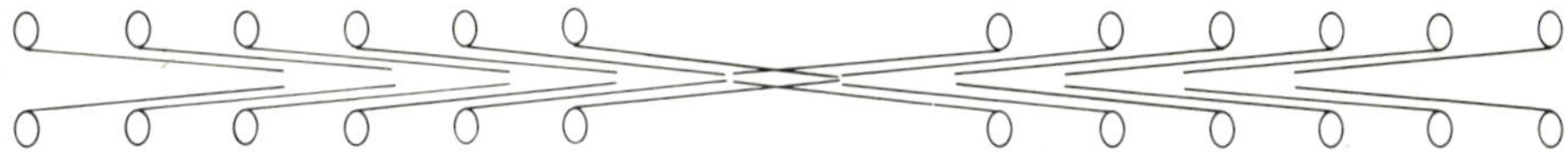

Figure 2 Arrangement of myosin molecules in the thick filaments. The backbone of the filaments is formed from the assembly of the 'tails' of the molecules, and the molecules form two anti-parallel sets, one in each half of the filament, with a reversal of structural polarity at the centre

molecules are positioned with opposite structural polarity on either side of the midpoint of the filament, their 'tails' always pointing to the centre of the filament (Figure 2). Each thick filament in a vertebrate striated muscle contains about 150 myosin molecules in each half of the filament, helically arranged with a subunit repeat of 143 Å (with probably three myosins per repeat) and a helical repeat of 3×143 Å $= 429$ Å.

In a striated muscle, both the actin and myosin filaments have sharply defined lengths and are arranged in partially overlapping arrays which give rise to the characteristic band-pattern. The thick filaments are about 1.6 μm in length and make up the A bands while the thin filaments are about 1 μm in length and are attached, with opposite structural polarity, on either side of the Z line. Contraction is brought about by a process in which the thin actin-containing filaments are drawn further into the array of myosin filaments in the A bands. During this sliding movement the filament overlap may change by up to 1 μm. It is now generally believed that the sliding movement is brought about directly by an interaction between the myosin cross-bridges and the actin units in the thin filaments. It is supposed that a repetitive cycle of cross-bridge attachment to a given actin monomer, followed by tilting of the attached cross-bridge, pulling the actin filament along a short distance (of the

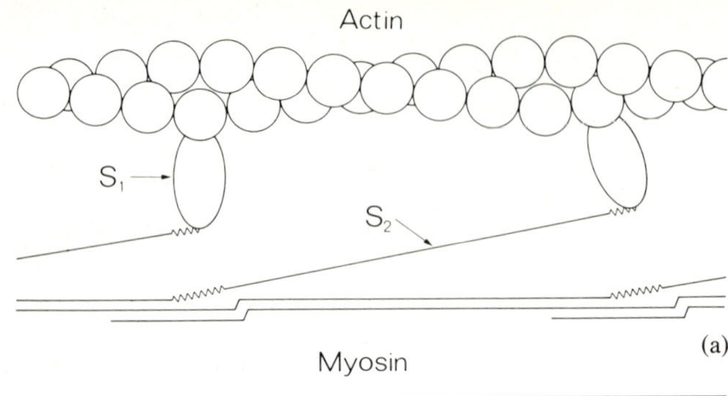

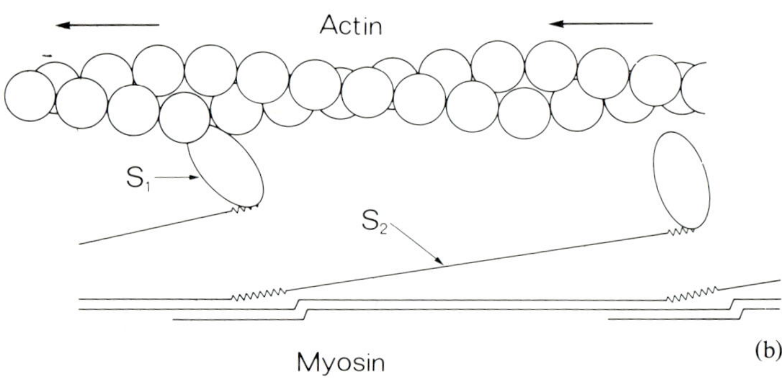

Figure 3 Active change in angle of attachment of cross-bridges (S_1 subunits) to actin filaments could produce relative sliding movement between filaments maintained at constant lateral separation (for small changes in muscle length) by long-range force balance. Bridges can act asynchronously since subunit and helical periodicities differ in the actin and myosin filaments. (a) Left-hand bridge has just attached; other bridge is already partly tilted; (b) left-hand bridge has just come to end of its working stroke; other bridge has already detached, and will probably not be able to attach to this actin filament again until further sliding brings helically arranged sites on actin into favourable orientation

order of 70 Å), followed by detachment of the cross-bridge, takes place (Figure 3); the mechanical movement cycle is coupled to the splitting of probably one (or two) molecules of ATP by the myosin cross-bridge. The cross-bridge can then attach to an actin monomer further along the actin filament, and the repetitive and asynchronous action of all the cross-bridges together produces a steady sliding force and the performance of mechanical work. Basically, muscle is a structure which integrates the activity of about 10^{17} molecules of myosin per cm^3 so that all the elements of force produced are lined up in the same direction and a large force (2–4 kg per cm^2) and a high velocity of movement (several cm/s per cm of muscle length) can be produced.

It may be useful to note some of the essential characteristics of this mechanism, in order to compare and contrast it with other types of motility. Structurally, the salient features are: (1) the actin filaments have a characteristic structural polarity which can be readily demonstrated by the angled attachment of heavy meromyosin or subfragment 1 giving rise to 'arrowheads' which always point away from the Z lines, and in the direction the actin will tend to move; (2) the actin filaments are attached to a specific structure at the Z lines, which contains the protein α-actinin; (3) actin interacts with a parallel assembly of myosin molecules, so that although the links are making and breaking asynchronously, links are always present between adjoining filaments at any given moment so that a continuous force is generated; (4) the myosin filaments are bipolar so that two sets of actin filaments are drawn towards each other; (5) the energy for contraction comes from the enzymatic splitting of ATP by myosin, which requires combination with actin to activate it by allowing a given cycle of dephosphorylation to be completed; hence energy release is automatically coupled to the performance of work; (6) the contractile machinery is switched off by blocking (probably sterically by tropomyosin movement linked to loss of calcium by troponin) the attachment of myosin cross-bridges to actin. Thus live, inactive muscle (i.e. containing ATP) is in the relaxed state, with the actin and myosin filaments free to slide past each other under a very small force.

In this mechanism, the actual form of the myosin filaments, i.e. their diameter and length, could vary within wide limits, and indeed a good deal of variation is found between muscles from different species. However, in a given muscle the filaments almost always have a constant length, and there is no evidence that depolymerization of either myosin or actin is involved in the normal activity of the muscle cell (with the possible exception of the strange behaviour of the myosin filaments in *Limulus* (horseshoe crab) muscle. It should also be noted that tropomyosin may also play a structural role in strengthening the thin filaments, for it is present even in muscles which lack troponin and where regulation takes place directly on myosin, and does not need tropomyosin, e.g. scallop striated muscle[3].

NON-MUSCLE MOTILE SYSTEMS

It has been recognized for several years (see, for example, reviews by Huxley[4], Pollard[5] and the Cold Spring Harbor Symposium on Cell Motility in 1975[2]) that proteins closely analogous to actin and myosin in their structure and interactions are present in a wide variety of cell types, especially where one or more forms of motile activity are taking place. In several cases (e.g. nerve growth cones) actin may be present in large amounts (20% of the soluble protein). These cellular actins are remarkably similar in amino acid sequence to muscle actin, but do show definite differences between different types of cells and between species, in a small proportion of residues. They form two stranded filaments indistinguishable from those of muscle actin, activate myosin ATPase, give closely similar 'arrowhead' structures when 'decorated' with myosin S_1, combine with tropomyosin, and have their ATPase-activating ability calcium-regulated by the tropomyosin–troponin system from muscle, though evidence about how their activity is regulated *in vivo* is scanty at present. They seem to differ from muscle actin, however, in the conditions under which they will polymerize and depolymerize, and in general seem much more liable to be in the depolymerized form. Thus Pollard[6] has shown that the large changes in the viscosity of extracts of *Acanthamoeba* cytoplasm which occur as a function of temperature are due to differences in the extent of polymerization of the actin they contain, and can be more or less duplicated even by quite highly purified *Acanthamoeba* actin preparations. Bray and Thomas[7] have also shown that a proportion of actin from nerve growth cones can remain in the unpolymerized form although it is not denatured, and Taylor, Condeelis, Moore and Allen[8] have described observations of the properties of other types of amoeba in which they believe that changes in the extent of polymerization of both actin and myosin play a part.

The myosins from non-muscle cells also display structural characteristics remarkably similar to muscle myosin; they show the same 'head and long tail' structure on electron microscopy and in many cases have the ability to form bipolar filaments closely similar to those of muscle myosin. However, in several cases their solubility is different from that of muscle myosin; for example, slime-mould myosin requires the presence of millimolar concentrations of calcium for filament formation[9].

Despite these differences from the muscle proteins, the extraordinarily close similarities found in the actins and myosins from non-muscle cells are so striking that it is difficult not to believe that important aspects at least of the contractile mechanism are closely analogous to those of muscle. Since the latter depend on a relative sliding motion between actin filaments and assemblies of myosin molecules, one presumes that a basically similar mechanism plays a part in all motile phenomena where both actin and myosin-like proteins are present. Since these proteins in non-muscle cells are not ordered into the same kind of large-scale structures that are present in muscle, any sliding motion will necessarily be between small groups of filaments. There are very many examples where bundles of actin filaments have been observed (see reviews already mentioned) and in which the type of movement taking place could be accounted for by a shearing force between those actin filaments and small groups of myosin molecules either in solution[10] or attached to other cell organelles either directly or via another group of actin filaments[4,11] (Figure 4).

The myosin moiety has been much more difficult to observe, no doubt because of its lability, and the much smaller amounts present, but its association with actin in cells has been clearly demonstrated by fluorescent antibody labelling methods[12] and in some instances by electron microscopy, for

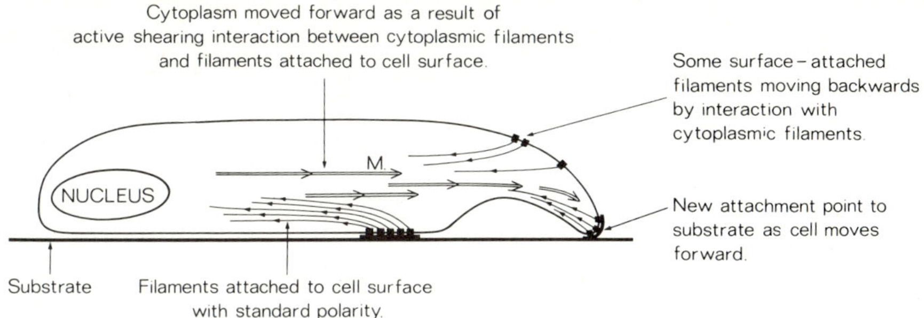

Figure 4 Diagrammatic representation of a mechanism by which active shearing forces developed between two sets of filaments could produce cytoplasmic streaming and cell movement

example in intestinal brush border[13]. It seems highly probable that myosin in non-muscle cells is usually present in the form of very much smaller filamentous aggregates than in muscle, but as explained earlier, even a very few myosin molecules can function in essentially the same way as they do in the thick filaments of muscle, albeit producing a much smaller force, provided they form a bipolar array. And even a unipolar array could move itself along an actin filament, and could move other structures to which it was attached.

OTHER PROTEINS

Tropomyosin has been identified in a number of non-muscle cells (e.g. in platelets[14], fibroblasts in tissue culture[15,16], intestinal brush border[13]), but so far identification of the troponin components seems much less certain, and the control mechanisms have not been established.

The protein α-actinin, which is a constituent of the Z lines in striated muscle, has also been identified by antibody labelling techniques in some non-muscle cells. Thus it appears to be present at the points where the actin filaments in intestinal microvilli abut against the cell membrane[17] possibly providing an attachment site; and α-actinin has also been found at sites along the bundles of actin filaments in tissue culture cells[16].

CONCLUSIONS

There are a very helpful number of correspondences between many of the proteins associated with contraction in muscle and the ones associated with several different forms of motility, including cases in which interactions with membranes are involved. However, it would be short-sighted to assume that all the systems function essentially exactly alike. Attention has already been drawn to the variable degree to which actin may be polymerized, and the effect this may have on the general consistency of the cytoplasm in different parts of a cell. In addition, very rapid actin polymerization is responsible for the extrusion of the acrosomal process in certain echinoderm sperm[18]. A further possibility might very well be, therefore, that the generation of microspikes and membrane ruffles on a wide variety of cells is also brought about by rapid polymerization of actin at local regions of the cell membrane which are thereby pushed outwards by the reaction of the growing actin filaments against other components of the cell. Once the microspike had formed, then actin–myosin interactions would either pull the process back into or along the cell, or start to pull the cell forward if the microspike or ruffle had attached to substrate or another cell, in the way outlined previously[4]. The function of actin near the membranes of liver cells is not known so far as I am aware, but clearly an agent such as phalloidin which alters the polymerization properties of actin and holds it in the polymerized form could interfere drastically with any phenomena dependent on the reversibility of this polymerization.

Thus a very rich field is opening up not only in the understanding of the basis of many more of the normal attributes of cells, but, as several of the papers later in this volume will show, in beginning to understand the basis of some of the important defects in cells under different pathological conditions.

References

1. Cold Spring Harbor Symposium on Quant. Biol. (1972). *The Mechanism of Muscular Contraction*, **37**
2. Cold Spring Harbor Meeting. (1975). *Cell Motility*. (In press)
3. Kendrick-Jones, J., Lehman, W. and Szent-Györgyi, A. G. (1970). Regulation in molluscan muscles. *J. Mol. Biol.,* **54,** 313
4. Huxley, H. E. (1973). Muscular contraction and cell motility. *Nature* (*London*), **243,** 445
5. Pollard, T. D. and Weihing, R. R. (1974). Actin and myosin and cell movement. *C.R.C Critical Rev. Biochem.,* **2,** 1
6. Pollard, T. D. (1976). The role of actin in the temperature-dependent gelation and contraction of extracts of *Acanthamoeba. J. Cell Biol.,* **68,** 579
7. Bray, D. and Thomas, C. (1976). Unpolymerized actin in fibroblasts and brain. *J. Mol. Biol.,* **105,** 527
8. Taylor, D. L., Condeelis, J. S., Moore, P. L. and Allen, R. D. (1973). The contractile basis of amoeboid movement. I. The chemical control of motility in isolated cytoplasm. *J. Cell Biol.,* **59,** 378
9. Nachmias, V. T. (1972). Filament formation by purified *Physarum* myosin. *Proc. Natl. Acad. Sci. USA*, **69,** 2011
10. Huxley, H. E. (1963). Natural and synthetic protein filaments from muscle. *J. Mol. Biol.,* **37,** 507
11. Bray, D. (1973). Model for membrane movements in the neural growth cone. *Nature* (*London*), **244,** 93

12. Weber, K. and Groeschel-Stewart, U. (1974). Antibody to myosin: the specific visualization of myosin-containing filaments in nonmuscle cells. *Proc. Natl. Acad. Sci. USA*, **71,** 4561
13. Mooseker, M. S. and Tilney, L. G. (1975). Organization of an actin filament–membrane complex. Filament polarity and membrane attachment in the microvilli of intestinal epithelial cells. *J. Cell Biol.*, **67,** 725
14. Cohen, I. and Cohen, C. (1972). A tropomyosin-like protein from human platelets. *J. Mol. Biol.*, **68,** 383
15. Lazarides, E. (1975). Tropomyosin antibody: the specific localization of tropomyosin in nonmuscle cells. *J. Cell Biol.*, **65,** 549
16. Lazarides, E. (1976). Actin–α-actinin, and tropomyosin interaction in the structural organization of actin filaments in nonmuscle cells. *J. Cell Biol.*, **68,** 202
17. Schollmeyer, J. V., Goll, D. E., Tilney, L. G., Mooseker, M., Robson, R. and Stromer, M. (1974). Localization of α-actinin in nonmuscle material. *J. Cell Biol.*, **63,** 304a
18. Tilney, L. G., Hatano, S., Ishikawa, H. and Mooseker, M. S. (1973). The polymerization of actin: its role in the generation of the acrosomal process of certain echinoderm sperm. *J. Cell Biol.*, **59,** 109

16
The Immunology of Contractile Proteins

E. J. HOLBOROW

Antibodies reactive by immunofluorescence with smooth muscle (SMA) occur persistently and usually at high titre in chronic active hepatitis patients[1], at lower titre in about 45% of patients with primary biliary cirrhosis[2], and transiently in most patients with acute viral hepatitis[3]. Recently a significantly raised incidence of SMA has been reported in alcoholic cirrhosis[4]. SMA are not an exclusive feature of liver autoimmunity, however, but are also found with high frequency in some other viral infections, in *Mycoplasma pneumoniae* infection, in some malignancies not necessarily involving the liver, and also in some healthy people (Table 1).

Table 1 Conditions in which smooth muscle antibodies are found

Chronic active hepatitis
Primary biliary cirrhosis
Viral hepatitis
Cytomegalovirus infection[5]
Infectious mononucleosis[6]
Wart virus[7]
M. pneumoniae[8]
Some malignancies[9,10]
Some healthy persons[1]

Many SMA-positive sera react also with non-muscle cells in cryostat tissue sections. Often such sera stain liver cell membranes[11] and renal glomeruli[12], and they may also stain thyroid epithelial cell membranes[13], the brush borders of renal tubular and intestinal epithelial cells[14], the membranes of bile ductular epithelial cells[15], and the cytoplasm of so-called myofibroblasts in healing wounds[16]. A characteristic filamentous staining of cells actively spreading in tissue culture is also seen[11], and the microvilli of lymphocytes stimulated with PHA, or of lymphoblastoid cells in culture are likewise stained prominently[17]. The inhibitory effect on staining of pretreatment of these cells with cytochalasin B, and a similar effect on staining of platelets, indicates that the antigens against which at least some SM antibodies are directed are constituents of the microfilaments[18]. The identification of actin, myosin and tropomyosin in association with microfilament bundles[15,19], and the ability of antisera raised in rabbits against actin, myosin, tropomyosin and other muscle proteins to give all the fluorescent staining patterns of different tissues seen with human SMA-positive sera[20], contribute to our belief that the antigenic constituents with which SMA specifically reacts are components of the intracellular system of contractile proteins ubiquitously distributed among the different cell types of body tissues.

The implication of these observations is that although SMA are detected in practice by their ability to stain smooth muscle tissue by immunofluorescence, they must arise *in vivo* under conditions in which contractile proteins of non-muscle cells have become immunogenic, presumably as a result of cellular changes associated with the viral, malignant and inflammatory disorders mentioned.

SPECIFICITY OF SMA

The SMA characteristic of sera of patients with chronic active hepatitis (CAH) react with both smooth and skeletal muscle. They give striational staining of the myofibrils of the latter[14]. Since both reactions are prevented by absorption with skeletal muscle actin (10 mg/0.1 ml serum) it is thought that they are both attributable to antiactin antibody cross-reactive with determinants common to the actins of smooth and skeletal muscle[21]. Using much less smooth muscle actin (100 μg/0.1 ml serum) we absorbed smooth muscle staining from only about 50% of CAH sera. Table 2 records the effects

Table 2 Effect of absorbing four different human SMA positive sera with extracts of smooth or skeletal muscle

Serum	*Unabsorbed staining pattern*	*Staining after absorption with SM extract*	*Staining after absorption with Sk extract*	*Antibodies*
RID	SM	0	SM	anti-SM
	Sk	0	0	anti-(SM,Sk)
PAT	SM	0	SM	anti-SM
BOW	SM	0	0	anti-(SM,Sk)
	Sk	0	0	
ANN	SM	0	SM	anti-SM
	Sk	Sk	0	anti-Sk

SM = Smooth muscle
Sk = Skeletal muscle
(Data from reference [23])

of absorption with crude extracts of smooth muscle or skeletal muscle on the staining patterns of four different SMA positive sera, and shows that the latter may contain (i) only cross-reactive antibody (presumably antiactin), (ii) only specific antismooth muscle antibody, (iii) mixtures of both, or (iv) mixtures of specific antismooth muscle and specific antiskeletal muscle antibodies.

Antibodies reactive with contractile proteins of smooth muscle other than actin are found mostly in patients with disorders not primarily affecting the liver. In absorption experiments on sera from, for example, rheumatoid arthritis and other forms of chronic inflammatory arthritis, myosin, heavy meromyosin, and a newly identified smooth muscle protein, janin[23], individually or in combination absorb out SMA, while actin is ineffective.

ARE CONTRACTILE PROTEINS CONSTITUENTS OF THE CELL MEMBRANE?

Microfilaments, as well as performing important muscle-like functions intracellularly, play a part in controlling the mobility and distribution of cell surface receptors[24]. Electron microscopy shows that microfilaments underlie the cell surface[25], and immunoperoxidase staining with rabbit antiactin serum shows actin in these submembranous filaments[15]. Recent work by Barber and Crumpton[26] demonstrates that a protein showing a high degree of homology with actin on peptide mapping is a major component of pig and human lymphocyte cell membranes. This protein, like actin, inhibits DNAse I activity.

Despite the evidently close association of at least one contractile protein with cell membranes, all attempts so far to demonstrate specific reactivity between SMA or antiactin antibodies, and the surface of living intact cells at any stage of activity have failed. The immunogenicity of contractile proteins in the conditions listed in Table 1 thus seems unlikely to be due to changes in contractile proteins already accessible at the cell surface, but rather to the accompanying cell abnormality either releasing intracellular actin (for example) or bringing it to the cell surface in immunogenic form. SMA occurs in conditions in which cellular necrosis and cellular proliferation are both prominent. In experiments in the rat we have found[27] that administration of hepatotoxins such as tannic acid and carbon tetrachloride does not lead to SMA production, but that acute liver damage surgically induced with a cryoprobe leads to rapid production of SMA with antiactin and antimyosin activity; these SMA, furthermore, are the only detectable tissue antibodies produced. We do not know whether such cryosurgical damage perhaps favours release of actin in the F-form, or indeed whether the F-form is more immunogenic than the G-form (see Figure 1 in Huxley's paper, chapter 15 in this book). Residual liver tissue, proliferating in rats after partial hepatectomy, shows a transient but marked loss of immunofluorescent submembranous liver cell staining by anticontractile protein antisera[28]. Again, this may reflect escape of potentially immunogenic protein from the proliferating cells, or its disassembly within them.

SMA AND VIRUS INFECTIONS

It is also not known whether intracellular virus replication endows membrane-associated contractile proteins with immunogenicity. Relevant to this possibility is the recent identification[29] of actin in enveloped viruses; for example, in the highly purified virions of oncornaviruses and paramyxoviruses. The contractile proteins of the host cell may be incorporated into viral proteins during their maturation, and viruses may preferentially synthesize polypeptides closely resembling contractile proteins, as has been shown for the actin-like polypeptides synthesized by Sendai virus growing in chick embryo fibroblasts[30]. The specificity of SMA in chronic active hepatitis for actin, and the wider reactivity of SMA in some viral diseases, might reflect modes of virus-induced autoimmunity to cellular contractile proteins. Since SMA, unlike many other autoantibodies, occur independently of age and sex[31], environmental factors likely to be important in stimulating their production, and a viral hypothesis to explain their occurrence is attractive. So far, however, actin or other contractile proteins have not been implicated in the processes of replication of the viruses known to lead to SMA production in the infected subject.

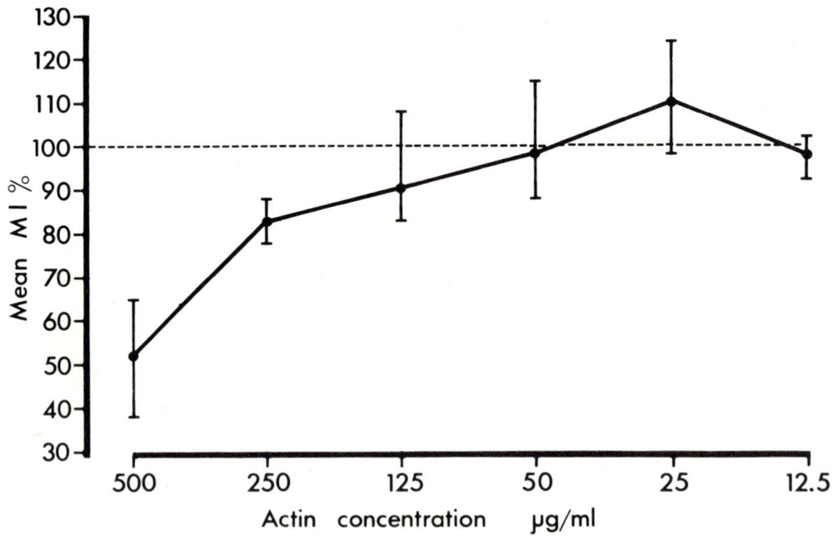

Figure 1 Mean migration index of normal human buffy coat cells at different concentrations of purified smooth muscle actin (T. Agoglu—unpublished data)

Whatever the environmental stimulating factor, a constitutional factor probably also influences SMA production, because of the raised incidence of the histocompatibility factor HLA-B8 in both chronic active hepatitis and alcoholic cirrhosis[4], and it has been suggested[32] that this antigen is linked to genes that promote abnormally raised and prolonged antibody responses.

Finally, Dr Tevfik Agoglu found recently in my laboratory that addition of purified actin to the culture medium at 50 μg/ml inhibits migration of both normal human buffy coat leukocytes and the mononuclear fraction of such cells, and likewise inhibits their ability to respond to stimulation by phytohaemagglutinin (Figure 1). This suggests that the lymphocytic and mononuclear cell infiltration of the liver parenchyma in chronic active hepatitis may be attributable in part to release of actin from damaged liver cells.

SUMMARY

The smooth muscle autoantibodies (SMA) found at high titres in some chronic active hepatitis patients and at lower titres in a number of other liver diseases and viral infections are directed at contractile proteins of the actomyosin group, which are constituents of ubiquitous intracellular microfilaments.

Although apparently not accessible at the cell surface, contractile proteins are functionally implicated in the mobility and distribution of cell surface receptors, and recent work has identified actin as an integral major component of lymphocyte cell membranes. Where SMA occur in human disease, whether affecting the liver or not, they presumably arise under conditions where the contractile proteins of non-muscle cells become immunogenic as the result of a certain type of cellular change associated with viral or other disorders.

References

1. Holborow, E. J. (1972). Smooth-muscle autoantibodies, viral infections and malignant disease. *Proc. R. Soc. Med.*, **65**, 481
2. Doniach, D., Roitt, I. M., Walker, J. G. and Sherlock, S. (1966). Tissue antibodies in primary biliary cirrhosis, active chronic (lupoid) hepatitis, cryptogenic cirrhosis and other liver diseases and their clinical implications. *Clin. Exp. Immunol.*, **1**, 237
3. Farrow, L. J., Holborow, E. J., Johnson, G. D., Lamb, S. G., Stewart, J. S., Taylor, P. E. and Zuckerman, A. J. (1970). Autoantibodies and the hepatitis-associated antigen in acute infective hepatitis. *Br. Med. J.*, **2**, 693
4. Bailey, R. J., Krasner, N., Eddleston, A. L. W. F., Williams, R., Tee, D. E. H., Doniach, D., Kennedy, L. A. and Batchelor, J. R. (1976). Histocompatibility antigens, autoantibodies and immunoglobulins in alcoholic liver disease. *Br. Med. J.*, **2**, 727
5. Andersen, P. and Andersen, H. K. (1975). Smooth muscle antibodies and other tissue antibodies in cytomegalovirus infection. *Clin. Exp. Immunol.*, **22**, 22
6. Holborow, E. J., Hemsted, E. H. and Mead, S. V. (1973). Smooth muscle autoantibodies in infectious mononucleosis. *Br. Med. J.*, **3**, 323

7. McMillan, S. A. and Haire, M. (1975). Smooth muscle antibody in patients with warts. *Clin. Exp. Immunol.*, **21,** 339
8. Biberfeld, G. and Sterner, G. (1976). Smooth muscle antibodies in *Mycoplasma pneumoniae* infection. *Clin. Exp. Immunol.*, **24,** 287
9. Whitehouse, J. M. A. and Holborow, E. J. (1971). Smooth muscle antibodies in malignant disease. *Br. Med. J.*, **4,** 511
10. Wasserman, J., Glas, U. and Blomgren, H. (1975). Autoantibodies in patients with carcinoma of the breast. *Clin. Exp. Immunol.*, **19,** 417
11. Farrow, L. J., Holborow, E. J. and Brighton, W. D. (1971). Reaction of human smooth muscle antibody with liver cells. *Nature New Biology*, **232,** 186
12. Whittingham, S., Mackay, L. R. and Irwin, J. (1966). Autoimmune hepatitis. Immunofluorescence reactions with cytoplasm of smooth muscle and renal glomerular cells. *Lancet,* **ii,** 1383
13. Biberfeld, G., Fagraeus, A. and Lenkei, R. (1974). Reaction of human smooth muscle antibody with thyroid cells. *Clin. Exp. Immunol.*, **18,** 371
14. Gabbiani, G., Ryan, G. B., Lamelin, J. P., Vassalli, P., Majno, G., Bouvier, C. A., Cruchaud, A. and Luscher, E. F. (1973). Human smooth muscle autoantibody. *Am. J. Pathol.*, **72,** 473
15. Holborow, E. J., Trenchev, P. S., Dorling, J. and Webb, J. (1975). Demonstration of smooth muscle contractile protein antigens in liver and epithelial cells. *Ann. N.Y. Acad. Sci.*, **254,** 489
16. Gabbiani, G., Hirschel, B. J., Ryan, G. B., Statkov, P. R. and Majno, G. (1972). Granulation tissue as a contractile organ. A study of structure and function. *J. Exp. Med.*, **135,** 719
17. Fagraeus, A., Lidman, K. and Biberfeld, G. (1974). Reaction of human smooth muscle antibodies with human blood lymphocytes and lymphoid cell lines. *Nature* (*London*), **252,** 246
18. Norberg, R., Lidman, K. and Fagraeus, A. (1975). Effects of cytochalasin B on fibroblasts, lymphoid cells and platelets revealed by human anti-actin antibodies. *Cell*, **6,** 507
19. Lazarides, E. and Weber, K. (1974). Actin antibody: the specific visualization of actin filaments in *non-muscle* cells. *Proc. Natl. Acad. Sci. USA*, **71,** 2268
20. Trenchev, P., Sneyd, P. and Holborow, E. J. (1974). Immunofluorescent tracing of smooth muscle contractile protein antigens in tissues other than smooth muscle. *Clin. Exp. Immunol.*, **16,** 125
21. Lidman, K., Biberfeld, G., Fagraeus, A., Norberg, R., Torstensson, R., Utter, G., Carlsson, L., Luca, J. and Lindberg, U. (1976). Anti-actin specificity of human smooth muscle antibodies in chronic active hepatitis. *Clin. Exp. Immunol.*,
22. McDonald, B. L., Dawkins, R. L. and Holborow, E. J. (1976). Evidence for immunological cross-reactivity between smooth and skeletal muscle. *Clin. Exp. Immunol.* (In press)
23. Trenchev, P. (1976). Antigenicity of janin, a new protein from smooth muscle. *Clin. Exp. Immunol.* (In press)
24. Nicolson, G. L. (1976). The cancer cell: dynamic aspects and modification in cell-surface organization. *N. Engl. J. Med.*, **295,** 197
25. Goldman, R. D., Lazarides, E., Pollack, R. and Weber, K. (1975). The distribution of actin in non-muscle cells. *Exp. Cell Res.*, **90,** 333
26. Barber, B. H. and Crumpton, M. J. (1976). Actin associated with purified lymphocyte plasma membrane. *FEBS Lett.* (In press)
27. Li, A. K. C., Trenchev, P. S., Holborow, E. J., Newsome, C. and Wynne, A. T. (1976). Experimental smooth muscle antibodies. *Clin. Exp. Immunol.* (In press)
28. Lampert, I. A., Trenchev, P. and Holborow, E. J. (1974). Contractile protein changes in the regenerating rat liver. *Virchows Arch. Abt. B. Zellpathol.*, **15,** 351
29. Wang, E., Wolf, B. A., Lamb, R. A., Choppin, P. W. and Goldberg, A. R. (1975). The presence of actin in enveloped viruses. *J. Cell Biol.*, **67,** 445a (abstract)

30. Lamb, R. A., Mahy, B. W. J. and Choppin, P. W. (1976). The synthesis of Sendai virus polypeptides in infected cells. *Virology*, **69,** 116
31. Hooper, B., Whittingham, S., Mathews, J. D., Mackay, I. R. and Curnow, D. H. (1972). Autoimmunity in a rural community. *Clin. Exp. Immunol.*, **12,** 79
32. Galbraith, R. M., Eddleston, A. L. W. F., Williams, R., Webster, A. D. B., Pattison, J., Doniach, D., Kennedy, L. A. and Batchelor, J. R. (1976). Enhanced antibody responses in active chronic hepatitis: relation to HLA-B8 and HLA-B12 and porto-systemic shunting. *Lancet*, **i,** 930

17
Endocytosis, Exocytosis and Vesicle Translocation

A. C. ALLISON

Extracellular materials can enter cells in two ways. They can penetrate through the plasma membrane by a specific transport mechanism, passive flux or some other process. Alternatively, they can be taken up within invaginations of the plasma membrane which become detached to form membrane-limited vacuoles or microvesicles that move into the cell. These often fuse with lysosomes and their contents are digested. If the products of digestion are small enough, they pass through the membrane surrounding the vacuole and enter the cytoplasm. Uptake of extracellular materials within membrane-limited vacuoles or microvesicles is termed endocytosis, three varieties of which can be distinguished. One is phagocytosis ('cell eating'), which for practical purposes can be defined as the ingestion of particles with diameter $\gg 1$ μm. This is a widespread mechanism observed in all classes of animals from protozoa to mammals. Most protozoa are phagocytic, and in metazoa specialized cells carry out this function: in mammals it is performed by cells of the mononuclear phagocyte system and polymorphonuclear leukocytes.

The intake of extracellular fluid within membrane-lined vesicles is termed pinocytosis ('cell drinking'), of which two main varieties can be distinguished. One is macropinocytosis, the formation of relatively large vacuoles visible by light microscopy (usually 0.3–2 μm in diameter). This is, for example, the mechanism by which colloid is ingested by thyroid epithelial cells and some fluid is taken up by mononuclear phagocytes. The second is micropinocytosis, the formation of microvesicles which in a variety of cell types have a remarkably uniform diameter of about 70 nm and are demonstrable only by electron microscopy. The first part of this paper summarizes what is known of these endocytic processes and the underlying mechanisms.

The term exocytosis is used for the release of secretory products from cells within membrane-limited vesicles or granules. Some products are secreted without any specific environmental stimulus, although factors such as the position of the cell in the mitotic cycle can influence secretion. This appears to be true, for example, of the release of immunoglobulin from plasma cells[1]. When a human or experimental animal bears a plasma cell tumour the immunoglobulin product accumulates in the serum as a characteristic monoclonal band, or spike. This type of secretion can be termed *stimulus-independent*.

In contrast, many secretory products are not continuously released but accumulate within secretory granules in the cytoplasm. They are discharged in response to a specific stimulus, which varies from cell to cell. In the β-cells of the pancreas, for instance, insulin-containing granules are released when the concentration of glucose in the blood rises. Appropriate agonists likewise trigger the release of catecholamine-containing granules from the adrenal medulla and of other hormones, as well as amylase from salivary glands and other secretions. Similarly, presynaptic nerve impulses bring about release of neurotransmitters from nerve endings, and cell-bound antibody and antigen stimulate the release of histamine from mast cells. Such modes of secretion can be termed *stimulus-coupled*, and their importance in physiology and pharmacology is evident. The second part of this paper is a summary of what is known about mechanisms of exocytosis.

In some types of endocytosis and exocytosis cytoplasmic contractile systems appear to play a major role. Two cellular contractile systems are known. Microtubules have been observed not only in systems that are obviously contractile, such as flagella, cilia and the mitotic spindle, but also in a wide range of dispositions in the cytoplasm of other cells[2,3]. The second system consists of actin and myosin-like proteins, which are not confined to muscle cells but are present also in many other cell types[4,5]. Both these systems are widespread constituents of living tissues. Probably their relative importance varies from one cell type to another and in the same cell at different times, and their interactions require prolonged and detailed investigation. The first attempts at such an analysis have depended on the use of compounds thought to interact more or less specifically with either microtubules or microfilaments, namely colchicine and vinca-alkaloids on the one hand, the cytochalasins on the other. A brief discussion of the mode of action and specificity of these compounds is therefore included in this paper.

EFFECTS OF CYTOCHALASINS ON MICROFILAMENTS

Earlier evidence that cytochalasins inhibit microfilament-mediated contractions has been reviewed elsewhere[5]. The other main action of cytochalasin B is competitive inhibition of facilitated diffusion of hexoses across the plasma membrane[6]. However, cytochalasin D has no significant effect on hexose uptake in Vero and MDBK cells, whereas the effects on cell shape and movement are very similar to those produced by cytochalasin B[7].

Evidence is accumulating that the effect of cytochalasin B on hexose transport is due to a relatively small number of sites with high affinity for the drug, whereas the effects on the microfilament system are due to the presence of a larger number of binding sites of relatively low affinity. The activation energies for binding of cytochalasin D are described by Tannenbaum *et al.*[8] Another useful approach has been the development of cell lines resistant to cytochalasin B as described by Kelly *et al.*[9] Some of these have lost the high-affinity plasma membrane-binding sites but retain the binding sites allowing for effects on the microfilament system.

MECHANISMS OF ENDOCYTOSIS

Phagocytosis

This process has been extensively studied in mammalian phagocytes. Two phases can be distinguished: a phase of attachment of a particle to the plasma membrane and a phase of ingestion. The attachment stage can be studied quantitatively using radioactive bacteria or other particles under conditions when internalization is inhibited[10]. Attachment can be non-specific, due to interaction of hydrophobic surfaces, for example the attachment of polystyrene latex particles or formalin-treated erythrocytes to macrophages[11]. However, mononuclear phagocytes have on their surfaces receptors for the Fc region of immunoglobulins of the G class (human IgG1, IgG3). Erythrocytes coated with IgG antibody are readily attached to mononuclear phagocytes and this is followed by phagocytosis[12]. Erythrocytes to which the third component of complement (C3) is bound become attached to mononuclear phagocytes but this is not followed by phagocytosis. Thus some attachment sites trigger endocytosis while others do not. Phagocytosis is blocked by inhibitors of glycolysis, such as high concentrations of 2-deoxyglucose, iodoacetate or fluoride[13]. Cytochalasins B and D in concentrations of 1 to 5 μg/ml have no demonstrable effect on the attachment stage but completely and reversibly block phagocytosis[10].

We conclude that phagocytosis requires a cooperative response of the plasma membrane and subjacent microfilament network. The network contracts, pleating the plasma membrane to which it is attached; the two layers of membrane in the pleat move over each other as described by Bray[14]; this results in eversion of the membrane around the particle, fusion of the lips of membrane and eventually completion of the endocytic process. While it is in contact with the plasma membrane, the vacuole is surrounded by a microfilament network which limits access of lysosomes, but when the vacuole has moved towards the nucleus (perhaps propelled by microfilaments and guided by microtubules) the microfilament network is less dense, and lysosomes have access to the phagocytic vacuole and often fuse with it.

In phagocytic cells treated with cytochalasin B, lysosomes approach the plasma membrane close to sites of attachment of bacteria and are also seen in the extracellular medium. This is the morphological counterpart of the selective release of lysosomal enzymes into the medium found by biochemical analyses[10, 15].

Macropinocytosis

The uptake of droplets of fluid by cells in culture was observed by Lewis[16], who introduced the term pinocytosis. Analogous processes have since been observed in many different types of cultured cells, including macrophages[13], tumour cells[17, 18] and fibroblasts[19]. *In vivo* uptake of fluid into membrane-lined vacuoles large enough to be observed by light microscopy occurs in several different cell types, including thyroid epithelial cells ingesting colloid[20], proximal convoluted tubule cells reabsorbing haemoglobin and other proteins[21] and in intestinal cells of fetal and newborn animals ingesting immunoglobulins and other gut contents[22]. These processes appear different in several respects from the formation of submicroscopic vesicles in membranes; the two types of process can be termed macropinocytosis and micropinocytosis, respectively.

Several non-specific and specific factors stimulate macropinocytosis. In macrophages a potent stimulator raises the concentration of bovine serum in the medium; the serum contains a macroglobulin antibody which reacts with the plasma membrane of the macrophage[13]. Certain nucleotides and acidic proteins (which may form complexes with serum constituents) are also effective[13]. In amoebae, macropinocytosis is stimulated by basic proteins or dyes or by raising the ionic strength of the medium[23]. In tumour cells uptake of labelled proteins is promoted by polycations[17]; whether this is due to macropinocytosis or micropinocytosis has not been established. Ingestion of colloid by thyroid epithelial cells is stimulated by thyrotrophic hormone[20], and in intestinal cells of newborn mammals macropinocytosis is stimulated by serum albumin. How these factors initiate macropinocytosis is not known, but it is again tempting to speculate that cross-linking of plasma membrane acidic glycoproteins may be involved. Basic proteins or polypeptides or anticellular antibodies could certainly have this effect, as would receptors for other effectively bifunctional agents.

The metabolic requirements for macropinocytosis in macrophages have been analysed by Cohn[13]. A variety of inhibitors of glycolysis or oxidative phosphorylation block macropinocytosis, from which it seems likely that the process requires metabolic energy. All forms of macropinocytosis so far investigated are reversibly inhibited by cytochalasin B in low concentrations, from which it again seems likely that the subjacent microfilament network is required for the envelopment of a vacuole containing fluid where ruffle membrane activity is occurring. Macropinocytic vacuoles are rapidly transported from this region towards the nucleus, and fuse with lysosomes. In colchicine-treated cells macropinocytic vacuoles are formed but fail to move towards the centrosomal region in a regular saltatory fashion[24]. This suggests that microtubules are not required for endocytic vacuole formation but guide the directional movement of vacuoles within the cell.

Micropinocytosis

Microvesicles, 70–100 nm in diameter, arise from the plasma membrane in many different cell types. They have been studied particularly in endothelial cells, where they have been shown to be involved in the slow passage of large molecules through cellular barriers[25–27]. For example, the slow leakage of proteins out of blood vascular system, even where there are non-fenestrated capillaries, may occur through microvesicles. This is more marked in some tissues, e.g. muscle, than in others, e.g. brain. By contrast, the rapid passage of water, small molecules and ions almost certainly takes place through intercellular junctions[28,29]. Micropinocytic vesicles have also been observed in peritoneal mesothelial cells[30], macrophages[31,32], neurosecretory terminals[33] and nerve cell terminals[34,35].

Casley-Smith[31] and Green and Casley-Smith[36] have pointed out that the formation of microvesicles does not require metabolic energy, in contrast to the formation of phagocytic or pinocytic vacuoles. In macrophages[32] and BHK-21 cells[38] uptake of ferritin or radioactive colloidal gold into microvesicles is not inhibited by cytochalasin B in concentrations that block macropinocytosis and phagocytosis. The analogy with the appearance of microvesicles in mammalian erythrocytes treated with aminoquinoline drugs[38] is interesting. All these findings suggest that the formation of microvesicles and their transport across the cytoplasm of cells takes place by Brownian motion and does not depend on a contractile system requiring metabolic energy.

Micropinocytosis may well represent an important mechanism for the uptake of neurotransmitters and other pharmacologically active substances, as well as for the reutilization of membrane following release of neurotransmitters or other products from neuronal, neurosecretory and other terminals. In neurosecretory and nerve terminals coated and smooth microvesicles are formed by invagination and vesiculation of the plasma membrane following exocytosis; the vesicles incorporate horseradish peroxidase from the extracellular fluid[33–35]. In nerve terminals the microvesicles may be used for repackaging of neurotransmitters, so that recycling of synaptic vesicles occurs[39]. The coating is thought by Kanaseki[39] and others to facilitate distortion of the membrane during the formation of micropinocytic vesicles, and smooth vesicles probably arise from the coated forms by shedding of the coats; only smooth vesicles are seen to fuse again with the plasma membrane. Increase of plasma membrane area following exocytosis is thus one factor stimulating micropinocytosis.

Radioactive colloidal gold can be used to quantitate both macropinocytosis and micropinocytosis, so that the way is now open to investigate factors selectively stimulating or inhibiting these processes[32,40]. It is already clear that both are blocked at low temperature. Since micropinocytosis is resistant to inhibitors of oxidative phosphorylation, the blocking at 4 °C may result from loss of fluidity of the membrane rather than from a requirement for metabolic energy. In cells so far investigated more fluid is taken up by micropinocytosis than by macropinocytosis. A major proportion of ^{198}Au uptake by mouse peritoneal macrophages is resistant to the effect of cytochalasin B. Even at the highest concentration of serum there is only 40% distribution of ^{198}Au uptake, and not all of this represents macropinocytosis since cells treated with cytochalasin B are rounded and consequently have less membrane available for pinocytic activity. In hepatoma cells there is even less inhibition of ^{198}Au uptake by cytochalasin B, suggesting that the great majority, perhaps all, of the medium taken into the cells enters by micropinocytosis. In liver cells[41] micropinocytosis is prominent at the vascular pole of the cells and no macropinocytosis is demonstrable.

Micropinocytic vesicles probably have various fates. Conceivably there is some fusion with the plasma membrane from which they arise, although two factors reduce the likelihood of this occurring. One is the fact that the microvesicles are released from stalks[26], and the second is the coating of the microvesicles at the time of their formation in at least some cells; loss of the coat appears to precede refusion. Electron microscopic observations show that microvesicles containing tracers can traverse the cytoplasm of endothelial and mesothelial cells by Brownian motion and fuse with the plasma membrane on the other side[26,27]. Recycling of microvesicles in nerve terminals has already been mentioned. Microvesicles can also fuse with one another to form small vacuoles or fuse with larger vacuoles of the lysosomal system[26,32].

The uptake of drugs into the lysosomes of living cells[42] may well occur by micropinocytosis, since it does not require serum in the medium and is inhibited at 4 °C. Incorporation of pharmacologically active agents into microvesicles may be a common and important process. Because of their small size, microvesicles may be able to penetrate into parts of the cell not accessible to larger vacuoles, e.g. movement through the cell to the plasma membrane on the other side, or to the nucleus. Once the microvesicles have fused with lysosomes the subsequent fate of their contents is presumably indistinguishable from that of materials taken into cells by other endocytic mechanisms. For penetration through the plasma membrane of the erythrocyte, and presumably the plasma membranes of other cells, drugs must be lipophilic and uncharged[43]. A much wider range of substances can be taken up by endocytosis, since the process does not involve penetration through a membrane. Some of the drugs or hormones taken up in this way are bound to macromolecules, from which they can be released by digestion.

DIGESTION OF MATERIAL TAKEN INTO CELLS BY ENDOCYTOSIS

Within secondary lysosomes macromolecules are digested to monosaccharides, dipeptides and amino acids[44]. The size of molecules that can traverse the lysosomal membrane has been analysed by following the fate of non-digestible sugars and peptides taken into cells by pinocytosis. If these pass through the lysosomal membrane, the vacuoles become smaller and disappear; otherwise the entrapped compounds exert osmotic effects and large vacuoles remain within the cells. As shown by Cohn[13] for macrophages and Nyberg and Dingle[19] for fibroblasts, monosaccharides can pass through the lysosomal membrane whereas sucrose, mannitol and other disaccharides cannot; certain dipeptides, such as $(D\text{-}Ala)_2$ can escape into the cytoplasm while others, such as $(D\text{-}Glu)_2$, and all tripeptides and higher peptides remain entrapped within the lysosomal system. The continuous digestion of body constituents in lysosomes is illustrated by the storage diseases resulting from the congenital absence of a lysosomal enzyme in humans[45]. Of special interest in this connection are the observations of Hickman and Neufeld[46] on correction of metabolic defects in cultured cells from individuals with two different storage diseases. The results show that some lysosomal hydrolytic enzymes are released from living cells into the extracellular fluid; they can be taken up by other cells and carry out digestion within their lysosomes.

STIMULUS–SECRETION COUPLING

The events coupling stimulation by electrical excitation or attachment of specific agonists to target cells with the release of packaged secretions have been extensively discussed. Two common events are depolarization of plasma membranes of target cells, not only in presynaptic nerve cells but also in pancreatic β-cells and adrenal medullary cells[47]; this increases the permeability of the plasma membrane and the level of Ca^{2+} in the cytoplasm. Substantial evidence that in several systems release of packaged secretions only occurs when Ca^{2+}, Ba^{2+} or Sr^{2+} are present in the extracellular fluid, whereas Mg^{2+} ions inhibit release, has been reviewed by Rubin[47] and Poste and Allison[48]. Only two recent observations supporting the involvement of Ca^{2+} in stimulus–secretion coupling need be cited. In the stimulated presynaptic nerve, studies with the luminescent protein aequorin have shown a rise in cytoplasmic Ca^{2+} concentration accompanying neurotransmitter release[49]. In rat mast cells Foreman, Mongar and Gomperts[50] have shown that exposure to cell-bound (reaginic) antibody and antigen raises the influx of $^{45}Ca^{2+}$ and releases histamine. Moreover, using a calcium ionophore (A23187), which increases the permeability of the plasma membrane to Ca^{2+}, they have found that raising the cytoplasmic concentration of Ca^{2+} (or Sr^{2+}) triggers an energy-dependent, cytochalasin-sensitive reaction discharging histamine from the cells.

Since the ions which are involved in stimulus–secretion coupling are the same as those triggering muscular contraction, many authors have speculated that an actomyosin system may participate in the discharge of packaged secretions. Others have favoured the view that contractile systems related to microtubules are involved in stimulus–secretion coupling[51,52]. Observations on the effects of cytochalasin B and of colchicine on these processes are therefore of interest.

Although the results are not open to simple interpretation, certain conclusions can be drawn. One is that stimulus-independent secretion, such as the synthesis and release of immunoglobulin from plasma cells, is resistant to even large doses of cytochalasin B and colchicine[1]. The immunoglobulin appears to be packaged in microvesicles which bud off the Golgi system and diffuse to the plasma membrane without the help of microtubules or microfilaments.

A second conclusion is that treatment of cells with cytochalasin B can increase the release of packaged secretions. Our own experience is largely with mammalian macrophages and neutrophil leukocytes[10,15]. When these are treated with cytochalasin, there is selective release of lysosomal but not cytoplasmic enzymes into the extracellular medium; in the presence of an agent which by itself induces exocytosis of lysosomal enzymes, such as phagocytosis of bacteria or treatment of cells with vitamin A, cytochalasin B augments the release. This is what might be expected if a microfilament network or 'terminal web' normally prevents access of the organelles to the plasma membrane. Analogous results have been obtained with insulin release from rat pancreatic β-cells[53].

However, in many situations cytochalasin B inhibits stimulus-coupled discharge of packaged secretions. For example, the discharge of histamine from rat peritoneal mast cells which is triggered by cell-bound antibody and antigen, the drug 48/80 or the calcium ionophore A23187, is strongly inhibited by cytochalasin but not by colchicine. The question arises how a microfilament network normally prevents release while contraction of the system allows release. The simplest model appears to be that the contracted system is in the form of bundles, with spaces in between them allowing access of the granules to the plasma membrane. Other models are, however, compatible with the observations[5].

There is no doubt that colchicine can also inhibit the release of certain cell products. In some cases, such as the release of thyroid [^{125}I]iodoprotein[54], inhibition is observed with low concentrations of colchicine. Analogues of the drug are active in proportion to their activity on microtubular systems, and vinblastine and 2-methyl-2,4-pentanediol also inhibit secretion. In such situations it seems probable that the inhibition is due to effects on microtubules rather than some unrelated action, and that microtubules participate at some stage in stimulated thyroid hormone secretion. This is a complex process involving endocytosis of colloid at one pole, digestion in lysosomes, and transport of products to the other pole of the cell from which they are secreted. Colchicine may affect cellular polarity through disorganization of microtubular structure.

SUMMARY

All cell types other than non-nucleated erythrocytes contain microfilaments composed of actin, myosin and the associated regulatory proteins. Cellular locomotion and ruffling movements of the plasma membrane depend on the activity of the microfilament system. During phagocytosis of particles or micropinocytosis of fluid into vesicles visible by light microscopy, ruffle-like extensions move over the particles and fuse to enclose the particles or fluid. These movements require metabolic energy and are inhibited by cytochalasins. They are initiated by specific triggers and mediated by active movements of the plasma membrane mediated by the microfilament system. In contrast, micropinocytosis, invagination of the plasma membrane to form microvesicles about 70 mm in diameter, does not require metabolic energy and is not inhibited by cytochalasins. Some secretory granules are separated from the plasma membrane by a network of microfilaments, and some secretory processes require intact microtubules. The most common trigger for secretion is an increase in the concentration of calcium in the cytoplasm. This can initiate contraction based on microfilament interactions, break down labile microtubules and promote fusion of the secretory granule membrane and the plasma membrane.

References

1. Parkhouse, R. M. E. and Allison, A. C. (1972). Failure of cytochalasin or colchicine to inhibit secretion of immunoglobulins. *Nature New Biology,* **235,** 220
2. Porter, K. R. (1966). Cytoplasmic microtubules and their functions. In: G. E. W. Wolstenholme and M. O'Connor (eds.). *Ciba Foundation Symposium on Principles of Biomolecular Organization*, pp. 308–326. (London: Churchill Livingstone)
3. Margulis, L. (1973). Colchicine-sensitive microtubules. *Int. Rev. Cytol.,* **34,** 333
4. Pollard, T. D. and Korn, E. D. (1972). The 'contractile' proteins of *Acanthamoeba castellani. Cold Spring Harbor Symp. Quant. Biol.,* **37,** 573
5. Allison, A. C. (1973). The role of microfilaments and microtubules in cell movement, endocytosis and exocytosis. *Locomotion on Tissue Cells: Ciba Foundation Symposium, Vol. 14*, pp. 109–148. (Amsterdam: Allied Science Publishers)
6. Plagemann, P. G. W., Zylka, J. H., Erbe, J. and Estensen, R. D. (1975). Membrane effects of cytochalasin B. Competitive inhibition of facilitated diffusion processes in rat hepatoma cells or cell lines and formation of functional transport sites. *J. Membr. Biol.,* **23,** 77

7. Miranda, A. E., Godman, G. C., Deitch, A. D. and Tannenbaum, S. W. (1974). Action of cytochalasin D on cells of established lines. 1. Early events. *J. Cell Biol.*, **61,** 481
8. Tannenbaum, J., Tannenbaum, S. W. and Godman, G. C. (1975). Subcellular localization of binding sites for cytochalasin D. Evidence from activation energies. *Biochim. Biophys. Acta,* **413,** 322
9. Kelly, F. *et al.* (1975). Variants of simian virus 40-transformed mouse cells resistant to cytochalasin B. *Cold Spring Harbor Symp. Quant. Biol.*, **39,** 345
10. Davies, P., Allison, A. C. and Haswell, A. D. (1973). Selective release of lysosomal hydrolases from phagocytic cells by cytochalasin B. *Biochem. J.*, **134,** 33
11. Rabinovitch, M. and De Stefano, M. (1970). Interactions of red cells with phagocytes of the wax-moth (*Galleria mellonella* L.) and mouse. *Exp. Cell Res.*, **59,** 272
12. Huber, H. and Fudenberg, H. H. (1970). The interaction of monocytes and macrophages with immunoglobulins and complement. *Sem. Haematol.*, **3,** 160
13. Cohn, Z. (1970). Endocytosis and intracellular digestion. In: R. van Furth (ed.). *Mononuclear Phagocytes*, pp. 121–132. (Oxford: Blackwell)
14. Bray, D. (1973). Model for membrane movements in the neural growth cone. *Nature* (*London*), **244,** 93
15. Davies, P., Fox, R. I., Polyzonis, M., Allison, A. C. and Haswell, A. D. (1973). The inhibition of phagocytosis and facilitation of exocytosis in rabbit PMN by cytochalasin B. *Lab. Invest.*, **28,** 16
16. Lewis, W. H. (1931). Pinocytosis. *Bull. Johns Hopkins Hosp.*, **49,** 17
17. Ryser, H. J. P. (1967). Studies on protein uptake by isolated tumor cells. 3. Apparent stimulations due to pH, hypertonicity, polycations, or dehydration and their relation to the enhanced penetration of infectious nucleic acids. *J. Cell Biol.*, **32,** 737
18. Wagner, R., Rosenberg, M. and Estensen, R. (1971). Endocytosis in Chang liver cells. Quantitation by sucrose-^{3}H uptake and inhibition by cytochalasin B. *J. Cell Biol.*, **50,** 804
19. Nyberg, E. and Dingle, J. T. (1970). Endocytosis of sucrose and other sugars by cells in culture. *Exp. Cell Res.*, **63,** 43
20. Williams, J. A. and Wolff, J. (1971). Thyroid secretion *in vitro*: multiple actions of agents affecting secretion. *Endocrinology*, **88,** 206
21. Ericsson, J. L. E. (1965). Transport and ingestion of hemoglobin in the proximal convoluted tubule. II. Electron microscopy. *Lab. Invest.*, **14,** 16
22. Orlic, D. and Lev. R. (1973). Fetal rat intestinal absorption of horseradish peroxidase from swallowed amniotic fluid. *J. Cell Biol.*, **56,** 106
23. Chapman-Andresen, C. (1965). The induction of pinocytosis in amoebae. *Arch. Biol.* (*Paris*), **76,** 189
24. Bhisey, A. N. and Freed, J. J. (1971). Altered movement of endosomes in colchicine-treated cultured macrophages. *Exp. Cell Res.*, **64,** 430
25. Bruns, R. R. and Palade, G. E. (1968). Studies on blood capillaries. I. General organization of blood capillaries in muscle. *J. Cell Biol.*, **37,** 244
26. Casley-Smith, J. R. and Chin, J. C. (1971). The passage of cytoplasmic vesicles across endothelial and mesothelial cells. *J. Microsc.*, **93,** 167
27. Simionescu, N., Simionescu, M. and Palade, G. E. (1973). Permeability of muscle capillaries to exogenous myoglobin. *J. Cell Biol.*, **57,** 424
28. Karnovsky, M. J. (1968). The ultrastructural basis of transcapillary exchanges. *J. Gen. Physiol.*, **52,** 645
29. Garlick, D. G. and Renkin, E. M. (1970). Transport of large molecules from plasma to interstitial fluid and lymph in dogs. *Am. J. Physiol.*, **219,** 1595
30. Fedorko, M. E. and Hirsch, J. G. (1971). Studies on transport of macromolecules and small particles across mesothelial cells of the mouse omentum. I. Morphological aspects. *Exp. Cell Res.*, **69,** 113

31. Casley-Smith, J. R. (1969). Endocytosis: the different energy requirements for the uptake of particles by small and large vesicles into peritoneal macrophages. *J. Microsc.*, **90,** 15
32. Wills, E. J., Davies, P., Allison, A. C. and Haswell, A. D. (1972). The failure of cytochalasin B to inhibit pinocytosis by macrophages. *Nature New Biology*, **240,** 58
33. Nagasawa, J., Douglas, W. W. and Sebutz, R. A. (1971). Micropinocytic origin of coated and smooth microvesicles ('synaptic vesicles') in neurosecretory terminals of posterior pituitary glands demonstrated by incorporation of horseradish peroxidase. *Nature (London)*, **232,** 341
34. Heuser, J. E. and Reese, T. S. (1973). Evidence for recycling of synaptic vesicle membrane during transmitter release at the frog neuromuscular junction. *J. Cell Biol.*, **57,** 315
35. Ceccarelli, B., Hurlbut, W. P. and Mauro, A. (1973). Turnover of transmitter and synaptic vesicles at the frog neuromuscular junction. *J. Cell Biol.*, **57,** 499
36. Green, H. A. and Casley-Smith, J. R. (1972). Calculations on the passage of small molecules across endothelial cells by Brownian motion. *J. Theor. Biol.*, **35,** 103
37. Goldman, R. D., Berg, G., Bushnell, A., Cheng-Ming C., Dickerman, L., Hopkins, N., Miller, M. L., Pollack, R. and Wang, E. (1973). Fibrillar systems in cell motility. *Locomotion of Tissue Cells: Ciba Foundation Symposium, Vol. 4*, pp. 83–107. (Amsterdam: Allied Science Publishers)
38. Ginn, F. L., Hochstein, P. and Trump, B. (1969). Membrane alternations in hemolysis: internalization of plasmalemma induced by primaquine. *Science*, **164,** 843
39. Kanaseki, T. and Kadota, K. (1969). The vesicle in a basket. A morphological study of the coated vesicle isolated from nerve endings in the guinea-pig brain, with special reference to membrane movements. *J. Cell Biol.*, **42,** 202
40. Davies, P., Allison, A. C. and Haswell, A. D. (1973). The quantitative estimation of pinocytosis using radioactive colloidal gold. *Biochem. Biophys. Res. Commun.*, **52,** 627
41. Bruni, C. and Porter, K. R. (1965). The fine structure of the parenchymal cell of the normal rat liver. General considerations. *Am. J. Pathol.*, **46,** 691
42. Allison, A. C. and Young, M. P. (1969). Vital staining and fluorescence microscopy of lysosomes. In: J. T. Dingle and H. B. Fell (eds.). *Lysosomes in Biology and Pathology*, **2,** pp. 600–624. (Amsterdam: North-Holland)
43. Schanker, L. S., Nafliotis, P. A. and Johnson, J. M. (1961). Passage of organic bases into human red cells. *J. Pharmacol. Exp. Ther.*, **133,** 325
44. Mego, J. L., Bertini, F. and McQueen, J. D. (1967). The use of formaldehyde-treated [^{131}I]albumin in the study of digestive vacuoles and some properties of these particles from mouse liver. *J. Cell Biol.*, **32,** 699
45. Hers, H. G. and van Hoof, F. (1969). Genetic abnormalities of lysosomes. In: J. T. Dingle and H. B. Fell (eds.). *Lysosomes in Biology and Pathology*, **2,** pp. 19–40. (Amsterdam: North-Holland)
46. Hickman, S. and Neufeld, E. (1972). A hypothesis for I—cell disease: defective hydrolases that do not enter lysosomes. *Biochem. Biophys. Res. Commun.*, **49,** 992
47. Rubin, R. P. (1970). The role of calcium in the release of neurotransmitter substances and hormones. *Pharmacol. Rev.*, **22,** 389
48. Poste, G. and Allison, A. C. (1973). The membrane fusion reaction. *Biochim. Biophys. Acta*, **300,** 421
49. Llinas, R., Blinks, J. R. and Nicholson, C. (1972). Calcium transport in presynaptic terminal of squid giant synapse: detection with aequorin. *Science*, **176,** 1127
50. Foreman, J. C., Mongar, J. L. and Gomperts, B. D. (1973). Calcium ionophores and movement of calcium ions following the physiological stimulus to a secretory process. *Nature (London)*, **245,** 249
51. Lacy, P. E., Howell, S. L., Young, D. A. and Fink, C. J. (1968). New hypothesis of insulin secretion. *Nature (London)*, **219,** 1177

52. Rasmussen, H. (1970). Cell communication, calcium ion and cyclic adenine macrophosphates. *Science,* **170,** 404
53. Van Obberghen, E., Somers, G., David, G., Vaughan, G. D., Malaisse-Lagae, F., Orci, L. and Malaisse, W. J. (1973). Dynamics of insulin release and microtubular microfilamentous system. I. Effect of cytochalasin B. *J. Clin. Invest.,* **52,** 1207
54. Williams, J. A. and Wolff, J. (1970). Possible role of microtubules in thyroid secretion. *Proc. Natl. Acad. Sci. USA*, **67,** 1901

18
Role of Microtubules in Hepatic Secretory Processes

B. JEANRENAUD, Y. LE MARCHAND and C. PATZELT

Since the work of these laboratories pertaining to microtubules has been reviewed recently, and some of the relevant literature cited[1,2], only a brief summary of our own work is given here.

'MITOTIC SPINDLE' INHIBITORS AND HEPATIC SECRETORY PROCESSES

Using an *in situ* perfusion technique[3] we have observed that the output of triglycerides into the perfusate (an index of very low density lipoprotein (VLDL) secretion) was markedly curtailed by the presence of colchicine or vincristine, the lowest effective dose of the drug being 10^{-6} M and 4×10^{-7} M, respectively[3]. Similar observations were made when protein secretion was investigated. Thus, in the presence of colchicine or vincristine, the secretion of globulins, albumin and even of small polypeptides was decreased[4]. That colchicine or vincristine brought about these effects without altering lipid or protein synthesis was suggested by the following experimental evidences: (1) The drugs did not alter basic hepatic functions such as maintenance of a normal ATP level, of glycogenolysis, urea production and oxygen consumption, or of fatty acid uptake, oxidation and esterification; (2) the drugs did not alter total lipogenesis or total protein synthesis as estimated, respectively, by the incorporation of labelled precursors into labelled lipids or labelled proteins of liver plus perfusate; (3) in the presence of either colchicine or vincristine, the observed decrease in triglyceride or protein secretion could be accounted for by an increase in intracellular accumulation of either triglycerides or proteins; (4) in the particular case of VLDL secretion, the lipid that accumulated intracellularly in the presence of colchicine or vincristine could

Table 1 Summary of metabolic alterations produced by colchicine or vincristine in perfused mouse livers

Parameter measured	*Result*
Triglyceride secretion	decreased
Albumin secretion	decreased
Globulin secretion	decreased
Small polypeptide secretion	decreased
Cellular ATP content	unchanged
Glucose production	unchanged
Urea production	unchanged
O_2 Consumption	unchanged
Fatty acid uptake	unchanged
Total lipogenesis*	unchanged
Total protein synthesis*	unchanged
Hepatic triglyceride content	increased
Hepatic labelled protein content	increased

* Measured by [^{14}C]precursors incorporated into hepatic plus perfusate lipids or proteins
Taken from data of references [3] and [4]

be shown, with the electron microscope, to be represented mostly by an accumulation (never seen in control liver) of clusters of vesicles containing VLDL-like particles and possibly other materials[3,5]. Taken as a whole these data[1–4] supported the concept, summarized in Table 1, of a blockade, brought about by colchicine or vincristine, of the intracellular movement and eventual release into the space of Disse of the vesicle-enclosed lipoproteins or proteins, the overall synthesis of these materials being unmodified. Such a blockade could well be due to an actual effect of the 'mitotic spindle' inhibitors on microtubular function.

HEPATIC MICROTUBULAR PROTEIN

In a subsequent attempt to understand the function of microtubules, the colchicine-binding activity of high-speed supernatants of liver was investigated since colchicine most likely binds to the microtubular protein, tubulin[6]. We found, as shown in Table 2, that most of the colchicine-binding activity of liver supernatants was in the soluble fraction, a possible indication of the presence of the microtubular system in the cytosol. The colchicine-binding activity of liver high-speed supernatants was temperature and time-dependent, little binding being seen at 0 °C, and equilibrium being reached after 15–20 min at 37 °C. When microtubular proteins from liver and brain were purified and analysed for colchicine binding, their respective binding activities were identical: colchicine binding reached a plateau at 10^{-4} M, half-maximal binding occurring at about 5×10^{-6} M[6]. Purified microtubular proteins from either liver or brain exhibited a fast-migrating band corresponding to the subunit,

Table 2 Distribution of colchicine-binding activity in the liver as measured by fixation to DEAE-Sephadex

	Bound [^{3}H]colchicine		
	pmol/mg protein	*pmol/ml*	*% of total*
Total homogenate	6.22	1258	100
1st supernatant	37.07	955	79
2nd supernatant	21.32	117	9
Pellet	5.35	297	23

Each value represents the mean of three experiments. Data taken from reference[6]

and other more slowly migrating fractions corresponding to dimeric and polymeric states. The fast-migrating fraction from either liver or brain could be split into α- and β-tubulin on addition of urea to the polyacrylamide gels[6]. Electron microscopic examination of purified microtubular proteins from both tissues revealed, after precipitation with vinblastine, a pattern of ordered structures that was almost identical for the two tissues[6] and very similar to those previously reported for vinblastine-treated brain tubulin[7]. In contrast to colchicine, vinblastine reacted not only with liver tubulin but also with several undetermined proteins that could be distinguished from microtubular protein on disc gel electrophoresis[6].

MEASUREMENT OF FREE TUBULIN AND POLYMERIZED TUBULIN IN THE LIVER: EXISTENCE OF AN ASSEMBLY–DISASSEMBLY CYCLE

Microtubules obtained from various tissues can be rapidly destroyed by exposure to low temperature[6]. Moreover, they can be stabilized by the addition of high concentrations of glycerol[6,8]. Making use of these properties, the respective amount of free tubulin, and of tubulin polymerized as microtubules, was measured in the liver. The preparation of liver homogenates was carried out either in the presence or in the absence of glycerol (4 M). The consecutive incubation of these two different homogenates at low temperature resulted in a complete destruction of microtubules in the sample without glycerol, while the homogenate with glycerol contained stabilized microtubules. After ultracentrifugation of the two samples, tubulin concentration in the supernatants was measured by a [^{3}H]colchicine-binding assay. The supernatant of glycerol-free homogenates contained, since no microtubules were left, the total amount of tubulin, while that of glycerol-containing homogenates contained only that portion of tubulin that was not originally present as formed microtubules. The difference in the colchicine-binding activities of the two supernatants was therefore a measure of that portion of tubulin assembled as microtubules

Table 3 Colchicine-binding activity in supernatants of liver homogenates prepared with or without glycerol

Experimental conditions	*Bound [^3H]colchicine*	
	pmol/mg protein	*%*
Without glycerol (total tubulin)	38.7* ± 6.4	100
With glycerol (4 M) (free tubulin)	23.2 ± 3.3	60

Each value represents the mean of eight experiments ± SD
* $p < 0.001$
Data taken from reference [6]

in the tissue[6]. When applying this method (Table 3), liver cells contained about 40% of total tubulin in the form of microtubules, the remaining 60% being free tubulin. Of further importance was the observation that colchicine or vincristine influenced the equilibrium between free and polymerized tubulin. In these experiments (Table 4), mouse livers were perfused for 60 min in the presence of vinblastine or colchicine added as a single dose to the perfusate. No interference between the initial exposure step to colchicine and the final [^{3}H]colchicine-binding assay was observed provided the perfusion was performed with labelled colchicine of specific activity identical to that used later for the binding assay. The total amount of tubulin remained essentially constant in the presence of colchicine but liver microtubules decreased (in a dose-dependent manner) in the presence of the drug, to eventually disappear at a colchicine concentration of 10^{-4} M (Table 4). In the presence of high enough concentrations of vinblastine, liver microtubules also eventually disappeared but the total amount of tubulin decreased in a similar fashion. Thus, in the

Table 4 Summary of the effect of colchicine or vinblastine on the state of the microtubular system in perfused mouse liver

Addition	*Total tubulin*	*Tubulin present as free tubulin*	*Tubulin present as polymerized microtubules*
	(*bound [^3H]colchicine d.p.m.-g tissue* $\times 10^{-5}$)		
None	4.6	2.9	1.7
Colchicine (10^{-4} M)	4.6	4.6	0
None	4.4	2.6	1.8
Vinblastine (10^{-4} M)	2.7	2.7	0

Each figure represents the mean of three experiments. Data taken from reference[6]
Only the highest concentrations of the drugs are represented

liver exists an assembly–disassembly cycle by which tubulin is polymerized to microtubules, and vice versa. Furthermore, this cycle is interfered with by colchicine or vinca-alkaloids. Colchicine is likely to interfere with the assembly–disassembly cycle of microtubular protein by its binding to free tubulin, thus preventing its polymerization to microtubules. The reduction of the total amount of tubulin by vinblastine is probably due to precipitation of the latter as paracrystals following binding of the drug with tubulin. It is also possible that vinblastine induces the precipitation of microtubules *per se*, thereby producing insoluble material no longer measurable by the present method.

ASSEMBLY–DISASSEMBLY CYCLE OF TUBULIN AND FUNCTION OF THE MICROTUBULAR SYSTEM

It is conceivable that the assembly–disassembly cycle of tubulin plays a role in the secretory process of the liver since secretory activity of the liver is curtailed[1,6] when it is interfered with by colchicine or vinblastine. However, the relationship between this cycle, the actual intracellular movement of vesicles containing secretory products and their discharge outside the cells remains unexplained. It can be speculated that the equilibrium which appears to exist between free tubulin and microtubules could be altered by various triggers such as trinucleotide availability, calcium concentration, etc.[2] It is also possible that the turnover rate of microtubules may be an important facet of their function. For example, one may assume that microtubules grow at one end while depolymerizing at the tail end, and that such a growth could be directional, i.e. towards the vascular pole of the hepatocytes[2]. One can further hypothesize that such growing microtubules could be associated, via ill-defined physicochemical reactions, to the vesicles containing secretory products in such a way that they would be 'pulled' (or 'pushed') towards the plasma membrane. This view considers the microtubular system not so much as a structural entity as a functional one. One should emphasize that this concept is highly speculative, and that several combined experimental approaches will be needed to understand the possible functional relationship between the assembly–disassembly cycle of tubulin and the intracellular movement of vesicle-enclosed secretory products.

Acknowledgements

This work has been supported by Grant No. 3.2180.73 of the Fonds National Suisse de la Recherche Scientifique, Berne, Switzerland.

References

1. Le Marchand, Y., Singh, A., Patzelt, C., Orci, L. and Jeanrenaud, B. (1975). *In vivo* and *in vitro* evidence for a role of microtubules in the secretory processes of liver. In: M. Borgers and M. de Brabander (eds.). *Microtubules and Microtubule Inhibitors*, p. 153. (Amsterdam: North-Holland/American Elsevier)
2. Patzelt, C., Singh, A., Le Marchand, Y. and Jeanrenaud, B. (1975). Characterization of hepatic microtubules: possible functional models. In: M. Borgers and M. de Brabander (eds.). *Microtubules and Microtubule Inhibitors*, p. 165. (Amsterdam: North-Holland/American Elsevier)
3. Le Marchand, Y., Singh, A., Assimacopoulos-Jeannet, F., Orci, L., Rouiller, C. and Jeanrenaud, B. (1973). A role for the microtubular system in the release of very low density lipoproteins by perfused mouse livers. *J. Biol. Chem.*, **248,** 6862
4. Le Marchand, Y., Patzelt, C., Assimacopoulos-Jeannet, F., Loten, E. G. and Jeanrenaud, B. (1974). Evidence for a role of the microtubular system in the secretion of newly synthesized albumin and other proteins by the liver. *J. Clin. Invest.*, **53,** 1512
5. Singh, A., Le Marchand, Y., Orci, L., and Jeanrenaud, B. (1975). Colchicine administration to mice: a metabolic and ultrastructural study. *Eur. J. Clin. Invest.*, **5,** 495
6. Patzelt, C., Singh, A., Le Marchand, Y., Orci, L. and Jeanrenaud, B. (1975). Colchicine-binding protein of the liver: Its characterization and relation to microtubules. *J. Cell Biol.*, **66,** 609
7. Bensch, K. G., Marantz, R., Wisniewski, H. and Shelanski, M. (1969). Induction *in vitro* of microtubular crystals by vinca-alkaloids. *Science,* **165,** 495
8. Shelanski, M. L., Gaskin, F. and Cantor, R. C. (1973). Microtubule assembly in the absence of added nucleotides. *Proc. Natl. Acad. Sci. USA*, **70,** 765

DISCUSSION

I. Sternlieb, New York (USA): The human liver, in addition to microtubules and filaments present peripherally, shows rich arrays of similar structures associated with the hepatocellular nuclei. The latter are easily seen in tangential sections of nuclei as a perinuclear network which seems to converge towards or emerge from nuclear pores[1]. In some sections they seem to branch out and reach into the cytoplasm (Figures 1 and 2). These observations in normal and diseased human liver cells suggest that: (a) transfer of substances from or to the nucleus may be mediated by a microtubular system; and, (b) an evaluation of the effects of colchicine or vinblastine on liver cell function or secretion should take into consideration the possible central and cellular effects of these drugs, in addition to those noted at the cell membrane.

Dr I. M. Arias, La Jolla (USA): Several studies suggest that membrane-bound polysomes may become free ribosomes and synthesize albumin in the cytoplasm rather than on membranes. Albumin accumulates in the cytoplasm and, for unknown reasons, is not released properly from the liver. Could coichicine and other drugs be acting in this manner rather than by affecting a tubulin-related transport process?

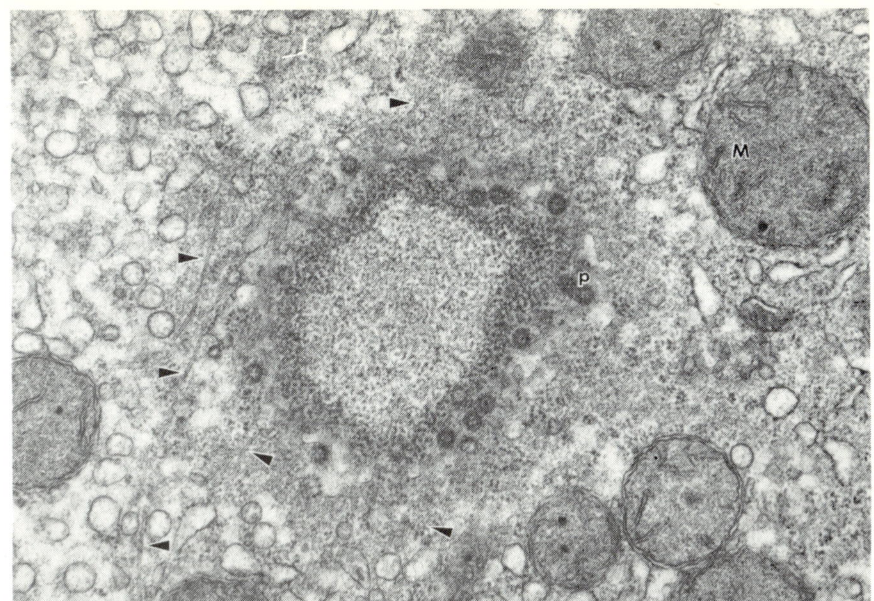

Figure 1 Nearly tangential section of nucleus of a hepatocyte from a liver biopsy specimen of a 17-year-old normal female subject. There are several microtubules indicated by arrowheads in the vicinity of the nucleus. M = mitochondrion; p = nuclear pores. Osmium tetroxide; epon; uranyl acetate and lead citrate; RCA-3H (× 32 400)

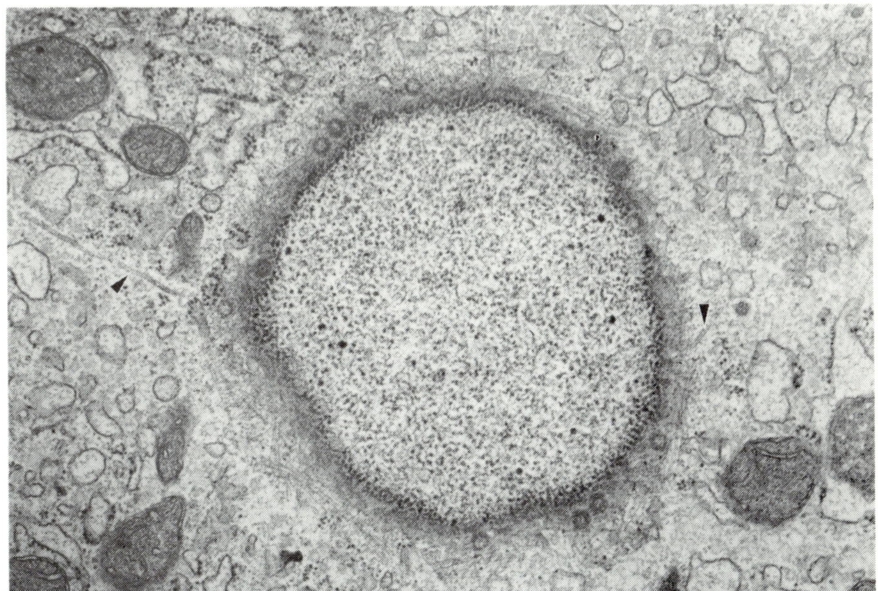

Figure 2 Numerous filaments surround this nucleus of a hepatocyte from the liver biopsy specimen of a 35-year-old man with chronic active hepatitis. At least two microtubules (arrowheads) radiating into the cytoplasm are seen in the plane of this section. Same processing as Figure 1. Siemens Elmiskop Ia (× 27 000)

Dr Jeanrenaud, Geneva (Switzerland): We have observed, in the presence of colchicine, a decrease in newly synthesized albumin (i.e. albumin labelled from labelled amino acids) secretion that can be accounted for completely by increase in intracellular accumulation of labelled albumin. However, we could not decide whether the accumulated albumin is free cytoplasmic or vesicle-enclosed albumin. But the decrease in protein secretion brought about by colchicine and reported by Redman *et al.*[2] was accompanied by an intracellular accumulation of this material within the Golgi complexes. We did not observe at the concentration of colchicine used (10^{-6} M), free ribosomes and the hepatic parenchyma of colchicine-treated livers was normally preserved. Although colchicine probably affects primarily the microtubular system, we cannot exclude other possibilities. The one you suggest might be one. Another one could be a membrane effect decreasing the output of secretory material without changing its synthesis.

Dr M. U. Dianzani, Torino (Italy): It was stated that in perfusion experiments, phalloidin did not affect lipoprotein 'secretion' by the liver. In studies with Drs Gravela and Poli on the influence of CCl_4 on the intracellular traffic of lipoproteins, colchicine prevented lipoprotein 'secretion' into the medium and the non-secreted material accumulated within the cells. With phalloidin, we obtained results somewhat different from those just reported. It inhibited both lipoprotein and protein 'secretion' by about 60% when added at 20 μM concentration. It produced also strong inhibition of polyribosomal profiles, but we do not consider the block in 'secretion' the primary event since cycloheximide, which practically completely blocks protein synthesis, did not block 'secretion' of lipoproteins during the first 30 min, only after 1 hour. By contrast, phalloidin inhibited by 60% already at 30 min. As cytochalasin B seems to be ineffective, this effect of phalloidin may be caused directly by alteration of the microfilament system. Possibly some additional action on the membrane (e.g. related to formation of blisters that may decrease the actively 'secreting' surface) may explain the results obtained.

Dr Jeanrenaud, Geneva (Switzerland): To clarify my message regarding the action of phalloidin upon isolated hepatocytes, observations currently obtained by Dr Marc Prentki in our laboratories can be summarized as follows: (1) Phalloidin does not influence only the secretion of proteins by hepatocytes but also protein synthesis from amino acids as well. The effect of phalloidin therefore cannot be ascribed exclusively (as is the case for colchicine) to an impairment of protein translocation *per se*. It is a combination of decreased translocation and decreased synthesis; (2) phalloidin does interfere with the translocation of lipoproteins. Thus, in the presence of phalloidin, [^{14}C]triglycerides derived from [^{14}C]palmitate are secreted into the incubation medium to a lesser extent than in control hepatocytes. The decrease in [^{14}C]triglyceride output observed in phalloidin-treated hepatocytes can be accounted for completely by an increased intracellular accumulation of [^{14}C]triglycerides. The relation of these phalloidin effects to the microfilamentous system is unsettled. Similarly, it is not understood why results are different for protein and lipoprotein secretion in the presence of phalloidin. We have also observed that the effects of phalloidin and of cycloheximide were not identical. Thus phalloidin effects are not simply related to microfilaments, and additional factors must come into play.

References

1. Sternlieb, I. (1965) Perinuclear filaments and microtubules in human hepatocytes and biliary epithelial cells. *J. Microsc.*, **4,** 551
2. Redman, C. M., Banerjee, D., Howell, K. and Palade, G. E. (1975). Colchicine inhibition of plasma protein release from rat hepatocytes. *J. Cell Biol.*, **66,** 42

19 Effect of Drugs on Hepatic Secretion and Degradation of Serum Lipoproteins

OLGA STEIN and Y. STEIN

SYNTHESIS AND SECRETION

The liver plays a central role in the synthesis and secretion as well as catabolism of serum lipoproteins. Different approaches have helped to elucidate this quite complex pathway. Using inhibitors of protein synthesis and various drugs, which were known to interfere with the delivery of the lipoproteins into the circulation, it became possible to determine at which stage of the pathway the block occurs. Some of the better known drugs and their possible site of action on the normal secretory pathway, depicted schematically in Figure 1, will be discussed presently. Biochemical and ultrastructural studies have contributed to the understanding of the secretory pathway and salient information was provided by the use of ethionine, orotic acid, cycloheximide (Figure 2) and colchicine (Figure 3). Thus, interruption of the synthesis of the apoprotein portion of lipoproteins by cycloheximide results in an almost complete inhibition of release of very low density lipoproteins from the liver, without formation of a fatty liver[1–3]. At the ultrastructural level only a few electron opaque particles are seen in the cisternae of the endoplasmic reticulum, indicating that inhibition of protein synthesis prevents the formation of the ultrastructural counterparts of serum very low density lipoproteins (VLDL). The lack of accretion of liver triglyceride is probably due in part to a fall in plasma free fatty acids (FFA) and an interference in the formation of triglycerides from diglycerides[4]. The administration of ethionine results in a much more complicated sequence of events, as both protein synthesis and the cellular ATP content are decreased and serum FFA are increased[5]. The accumulation

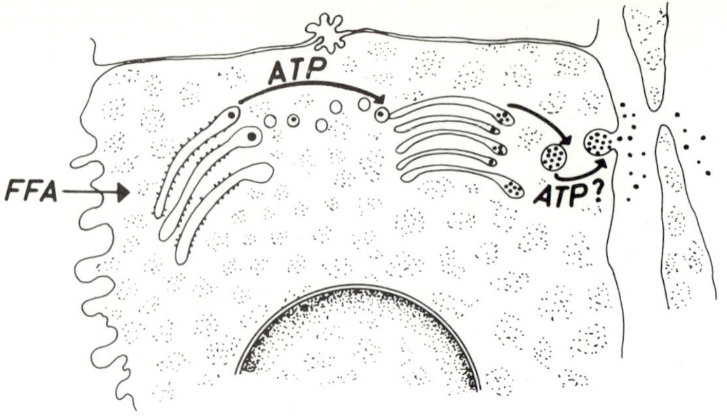

Figure 1 A schematic representation of synthesis, intracellular transport and secretion of very low density lipoproteins in a liver cell. The synthesis of the lipid and protein moieties occurs in the endoplasmic reticulum; the final glycosidation and formation of secretory vesicles takes place in the Golgi apparatus and the release of the lipoprotein particles into the space of Disse occurs after fusion between the membranes of the secretory vesicle and the plasma membrane

of VLDL particles in the cisternae of the endoplasmic reticulum 3–5 hours after administration of ethionine, and their absence from the Golgi apparatus, at a time when there is a significant fall in cellular ATP[5–7] suggests that this

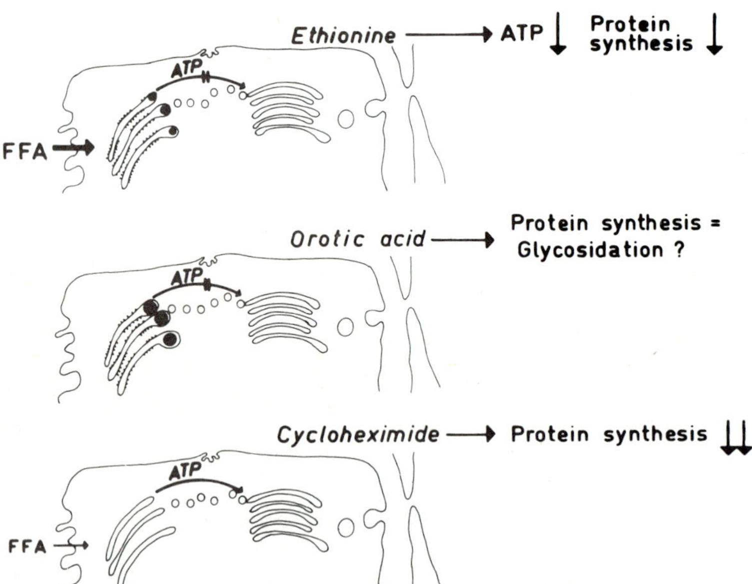

Figure 2 Schematic representation of the pathway of lipoprotein secretion and the sites at which it is interfered with by cycloheximide, orotic acid and ethionine

transfer is an energy-dependent step. Accumulation of intracisternal 'giant liposomes' occurs in livers of rats fed a purified diet containing orotic acid[8–10], without interference with protein synthesis. The apoprotein content of the giant intracisternal droplets was shown to comprise all the apoproteins normally present in serum VLDL, but was relatively deficient in *N*-acetylglucosamine, galactose and sialic acid[11]. As these carbohydrate moieties are added in the Golgi apparatus, and as no lipoprotein granules are seen in the Golgi apparatus after orotic acid treatment it seems that orotic acid interferes with the transport and subsequent glycosidation of the newly synthesized lipoprotein.

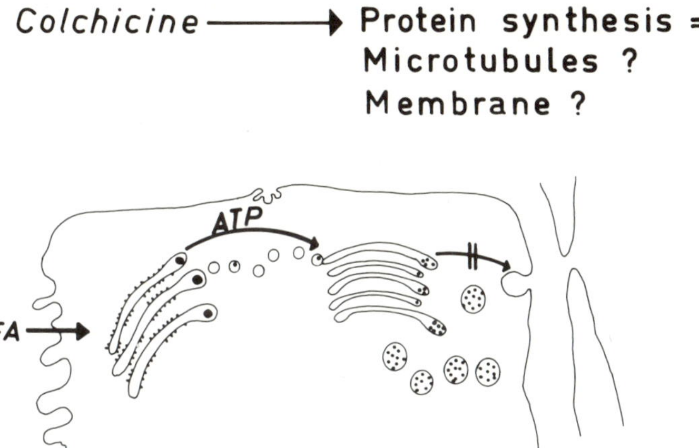

Figure 3 Schematic representation of the possible effect of colchicine on the last stages of lipoprotein secretion by the liver

The delivery of the VLDL from the intracellular site of formation is achieved by Golgi-derived secretory vesicles (Figure 1) which eventually fuse with the sinusoidal plasma membrane. The regulation of this step has been studied with the help of colchicine, which has been shown to interfere with the secretory process in various glands[12–15]. Injection of labelled fatty acid and leucine to rats which had been pretreated with colchicine, resulted in a marked fall in labelled triglyceride and labelled protein in the circulation, at a time when fatty acid esterification and protein synthesis were normal[16,17]. Using Triton WR 1339, which blocks triglyceride removal from the circulation it was possible to show a concomitant fall in serum labelled and non-labelled triglyceride (Figure 4). The reversibility of the colchicine effect depends on the dose used and restoration to normal occurred about 8 hours after injection of 0.05 mg colchicine/100 g body weight, but not after 0.5 mg (Figure 4).

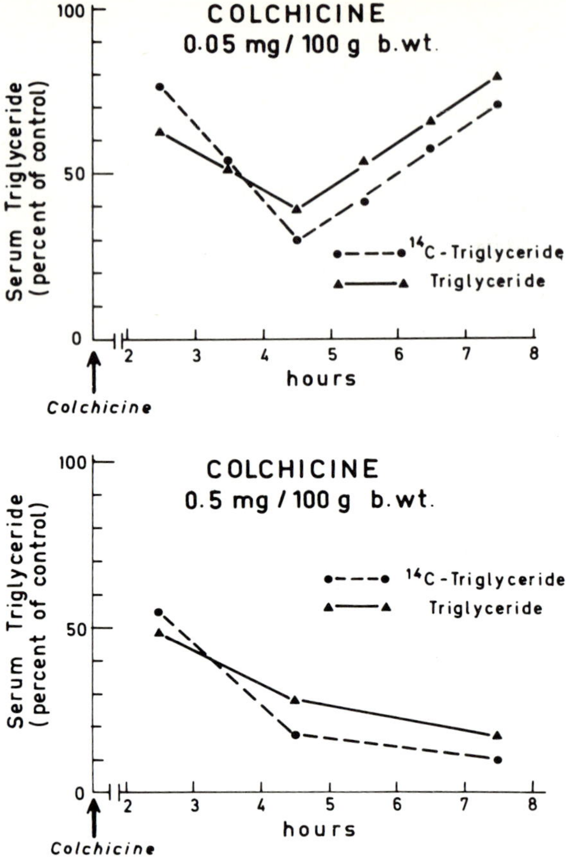

Figure 4 Effect of colchicine on the secretion of labelled triglycerides and the level of total triglycerides at different times after injection of the drug. For the details of the experiment see reference [17]

Colchicine prevented the release of VLDL, HDL and of albumin[18] and also had similar effects in perfused liver[19]. The morphological expression of colchicine treatment was the accumulation of Golgi-derived secretory vesicles filled with lipoprotein particles. Since in many areas signs of disintegration of the intravesicular particles were evident it seems that the drug did not affect the internal degradation by lysosomes (criniophagy) (Figure 5). The exact mechanism by which colchicine blocks the final stage of the secretory process has not been finally elucidated. Its well-known action on the microtubules has focused attention on the importance of this system. However, colchicine has been shown to affect cytomembranes, which leaves also the possibility of a more direct effect (Figure 3).

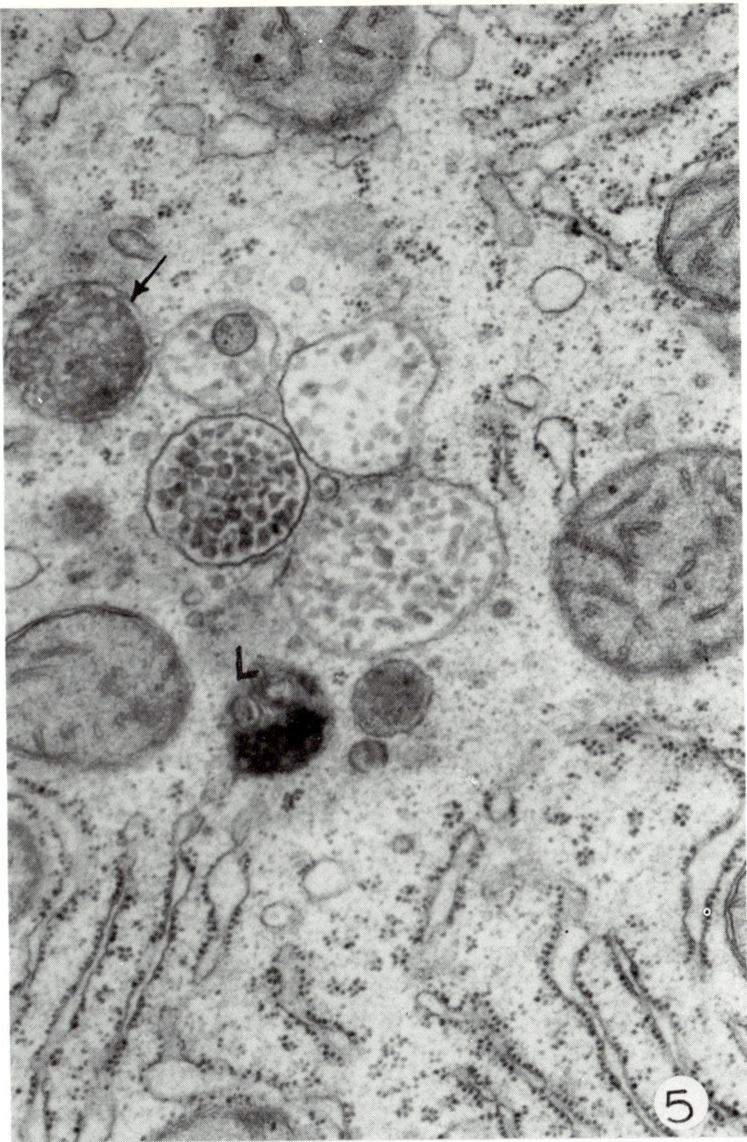

Figure 5 The secretory vesicles, containing very low density lipoprotein particles, are seen in the cytoplasm of the hepatocyte. In close proximity there is a secondary lysosome (L) and small vesicles, which could represent primary lysosomes. In one of the secretory vesicles (arrow), the particles have lost their distinct outlines, suggesting that they might be undergoing degradation. (From Stein *et al.*[17], reproduced with permission of the editor of *J. Cell Biol.*) (× 39 000)

CATABOLISM

In addition to being the source of serum lipoproteins, the liver is also the main site of degradation of certain serum lipoproteins. Serum VLDL is catabolized in a stepwise manner; following the loss of the bulk of triglyceride due to the action of lipoprotein lipase situated on the surface of muscle and fat capillaries, a remnant particle is formed which is cleared by the liver. Of the other two main serum lipoprotein classes, HDL is catabolized mainly by the liver[20,21], while the extent to which the liver participates in LDL metabolism is still not certain[22]. In various studies in which [^{125}I]-labelled VLDL[23] or HDL[21] had been used, concentration of label was found in relation to lysosomes of hepatocytes. Therefore it seemed of interest to elucidate the role of hepatic lysosomal enzymes in the final stages of catabolism of serum lipoproteins. In this approach chloroquine was used, as this drug has been shown to concentrate in the lysosomes of various cells in culture[24–27]. Studies with cultured cells have provided evidence that chloroquine interferes with the degradation of intracellular proteins[25] and mucopolysaccharides[26] and that it is an inhibitor of cathepsin B_1[25]. More recently inhibition of the protein and cholesterol ester moiety of low density lipoproteins has been shown to occur in human fibroblasts and in smooth muscle cells in culture[27,28].

To achieve an adequate concentration of chloroquine in the liver, rats were primed by two intraperitoneal injections and were injected with [^{125}I]-labelled VLDL of rat or human origin or with human LDL. The morphological changes observed in the liver are shown in Figure 6 and consist mainly of an increase in secondary lysosomes. Serum and liver protein and lipid radioactivity was determined at various time intervals after injection and the drug had no effect on the clearance of the lipoproteins from the circulation. However, the retention of labelled protein in the liver was significantly higher in the chloroquine-treated rats (Table 1). The same was true also for the labelled lipid.

Since chloroquine did not affect the disappearance rate of the lipoproteins from the circulation, it seems that the increased retention was due to an inhibition of protein degradation. This was further studied *in vitro* using the postnuclear supernatant of rat liver homogenates as the source of enzyme. The rate of hydrolysis was determined by the appearance of non-iodide TCA soluble material 2 hours after incubation with various [^{125}I]-labelled serum lipoproteins. The pH optimum was found to be 4.4 and the reaction could be saturated in the presence of 25–50 μg of lipoprotein–protein. Under these conditions addition of 100 μM chloroquine resulted in a complete inhibition of hydrolysis and individual lipoproteins showed different susceptibility to lower doses of chloroquine.

These results indicate that yet another facet of the hepatic metabolism of serum lipoproteins has been elucidated with the help of a drug. If chloroquine acts mainly by inhibiting lysosomal hydrolases the present findings support the notion that the end stages of serum VLDL-remnant degradation take place in the lysosomes of hepatocytes.

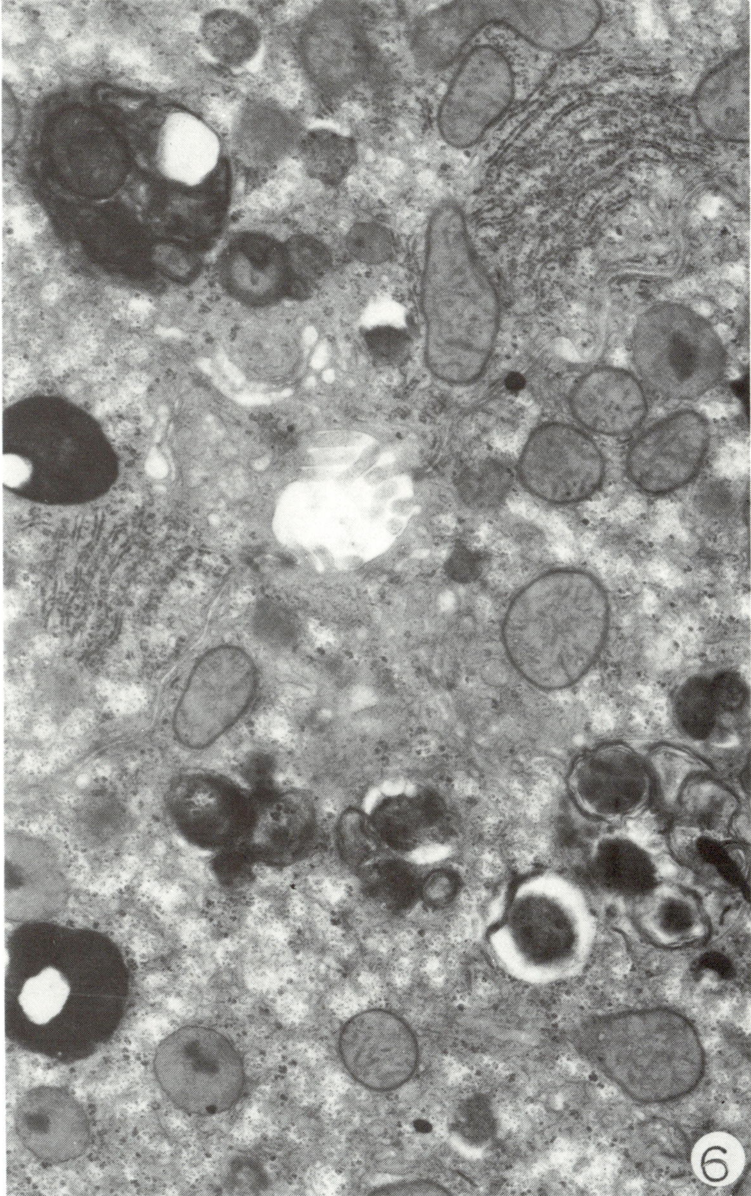

Figure 6 Section of rat liver 3 hours after injection of chloroquine (75 mg/kg body weight). Note the increase in autophagic vacuoles and secondary lysosomes (for further experimental details, see reference [30] (× 20 000)

Table 1 Effect of chloroquine on the retention of [^{125}I]-labelled protein in rat liver after injection of rat VLDL and human VLDL or LDL

Lipoprotein protein	*Treatment*	*Minutes*	*Percent of injected dose*	
			Liver	*Serum*
Rat VLDL	Saline	45	23.2 ± 085	46.5 ± 2.07
	Chloroquine	45	36.8 ± 2.09	44.3 ± 2.50
Human VLDL	Saline	45	18.7 ± 0.49	22.3 ± 0.69
	Chloroquine	45	49.1 ± 5.19	20.9 ± 1.08
Human LDL	Saline	240	4.4 ± 0.46	48.3 ± 3.20
	Chloroquine	240	10.4 ± 1.39	49.1 ± 3.41

Conditions: The rats were primed by two injections of chloroquine 75 and 50 mg/kg body weight 1 hour apart respectively. Labelled VLDL or LDL was injected 2 hours after the first injection. In the LDL study the rats received additional injections of chloroquine (50 mg/kg) 3 hours and 5 hours after the first injection; 100–300 μg of VLDL protein and 50 μg of LDL protein were injected and rats were killed 45 or 240 minutes thereafter. For the method of isolation and iodination of serum lipoproteins see reference[23]. Four to five rats were used in each experiment and values are means ± SE. Protein radioactivity was determined following TCA precipitation and lipid extraction according to Folch *et al.*[29]

In the case of rat or human VLDL more than 60% of the labelled protein which had disappeared from the circulation was recovered in the liver of chloroquine-treated rats while only 20% was found in the case of LDL. These findings were taken as an indication that the liver plays only a minor role in the degradation of serum LDL, even though the enzymes which can degrade LDL *in vitro* are present in the liver, and can be inhibited by addition of chloroquine *in vitro*. These results support also the thesis of Sniderman *et al.*[22] that serum LDL is catabolized mainly in extrahepatic tissues.

References

1. Verbin, R. S., Goldblatt, P. J., and Farber, E. (1969). The biochemical pathology of inhibition of protein synthesis *in vivo*. The effects of cycloheximide on hepatic parenchymal cell ultrastructure. *Lab. Invest.*, **20,** 529
2. Jazcilevich, S. and Villa-Trevino, S. (1970). Induction of fatty liver in the rat after cycloheximide administration. *Lab. Invest.*, **23,** 590
3. Riede, U. N., Seebass, C. and Dohr, H. P. (1971). Ultrastrukturell-morphometrische Untersuchungen an der Leberparenchymzelle der Ratte nach Hemmung der Proteinsynthese durch Cycloheximid. *Virchows Arch. Abt. B. Zellpathol.*, **9,** 16
4. Bar-On, H., Stein, O. and Stein, Y. (1972). Multiple effects of cycloheximide on the metabolism of triglycerides in the liver of male and female rats. *Biochim. Biophys. Acta,* **270,** 444
5. Farber, E. (1967). Ethionine fatty liver. *Adv. Lipid Res.*, **5,** 119
6. Baglio, C. M. and Farber, E. (1965). Reversal by adenine of the ethionine-induced lipid accumulation in the endoplasmic reticulum of the rat liver: a preliminary report. *J. Cell Biol.*, **27,** 591
7. Arstila, A. U. and Trump, B. F. (1972). Ethionine-induced alterations in the Golgi apparatus and in the endoplasmic reticulum. *Virchows Arch. Abt. B. Zellpathol.*, **10,** 344

8. Handschumacher, R. E., Creasey, W. A., Jaffe, J. J., Pasternak, C. A. and Hankin, L. (1960). Biochemical and nutritional studies on the induction of fatty livers by dietary orotic acid. *Proc. Natl. Acad. Sci. USA*, **46,** 178
9. Novikoff, A. B., Roheim, P. and Quintana, N. (1966). Changes in rat liver cells induced by orotic acid feeding. *Lab. Invest.*, **15,** 27
10. Novikoff, P. M., Roheim, P. S., Novikoff, A. B. and Edelstein, D. (1974). Production and prevention of fatty liver in rats fed clofibrate and orotic acid diets containing sucrose. *Lab. Invest.*, **30,** 732
11. Pottenger, L. A., Frazier, L. E., DuBien, L. H., Getz, G. S. and Wissler, R. W. (1973). Carbohydrate composition of lipoprotein apoproteins isolated from rat plasma and from the livers of rats fed orotic acid. *Biochem. Biophys. Res. Commun.*, **54,** 770
12. Lacy, P. E., Howell, S. L., Young, D. A. and Fink, C. J. (1968). New hypothesis of insulin secretion. *Nature* (*London*), **219,** 1177
13. Williams, J. A. and Wolff, J. (1970). Possible role of microtubules in thyroid secretion. *Proc. Natl. Acad. Sci. USA*, **67,** 1901
14. Malaisse-Lagae, F., Greider, M. H., Malaisse, W. J. and Lacy, P. E. (1971). The stimulus–secretion coupling of glucose-induced insulin release. IV. The effect of vincristine and deuterium oxide on the microtubular system of the pancreatic beta cell. *J. Cell Biol.*, **49,** 530
15. Pelletier, G. and Bornstein, M. B. (1973). Effect of colchicine on rat anterior pituitary gland in tissue culture. *Exp. Cell Res.*, **70,** 221
16. Stein, O. and Stein, Y. (1973). Colchicine-induced inhibition of very low density lipoprotein release by rat liver *in vivo*. *Biochim. Biophys. Acta*, **306,** 142
17. Stein, O., Sanger, L. and Stein, Y. (1974). Colchicine-induced inhibition of lipoprotein and protein secretion into the serum and lack of interference with secretion of biliary phospholipids and cholesterol by rat liver *in vivo*. *J. Cell Biol.*, **62,** 90
18. Redman, C. M. (1973). Colchicine and vinblastine inhibit the secretion of albumin by rat liver slices. *Ninth International Congress of Biochemistry, Stockholm,* **259**
19. Le Marchand, Y., Patzelt, C., Assimacopoulos-Jeannet, F., Loten, E. G. and Jeanrenaud, B. (1974). Evidence for a role of the microtubular system in the secretion of newly synthesized albumin and other proteins by the liver. *J. Clin. Invest.*, **53,** 1512
20. Roheim, P. S., Rachmilewitz, D., Stein, O. and Stein, Y. (1971). Metabolism of iodinated high density lipoproteins in the rat. I. Half-life in the circulation and uptake by organs. *Biochim. Biophys. Acta*, **248,** 315
21. Rachmilewitz, D., Stein, O., Roheim, P. S. and Stein, Y. (1972). Metabolism of iodinated high density lipoproteins in the rat. II. Autoradiographic localization in the liver. *Biochim. Biophys. Acta*, **270,** 414
22. Sniderman, A. D., Carew, T. E., Chandler, J. G. and Steinberg, D. (1974). Paradoxical increase in rate of catabolism of low density lipoproteins after hepatectomy. *Science*, **183,** 526
23. Stein, O., Rachmilewitz, D., Sanger, L., Eisenberg, S. and Stein, Y. (1974). Metabolism of iodinated very low density lipoprotein in the rat; autoradiographic localization in the liver. *Biochim. Biophys. Acta*, **360,** 205
24. Fedorko, M. E., Hirsch, J. G. and Cohen, Z. A. (1968). Autophagic vacuoles produced *in vitro*. II. Studies on the mechanism of formation of autophagic vacuoles produced by chloroquine. *J. Cell Biol.*, **38,** 392
25. Wibo, M. and Poole, B. (1974). Protein degradation in cultured cells. II. The uptake of chloroquine by rat fibroblasts and the inhibition of cellular protein degradation and cathepsin B_1. *J. Cell Biol.*, **63,** 430
26. Lie, S. O. and Schofield, M. (1973). Inactivation of lysosomal function in normal cultured human fibroblasts by chloroquine. *Biochem. Pharmacol.*, **22,** 3109
27. Goldstein, J. L., Brunschede, G. Y. and Brown, M. S. (1975). Inhibition of the proteolytic degradation of low density lipoprotein in human fibroblasts by chloroquine, Concanavalin A and Triton WR 1339. *J. Biol. Chem.*, **250,** 7854

28. Stein, O., Vanderhoek, J., Friedman, G. and Stein, Y. (1976). Deposition and mobilization of cholesterol ester in cultured human skin fibroblasts. *Biochim. Biophys. Acta,* **450,** 367
29. Folch, J., Lees, M. and Sloane-Stanley, G. H. (1957). Simple method for the isolation and purification of total lipids from animal tissues. *J. Biol. Chem.,* **226,** 497
30. Stein, Y., Ebin, V., Bar-On, H. and Stein, O. (1977). Chloroquine-induced interference with degradation of serum lipoproteins in rat liver, studied *in vivo* and *in vitro. Biochim. Biophys. Acta* (In press)

DISCUSSION

Dr R. Coleman, Birmingham (England): As colchicine did not affect the secretion of biliary lecithin, has it any effect on the secretion of other biliary constituents, especially bile salts, cholesterol and bilirubin?

Dr Stein, Jerusalem (Israel): Colchicine does not inhibit cholesterol secretion in the bile (see reference[7] in the paper) nor the excretion of bilirubin glucuronides into the bile (Gratz and Schwab (1976). *Cytobiologie*, **13,** 199).

20
Transmembrane Water and Ion Shifts Following Experimental Induced Endocytotic Fluid Uptake in the Isolated Perfused Rat Liver

W. JAHN

Liver parenchymal cells react to various injuries by forming endocytotic vacuoles. The simplest way to induce vacuolization in the liver is by congestion *in vivo*[1] or *in vitro*[2]. We have studied this mechanically induced endocytosis in a model system, the isolated perfused rat liver. The term 'endocytosis' will be used in this context, since endocytosis is generally understood as the process of uptake of any material by cells by invagination of the cell membrane and the subsequent forming of vacuoles[3].

METHODS

Livers of fed, male Wistar rats were perfused at 27 °C with buffer solutions containing 8% albumin or dextran but no red cells. The composition of the salt solution was: NaCl, 120 mM; KCl, 4 mM; $CaCl_2$, 1.4 mM; $MgCl_2$, 0.5 mM; NaH_2PO_4, 0.4 mM; $NaHCO_3$, 30 mM. The pH of the perfusion medium was 7.3–7.4. The perfusion rate was 3.5 ml/g · min. During perfusion with high posthepatic pressure the perfusion rate was reduced to about 1.5 ml/g · min, to avoid irreversible swelling. For perfusion with high posthepatic pressure a cannula was connected to the proximal end of the vena cava, such that the venous outflow pressure was 15 cm water.

Transmembrane water shifts between the intra- and extracellular space of the liver were monitored as changes in the dextran or albumin concentration of the perfusion medium, measured polarimetrically with allowance for the glucose present[10].

In experiments with a medium containing the optically inactive polyvinylpyrrolidone instead of dextran or albumin it was found that the change in optical rotation of the perfusion medium, resulting from the release of unknown optically active substances from the liver, was very low. The error arising from this source in the measurement of water shifts by polarimetric estimation of the dextran or albumin concentration was less than the deviation of individual measurements and could be neglected.

The sodium and potassium concentration of the perfusion medium was measured by flame photometry. Transmembrane potassium shifts were calculated from the changes in potassium concentration of the perfusion medium and the volume of the perfusion medium. It was not possible to estimate the transmembrane sodium shifts in the same way because of the high extracellular sodium concentration. Within the range of error of measurement the total concentration of alkali ions (Na^+ and K^+) remained constant during the experiments, because of osmotic equilibrium. The transmembrane sodium shifts were calculated on the basis of this finding and with regard to the observed transmembrane water shifts.

The extracellular space of the liver was estimated by perfusion of the liver outside the perfusion apparatus with a known amount of buffer solution and an estimation of the dextran concentration in this solution. The intracellular space (intracellular water) was then calculated from wet weight and dry weight of the liver[10].

Experimental data (usually $\bar{x} \pm$ S.D.) were calculated per g liver (wet weight).

WATER AND ION SHIFTS IN PERFUSED LIVER

Simple physical considerations indicate that not only the increased sinusoidal pressure forces water into cells during liver perfusion. When the posthepatic pressure is increased without altering the perfusion rate, the prehepatic pressure increases to the same extent. This raises the overall pressure in tissue, but does not give rise to any pressure gradient, which might be able to shift water into the cell. This is independent of any secondary change in vascular resistance (Figure 1).

When the posthepatic pressure in the liver — during perfusion at a constant rate — is increased by immersion of the organ in a flask filled with perfusion medium to a maximum of 15 cm for example, nothing happens. The only difference between this situation and the perfusion with an increased outflow pres-

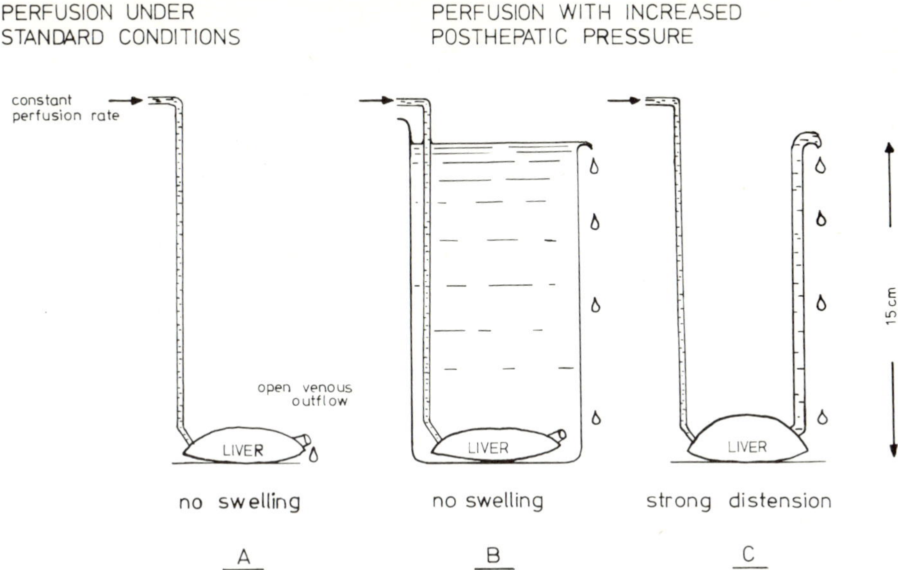

Figure 1 Schematic sketch for demonstration of the liver distension induced by the pressure gradient between the vascular system and the surroundings of the liver

sure alone is the pressure gradient which exists at the liver surface in the latter case. This pressure gradient between the vascular system and the surroundings of the liver is the only physical reason for the swelling of the organ.

When the liver swells the network of liver cells will be distended and this generates stretching forces on the individual cells, which can account for the increase in volume by vacuole formation. Changes in vascular resistance secondary to this distension will also occur as perfusion with posthepatic pressure of 15 cm for 5 min produce numerous vacuoles, up to 20 μm in diameter, in parenchymal cells (Figure 2). When after this treatment the liver is perfused under standard conditions — i.e. with open venous outflow — the vacuoles decrease in number and size (Figure 3). They contain a more or less electron-dense material; if the perfusion medium contained dextran instead of albumin, they appear to be empty[2].

The water taken up by endocytosis is at least partly released, while the colloids of the perfusion medium were concentrated inside the vacuoles. The endocytotic water shifts do not change the concentration of colloids in the perfusion medium, but in contrast the transmembrane water shifts can be monitored continuously by following polarimetrically the changes in the dextran or albumin concentration of the perfusion medium. Figure 4 shows the measurement of the transmembrane water movement in the distended liver. Both a 1 min and a 5 min period of distension induce a small transmembrane

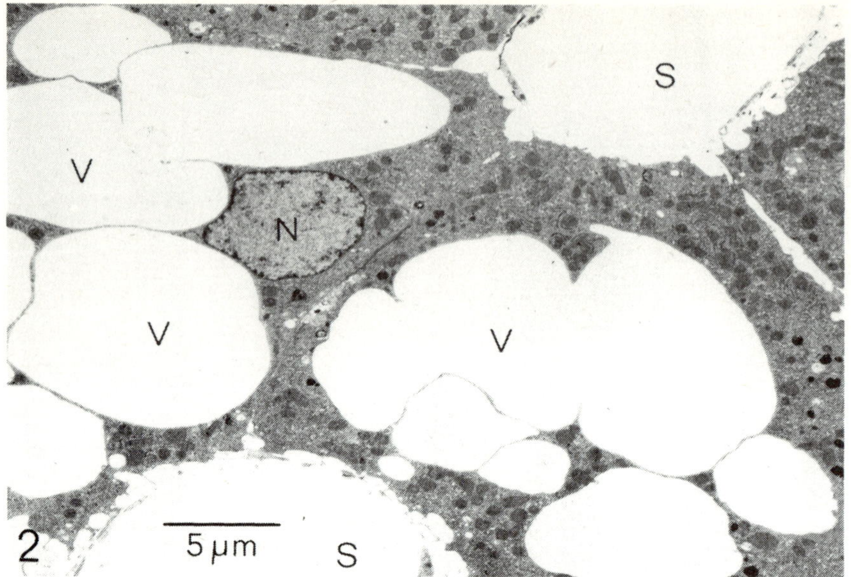

Figure 2 Electron micrograph of an isolated rat liver perfused for 5 min with a posthepatic pressure of 15 cm. The perfusion medium contained albumin. Many vacuoles (V) are present in the liver parenchymal cells (N = nucleus, S = sinus) (× 3000)

water uptake, followed by water release. The close agreement between experiments with albumin and dextran in the perfusion medium indicates that this reaction is essentially independent of the nature of the colloids in the perfusion fluid. All experiments described below contained dextran in the perfusion medium.

Any water transfer is associated with ion movements. In Figure 5 the transmembrane sodium and potassium shifts connected with the water shifts in the liver are shown after a 5 min period of distension. Surprisingly, the water release from the distended liver, which is mainly a release of sodium, is associated with a transient potassium loss. One element in this mechanism may be the transfer of sodium from the endocytotic vacuoles to the extracellular space without substitution of sodium for potassium, since otherwise it would be impossible to restore the original ion content of the cell and the original cell volume. The transient potassium loss may be a component in the mechanism of water release to overcome this problem.

EFFECTS OF DRUGS ON WATER/ION SHIFTS

In a search for drugs which might influence the mechanically induced endocytosis, we found that a group of substances which cause a potassium reuptake in the phalloidin-poisoned rat liver inhibit the reaction in the distended

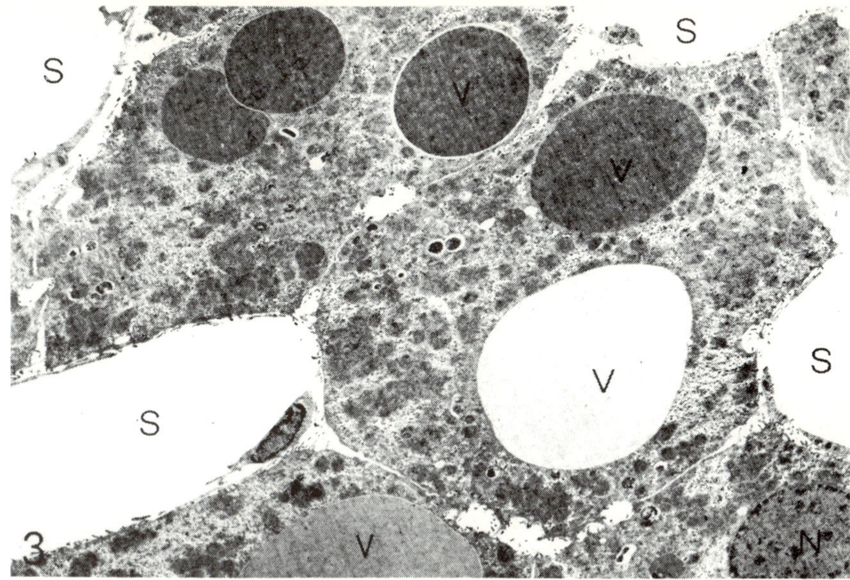

Figure 3 Electron micrograph of an isolated liver, which was perfused for 55 min with open venous outflow after a 5 min period of distension. There are a few vacuoles (V) in the parenchymal cells, containing a more or less electron-dense material (N = nucleus, S = sinus) (× 3000)

liver. These compounds[4] are coronary therapeutic agents like dipyridamole or hexobendine, but also simple chemical compounds like 4,7-phenanthroline. The latter substance, which causes no significant changes in water or ion content of the control liver, inhibits not only the water release (Figure 6) but also partly inhibits the endocytotic water uptake, when given before the distension. While in the control liver in a series of six experiments the intracellular space increased from 571 ± 32 μl/g to 814 ± 75 μl/g during a 5 min period of distension, the corresponding values in the phenanthroline-treated liver were 581 ± 19 μl/g and 707 ± 39 μl/g. The difference in the increase of the intracellular space is significant ($p = 0.01$, six experiments).

ACTION OF CYTOCHALASIN AND PHALLOIDIN

If a mechanical distension of liver cells induces endocytotic fluid uptake, agents which affect contractile properties of the cell should induce endocytosis in the isolated liver. Agents which act on contractile proteins include the cytochalasins and phalloidin.

Cytochalasin B induces the formation of endocytotic vacuoles in the isolated perfused rat liver[5]. The transmembrane water and ion shifts which accompany this endocytosis are very similar to those observed in the distended

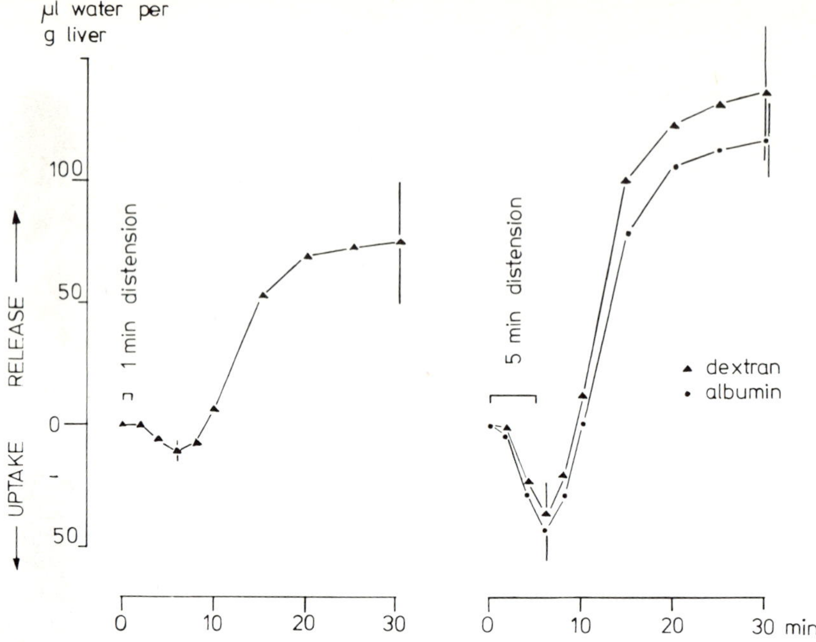

Figure 4 Transmembrane water shifts in the isolated rat liver, perfused for 1 or 5 min with a posthepatic pressure of 15 cm water. The perfusion medium contained albumin (●) or dextran (▲). Each curve represents six to eight experiments

liver (Figure 7). The action of cytochalasin on the isolated liver is greater at a high perfusion rate. This indicates that a mechanical factor is involved in the effect of this agent.

Regarding the similarity between the action of cytochalasin and the mechanically induced endocytosis, the effect of cytochalasin can be described as a lowering of the mechanical 'threshold' for the induction of the distension reaction, such that the low stretching forces present under control conditions due to the perfusion pressure alone, became sufficient to induce endocytosis.

There are also similarities between the liver distension and phalloidin poisoning. Phalloidin induces vacuolization in the liver parenchymal cells[6] and, with Lengsfeld, we could demonstrate that these vacuoles in the isolated perfused rat liver are of endocytotic origin[7]. The same was shown for the liver *in vivo* by Tuchweber *et al.*[8]

The rate of vacuolization in the phalloidin-poisoned rat liver depends largely on the perfusion pressure[9,10]. This pressure dependence indicates that also under these conditions endocytosis in the liver parenchymal cell depends on stretching forces.

However, striking differences exist between the phalloidin-poisoned and the distended liver. The potassium release from the phalloidin-poisoned liver[11],

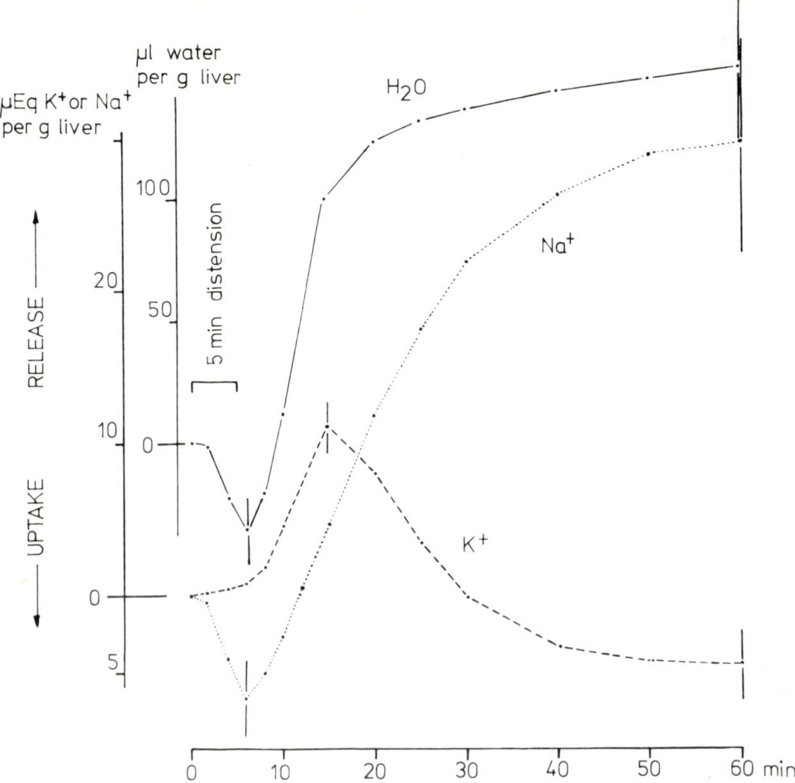

Figure 5 Transmembrane water, sodium and potassium shifts in the isolated liver, perfused for 5 min with a posthepatic pressure of 15 cm. The perfusion medium contained dextran (eight experiments)

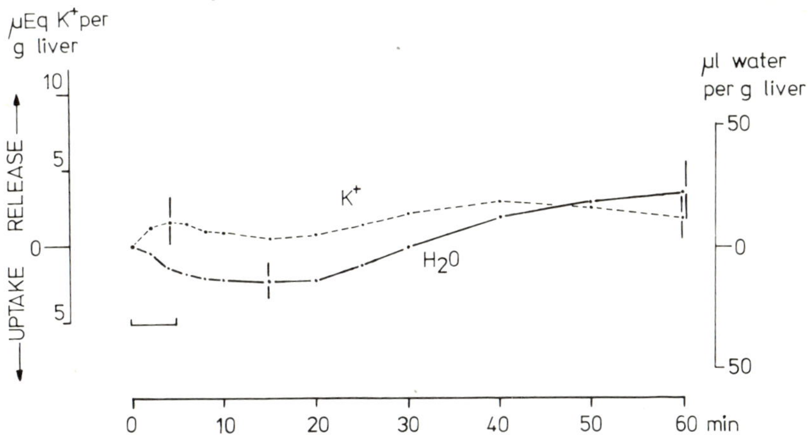

Figure 6 Distension of the isolated perfused rat liver in the presence of 4,7-phenanthroline (30 mg/50 ml), given 10 min before the distension (six experiments)

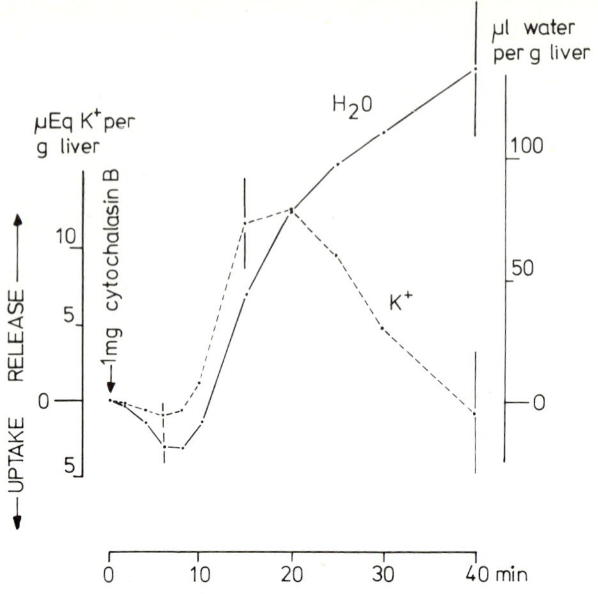

Figure 7 Transmembrane water and potassium shifts in the isolated perfused rat liver after application of 1 mg cytochalasin B. In these experiments the perfusion rate was 5 ml/g min (six experiments)

is not spontaneously reversible and the accompanying transmembrane water shifts are more complicated. There is a strong diphasic water uptake, which only in later stages of poisoning is followed by water release[12]. Only in a pharmacological altered state exist similar water and ion movements in the phalloidin-poisoned and the distended liver. A high extracellular Mg^{2+} concentration inhibits the potassium release from the phalloidin-poisoned liver but does not prevent the endocytotic fluid uptake[13]. In the presence of 20 mM Mg^{2+} phalloidin induces a spontaneously reversible potassium release accompanied by transmembrane water release which resembles the reaction of the distended liver (Figure 8). The action of phalloidin on the isolated rat liver, however, is not limited to the induction of endocytosis, but some symptoms of phalloidin poisoning may be a consequence of the induced endocytotic fluid uptake.

SUMMARY

Perfusion of the isolated rat liver with high posthepatic pressure leads to a distension of the liver tissue, which causes endocytotic fluid uptake in liver parenchymal cells. During subsequent perfusion with open venous outflow, most of the water taken up by endocytosis is released while the colloids of the

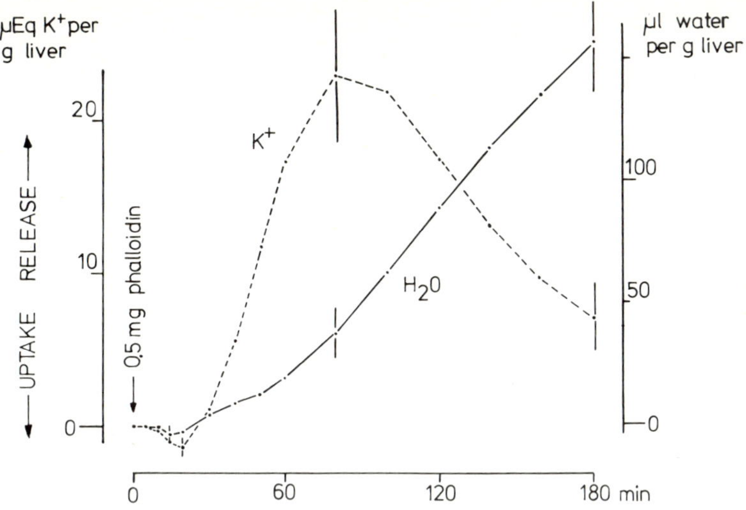

Figure 8 Transmembrane water and potassium shifts in the isolated perfused rat liver following phalloidin poisoning in the presence of 20 mM Mg^{2+} (six experiments)

perfusion medium were concentrated inside the endocytotic vacuoles. The water release is associated with a transient potassium loss.

Drugs which affect contractile properties of the parenchymal cells can also induce endocytosis in the isolated perfused rat liver. The transmembrane water and ion shifts accompanying the endocytotic water uptake after application of cytochalasin B are very similar to those observed in the distended liver. In contrast, the endocytosis induced by phalloidin is followed by irreversible potassium release, but in the presence of high extracellular Mg^{2+} concentrations phalloidin also induces transmembrane water release and a transient potassium loss.

It is concluded that the transient potassium loss following endocytotic fluid uptake is an element in the mechanism of water transfer from the endocytotic vacuoles to the extracellular space and that at least part of the potassium loss from the phalloidin-poisoned liver is a consequence of the endocytotic water uptake.

Acknowledgements

I thank Professor Dr Theodor Wieland for his support and his interest in these investigations.

References

1. David, H., Hecht, A. and Uerlings, I. (1965). Elektronenmikroskopische Befunde an der experimentellen Stauungsleber. *Acta Biol. Med. Germ.*, **15**, 513
2. Jahn, W. and Lengsfeld, A. (1974). Untersuchung der von den Perfusionsbedingungen abhängigen Aufnahme und Abgabe hochmolekularer Substanzen an der isolierten Rattenleber. *Naunyn-Schmiedeberg's Arch. Pharmacol.*, **281**, 241
3. Jacques, P. J. (1969). Endocytosis. In: J. T. Dingle and H. B. Fell (eds.). *Lysosomes in Biology and Pathology*, pp. 395–420. (Amsterdam: North-Holland)
4. Jahn, W. (1972). Hemmung der Phalloidinvergiftung *in vitro* durch 4,7-Phenanthrolin. *Naunyn-Schmiedeberg's Arch. Pharmacol.*, **272**, 182
5. Jahn, W. (1975). Endocytosis in the isolated perfused rat liver, induced by cytochalasin B. *Naturwissenschaften*, **62**, 445
6. Wieland, O. (1965). Changes in liver metabolism induced by the poisons of *Amanita phalloides*. *Clin. Chem.*, **11**, 323
7. Lengsfeld, A. and Jahn, W. (1974). Endocytose an der isoliert perfundierten Rattenleber nach Phalloidinvergiftung. *Cytobiologie*, **9**, 391
8. Tuchweber, B. Kovacs, K., Khandekar, J. D. and Garg, B. D. (1973). Peliosis-like changes induced by phalloidin in the rat liver. *J. Med.*, **4**, 327
9. Frimmer, M. (1972). The influence of physical conditions on swelling and K^+-release in perfused rat livers poisoned with phalloidin. *Naunyn-Schmiedeberg's Arch. Pharmacol.*, **275**, 393
10. Jahn, W. (1972). Aufnahme von Wasser und hochmolekularer Substanzen durch die isoliert perfundierte Leber nach Phalloidinvergiftung und nach Erhöhung des posthepatischen Druckes. *Naunyn-Schmiedeberg's Arch. Pharmacol.*, **275**, 405
11. Frimmer, M., Gries, J., Hegner, D. and Schnorr, B. (1967). Untersuchungen zum Wirkungsmechanismus des Phalloidins. *Naunyn-Schmiedeberg's Arch. Pharmacol. Exp. Pathol.*, **258**, 197
12. Jahn, W. (1974). Dependence of the transmembrane water and ion shifts in the phalloidin poisoned rat liver on pH of the perfusion medium. *Naunyn-Schmiedeberg's Arch. Pharmacol.*, **286**, 157
13. Jahn, W. (1976). Differential effect of raised Mg^{2+} concentration on K^+ ion movement and on swelling in the isolated perfused rat liver poisoned by phalloidin. *Naunyn-Schmiedeberg's Arch. Pharmacol.*, **293**, 197

21
Morphology of Phalloidin Intoxication

B. AGOSTINI and W. HOFMANN

The poisonous fungus *Amanita phalloides* contains, apart from the thermolabile and digestible phallolysins[1], two families of toxins, the amatoxins and the phallotoxins[2]. Humans die a few days after ingestion of the fungus with severe disorders of the liver, due to the lethal effect of amatoxins. The phallotoxins may not play any significant role in human intoxication, as they are ineffective when administered perorally to experimental animals[3]. The phallotoxins are very poisonous, however, when injected parenterally, and lead to death of adult animals within a few hours by a characteristic haemorrhagic necrosis of the liver[2]. In contrast to this, no significant alteration has yet been observed in other organs.

In this presentation the early morphological changes induced by phalloidin will be summarized as an approach to the discussion of the mechanism of the intoxication. Moreover, more recent studies on neonatal and tumour-bearing rats will be reported because of their importance in the pathogenesis of phalloidin poisoning.

THE MECHANISM OF PHALLOIDIN INTOXICATION AND THE MORPHOGENESIS OF LIVER ALTERATION

In the last few years it has been shown in this laboratory by studies with purified phalloidin[4,5] that an early link in the chain of events leading to the dramatic alteration of the liver is the interaction of the poison with the plasma membrane of the hepatocytes[6,7], resulting rapidly in a striking increase of actin filaments and in a peculiar endocytotic vacuolization within the cell[8]. These filaments have been demonstrated to be a target structure, i.e. receptor of phalloidin[9]. They have been identified as actin by decoration with heavy meromyosin[10] and by immunocytochemical studies with labelled antibodies[4,5,11].

Correlated biochemical, ultrastructural and immunocytochemical investigations[4,5] have clearly demonstrated that the phalloidin-induced proliferation of actin filaments occurs first close to the plasma membrane of the hepatocytes at the level of the so-called junctional complexes — mainly at the macula adherens and at the intermediate junction[12] — between contiguous hepatocytes and around bile canaliculi, i.e. in cell portions in which, also in control hepatocytes, a microfilamentous web is well developed and supposedly engaged in the various motile activities of the hepatocyte, is located[13,14].

The question why the hepatocytes are the only cells damaged by phalloidin in experimental animals has been discussed for several years. Studies with [^{3}H]desmethylphalloin on rats and mice demonstrated[15] that the toxin is almost completely taken up into the liver in 2 hours, where it exerts its poisonous effect, and that almost 60% of the injected poison is excreted unchanged in the urine within the first day. Since actin, i.e. the target structure of phalloidin, is contained in many other cells which are not altered by the poison, it is reasonable to assume that phalloidin is capable of penetrating only into the hepatocytes, so that it is only in these cells where its toxicological effect, which includes the early polymerization of G-actin to F-actin[10], can occur. In other words, only the plasma membrane of the hepatocyte would be able to bind the poison[6,7] and to transfer it to the target structure, the actin, which is located at the inner site of the cytoplasmic membrane. This is also strongly supported by the results of some more recent studies on the effect of phalloidin on newborn and very young rats[16,17] and on rats with autochthon and transplantable hepatomas[18,19].

STUDIES ON NEWBORN AND VERY YOUNG RATS

It is known that in contrast to adult animals, newborn and very young rats survive phalloidin poisoning[20] and die only when the poison is administered after the age of 3 weeks[21,22].

We injected three groups of 12 Wistar albino rats, aged 7, 11, and 15 days intraperitoneally with 10 mg/kg body weight of phalloidin, i.e. a dose 10-fold the LD_{50} for adult rats.

As reported also by Siess *et al.*[23], the examination of the liver a few hours after poisoning showed alterations similar to those of adult animals treated with sublethal doses, in which the foci of haemorrhagic necrosis are also irregularly scattered throughout the organ. On comparison with the control (Figure 1), the histological examination of intoxicated liver (Figure 2) initially shows strongly vacuolized cytoplasms with erythrocytes containing vacuoli in the hepatocytes and sinusoids engulfed with blood. Necrotic areas are usually surrounded by a modest reaction of fibroblasts and histiocytes and by liver tissue of normal appearance (Figure 3).

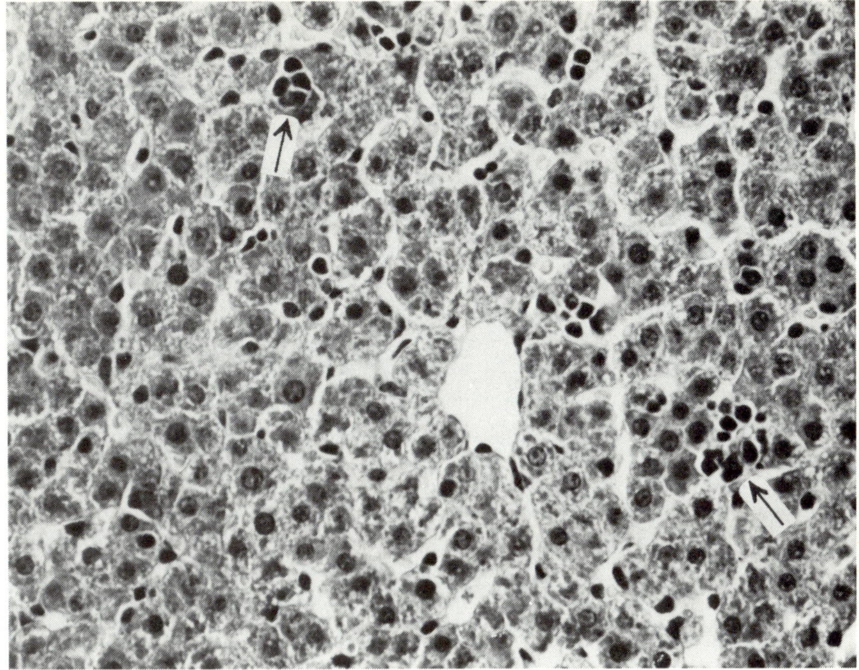

Figure 1 Paraffin section through the liver of a 15-day-old control Wistar rat injected intraperitoneally with a dose 10-fold the LD_{50}. Haemotoxylin and eosin stain. Midzone of a hepatic lobule with a centrolobular vein. Several small vacuoli, positive to fat staining in frozen sections can be observed. Arrows indicate haemopoietic cells in the sinusoids (× 800)

Also with the electron microscope[23] alterations to the hepatocytes, including the proliferation of actin filaments[17], appeared similar to those observed in adult animals. Figure 4 confirms the endocytotic pathogenesis of hepatocyte vacuoli since they contain erythrocytes and fibrin, whereas Figure 5 shows two enlarged bile canaliculi, with altered microvilli, containing bile and surrounded by a thickened filamentous web. As reported earlier[16,17], this, as well as the microfilamentous material contained in other cell portions, could be identified as actin by immunocytochemical studies with labelled antiactin antibodies from patients with severe chronic hepatitis.

These observations strongly suggest that the phalloidin tolerance of newborn and very young rats cannot depend on the absence of a target structure within the hepatocytes or on a detoxification mechanism. It is more probably related to some intrinsic capability of the fast-growing hepatocytes of neonatal liver to react against the effect of the poison, after penetration into the hepatocytes. Tolerance to doses of phalloidin higher than the LD_{50} was also observed in rats with regenerating liver following partial hepatectomy[24,25]. Moreover, preliminary results of autoradiographic studies with [^{3}H]thymidine on very young

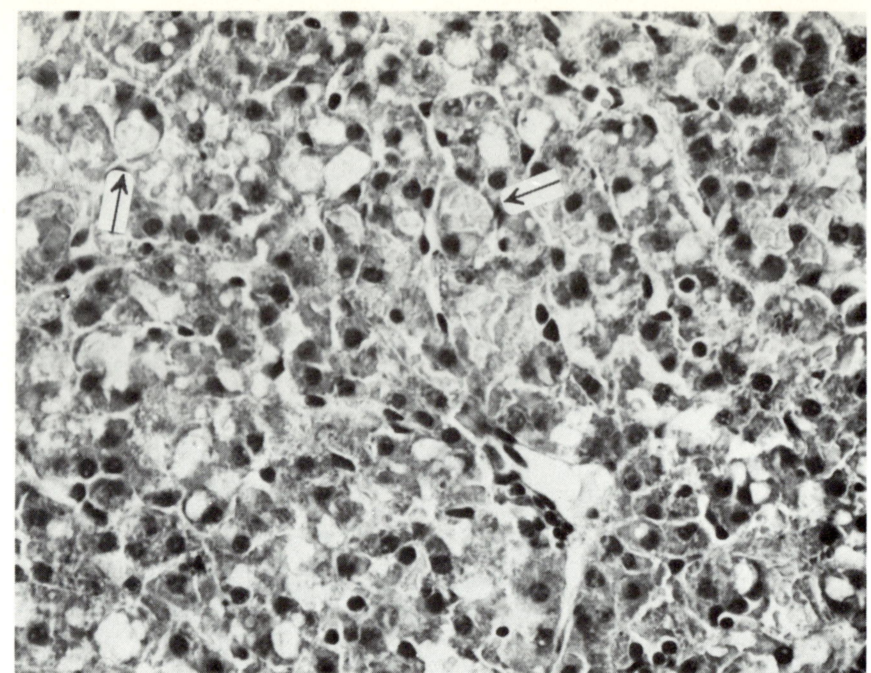

Figure 2 As Figure 1. Hepatic lobule of a 15-day-old rat killed 2 hours after poisoning with phalloidin. The rat was injected intraperitoneally with a dose 10-fold the LD_{50}. Note numerous variously sized vacuoli, many of which (arrows) contain erythrocytes, in the liver cells (× 800)

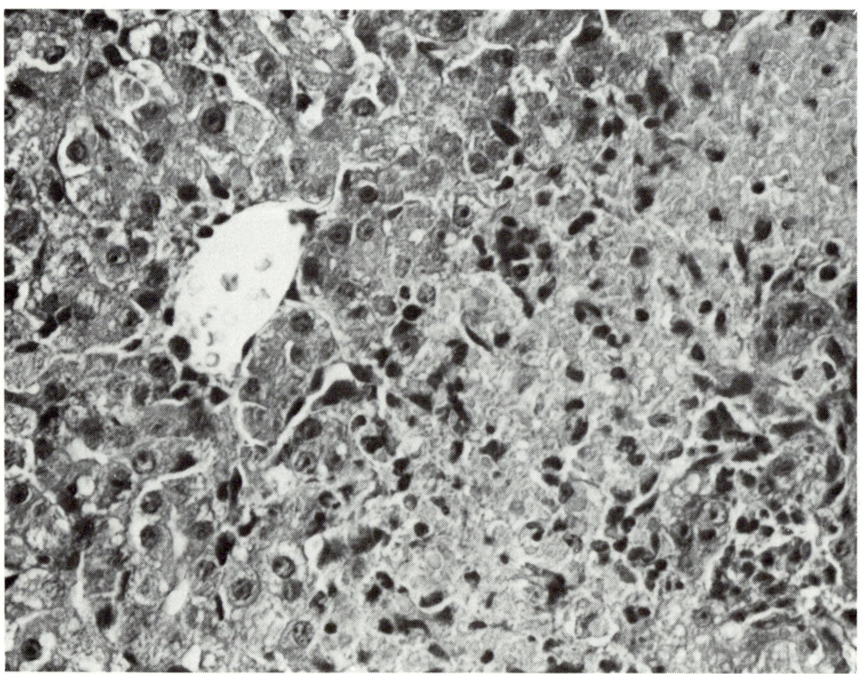

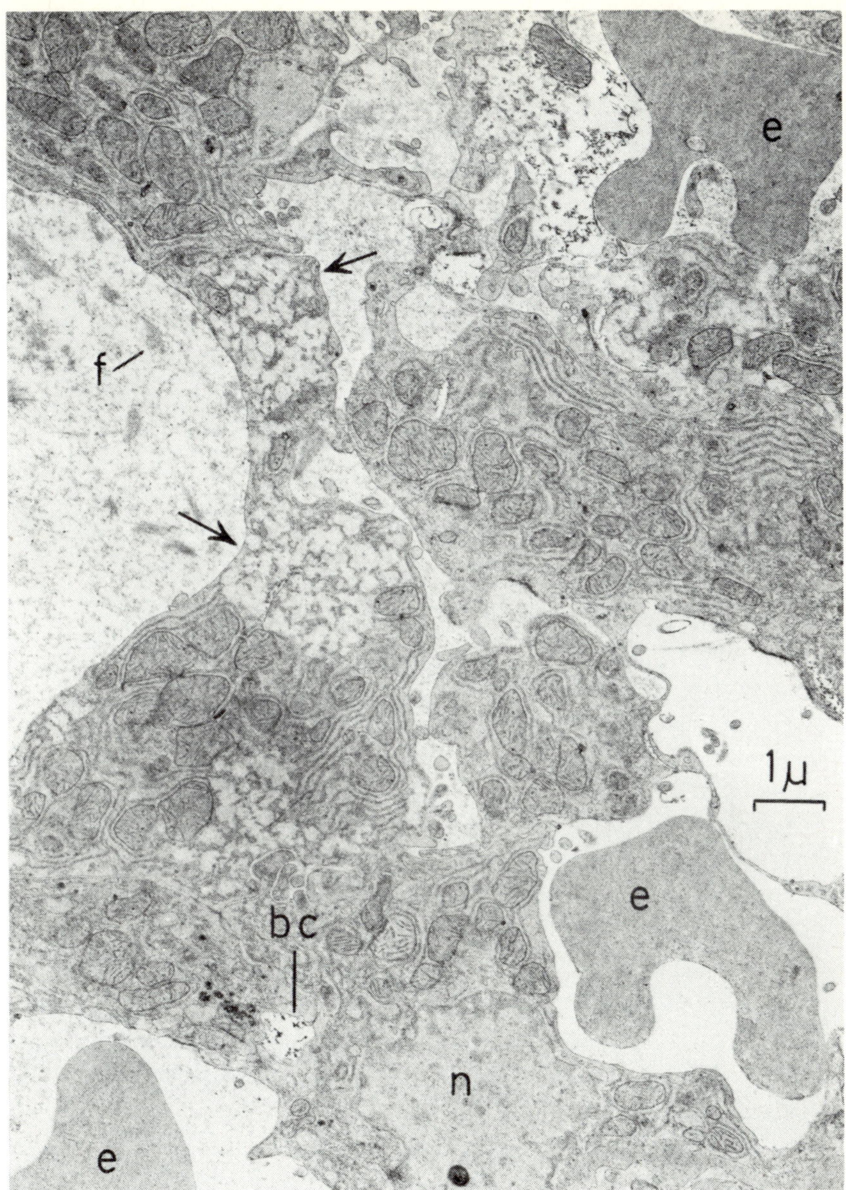

Figure 4 Electron micrographs of a thin section through the liver of a 15-day-old rat poisoned for 2 hours (*cf.* Figure 2). Double fixation with glutaraldehyde[33] and osmium[34], embedding in epon 812, and counterstaining with uranyl acetate[35] and lead citrate[36]. Erythrocytes (e) and fibrin (f) are contained in large vacuoli of the liver cells. Note the widening of the intercellular space. Arrows indicate pale areas of probable cytoplasmic necrosis. n: nucleus. bc: bile canaliculus (× 9000)

Figure 3 (left). As Figure 1. Hepatic lobule of a twin animal as in Figure 2 sacrificed 48 hours after poisoning. A large necrotic area is easily recognizable on the right. Note a few red cells in the centrolobular vein (× 800)

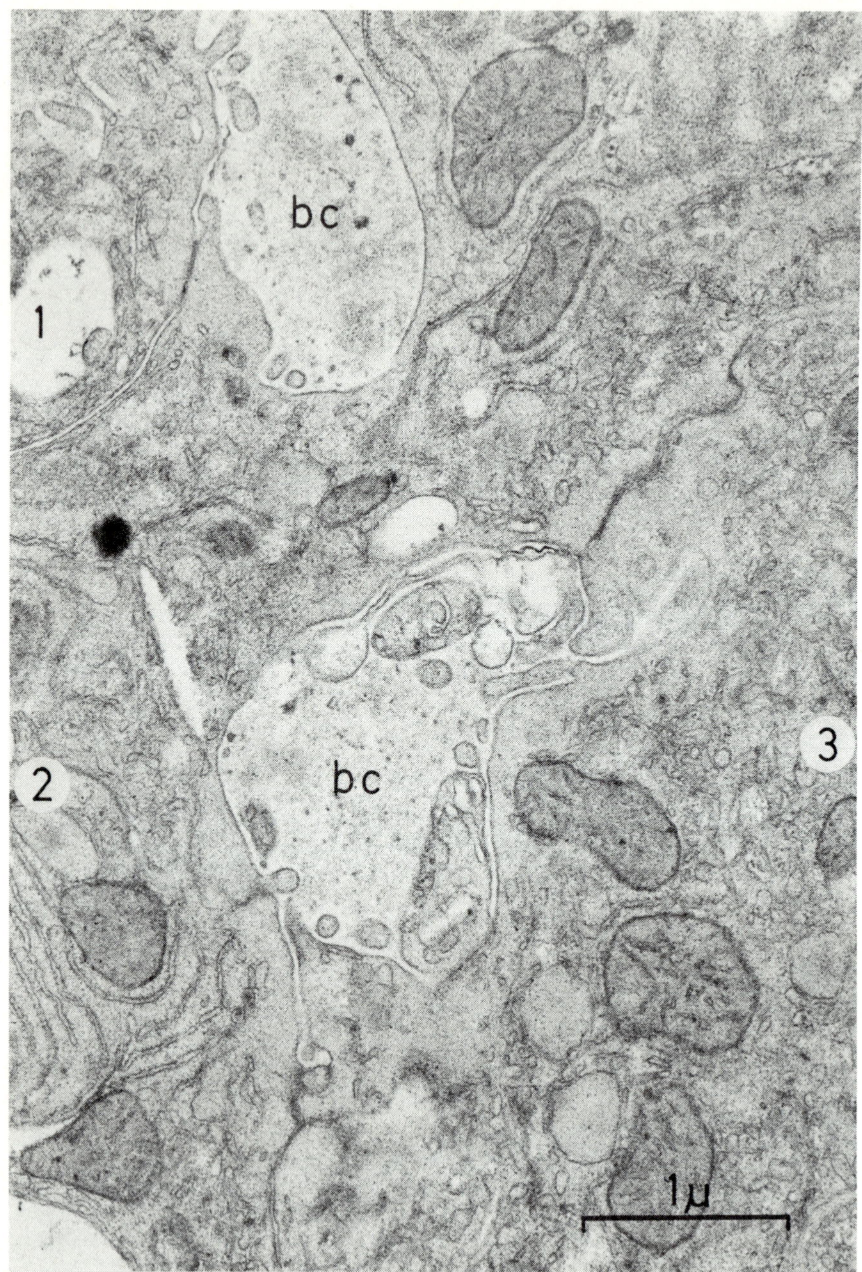

Figure 5 Thin section through the same sample as in Figure 4. Part of three liver cells (1, 2, and 3) and among them two enlarged bile canaliculi (bc), which have lost most of their microvilli and contain bile. Note the thickening of the pericanalicular filamentous web (× 27 000)

rats[26] demonstrate a rapid reparative regeneration of the hepatocytes which is characterized by maxima of the [^{3}H]index 24 hours after intoxication. The possiblity should be taken into account, however, that neonatal hepatocytes could have a lower phalloidin-binding capacity than those of adult animals.

STUDIES ON RATS WITH AUTOCHTHON AND TRANSPLANTABLE HEPATOMAS

These studies were performed to confirm the above hypothesis. We investigated first the effect of phalloidin on non-differentiated hepatocytes of three Wistar and three BD IX rats with primary liver carcinoma induced by diethylnitrosamine[18]. Two BD IX rats also with the same tumour served as a control.

Phalloidin was injected intraperitoneally in various amounts up to ten times the LD_{50}. None of the six animals died and no signs of haemorrhagic lesion could be found on histological examination of liver tissue not involved in the neoplastic process of two rats sacrificed 24 hours after phalloidin poisoning. Necrotic haemorrhagic areas, which are characteristic of diethylnitrosamine-induced carcinoma[27], could be observed in tumour nodules of poisoned as well as in those of untreated animals. In contrast to this, a very large amount of actin-like microfilamentous material could be seen by electron microscopy within normal and particularly in transformed hepatocytes (Figure 6).

For comparison we extended our studies to eight adult ACI rats bearing a fast-growing transplantable Morris hepatoma in the subcutaneous tissue of the back.

The animals were injected with various amounts of phalloidin 10–12 days after transplantation of the tumour, which at this time was always well developed up to a weight of 15–20 g.

As reported earlier[18], the animals did not survive the LD_{100} for control rats and like the latter developed within a few hours the lethal haemorrhagic lesion of the liver, while no significant differences could be observed between the hepatomas of poisoned and those of control animals, which also displayed necrotic haemorrhagic areas.

As illustrated in Figure 7, in cooperation with Professor Ivankovic, similar behaviour could be observed in five BD IX adult rats with secondary carcinomatous nodules which developed a few weeks after intraperitoneal inoculation of malignant cells from a homologous diethylnitrosamine-induced liver carcinoma[19].

It appears, therefore, that phalloidin is capable of damaging only normal hepatocytes. Transformed hepatocytes may also be affected by the poison, as it seems to be suggested by the very large amount of actin-like microfilaments in the neoplastic hepatocytes with diethylnitrosamine-induced carcinoma. These cells would not undergo, however, the alteration of phalloidin poisoning due to their intrinsic capability — as non-differentiated hepatocytes — to

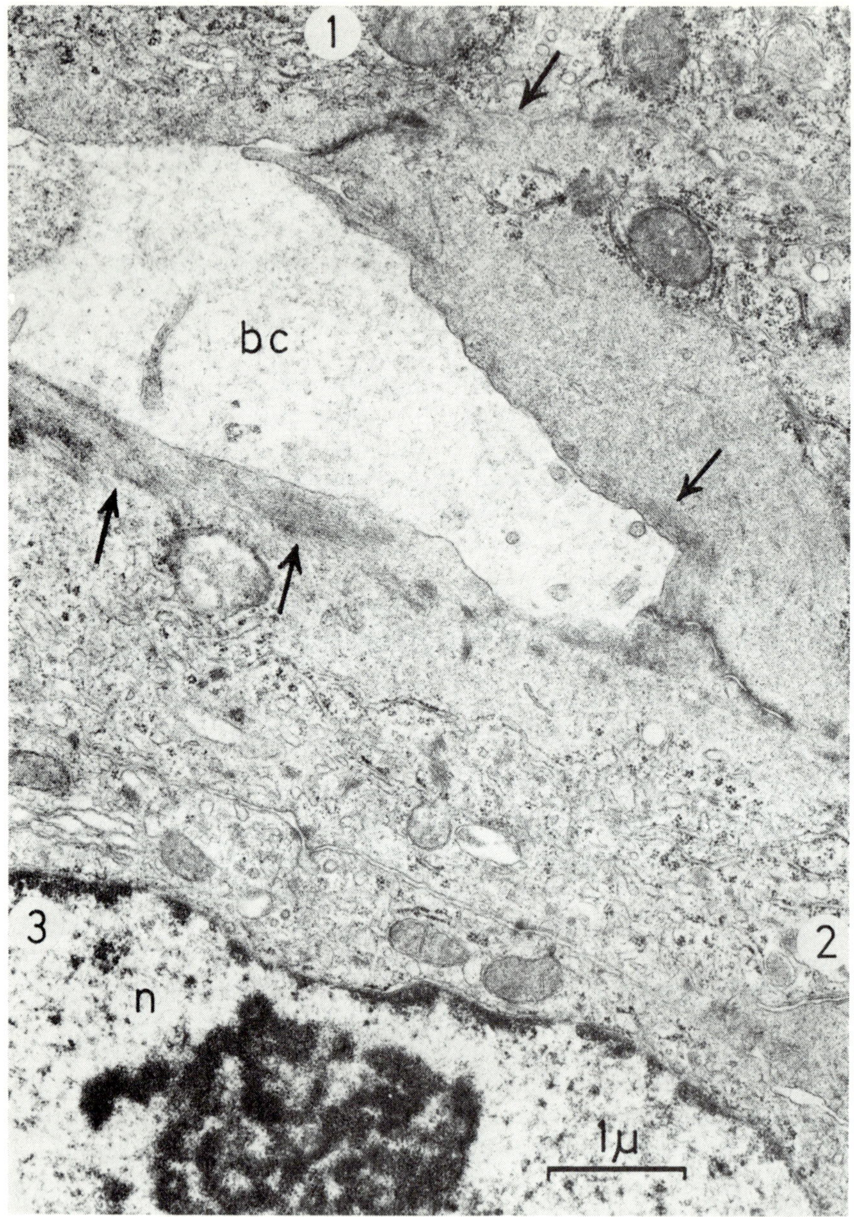

Figure 6 Electron micrograph of a section through three transformed hepatocytes (1, 2, and 3) of a BD IX rat with liver carcinoma induced by diethylnitrosamine which survived a 10-fold LD_{50} of phalloidin (section preparation as Figure 4). Note a dilated bile canaliculus (bc) surrounded by a well-developed filamentous web. Arrows indicate longitudinally cut microfilaments. n: nucleus with a prominent nucleolus (× 18 000)

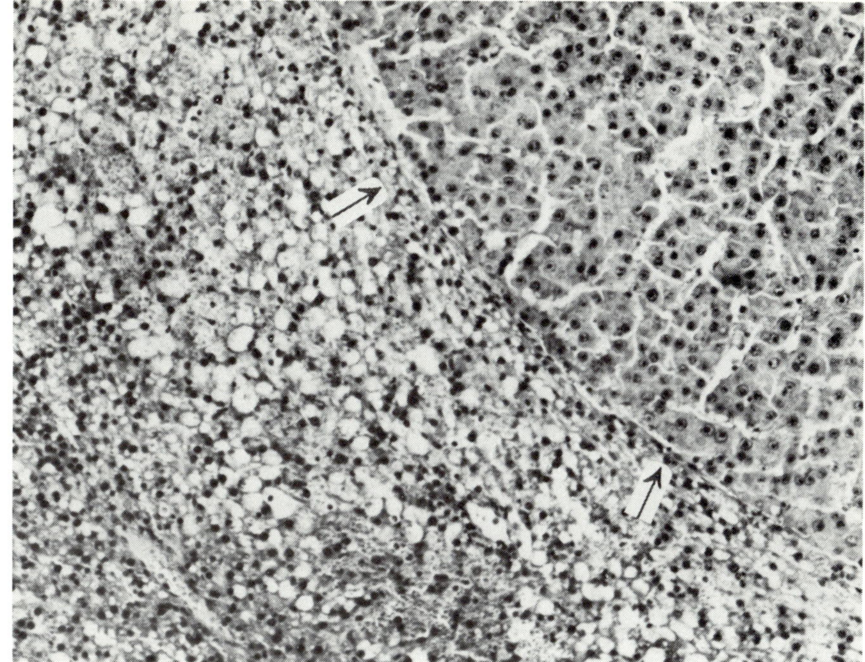

Figure 7 Paraffin section through the liver of an adult BD IX rat with a secondary carcinomatous nodule developed 6 weeks after intraperitoneal inoculation of cell from a diethylnitrosamine-induced carcinoma of a homologous liver[19]. The animal died 4 hours after injection of a LD_{50} of phalloidin. While no alterations can be observed in the neoplastic cells (arrows), the surrounding liver tissue consists of strongly vacuolized hepatocytes and sinusoids engulfed with blood, as it is characteristic after phalloidin poisoning (haemotoxylin and eosin) (× 480)

react against the effect of the poison, or depending also on some special relationship among adjoining neoplastic cells, which would avoid the endocytotic vacuolization and the subsequent engulfment, *ex vacuo*, of sinusoids and vacuoli with blood.

Alternatively, other transformed hepatocytes, like those of Morris hepatoma, may not be affected by phalloidin due to some chemical abnormality of their surface with a deficiency of certain critical functions[28] normally involved also in the binding of phalloidin and the transfer of it to its target structure, i.e. the actin, which is located also in the cells of Morris hepatoma intimately close to the inner leaflet of the plasma membrane. The possibility is strongly supported by the results of some experiments on Wistar rats with AH 103 Yoshida ascites hepatoma.

When these animals are injected with a LD_{100} of phalloidin, they develop in a few hours the characteristic lethal lesion of the liver, including the proliferation of actin filaments within the hepatocytes, without displaying, however, any alteration in the neoplastic ascites cells. These differences between hepatocytes and neoplastic cells are shown in Figures 8, 9 and 10.

Figure 8*a* represents as a control a cryostat section through the liver of a rat with well-developed ascites hepatoma challenged with antiactin antibodies prepared by Professor Seelig from a patient with severe chronic hepatitis. The well-known characteristic polygonal pattern of immunofluorescence staining of hepatocytes borders, due to the presence of actin close to the plasma membrane and some bright dots, probably corresponding to bile canaliculi surrounded by the microfilamentous web, are easily recognizable. When a liver

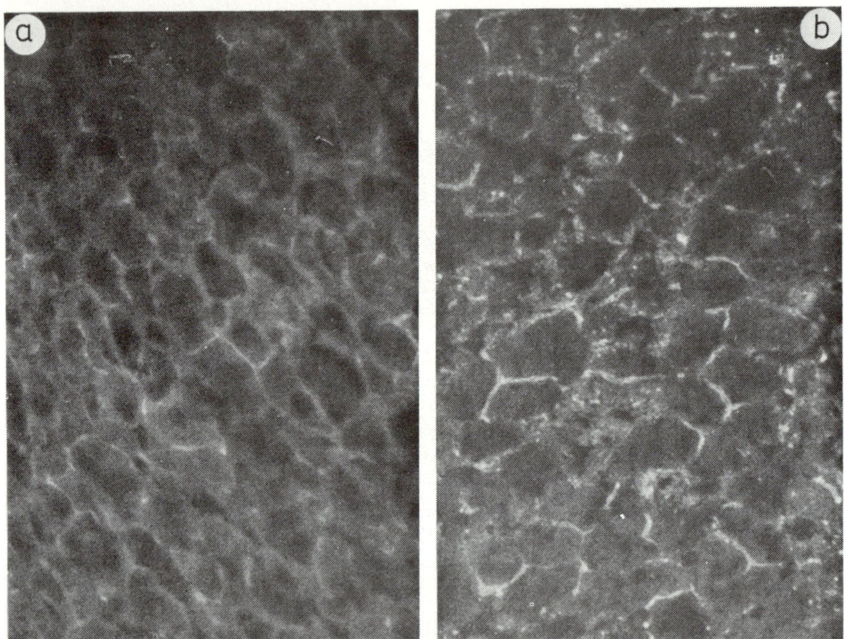

Figure 8a and b Epifluorescence micrographs of cryostat sections of the liver of two adult Wistar rats with a 10-day-old AH 103 Yoshida ascites hepatoma stained with labelled antiactin antibodies from a patient with liver hepatitis B. Leitz Ortolux epifluorescence microscope, filter BG 480 and K 510, lamp: Hg 200 W, Ilford film HP4 (27–29 DIN). (*a*) Liver of unpoisoned animal as a control. Note the characteristic polygonal pattern of hepatocyte borders (*cf.*[39]). (*b*) Animal sacrificed 3 hours after intraperitoneal injection of a LD_{100} of phalloidin. On comparison with the liver of the untreated animal (*a*) the immunofluorescent staining is increased (× 1100)

section of a twin animal poisoned with phalloidin is observed under the same conditions (Figure 8*b*), the immunofluorescence is increased along the hepatocyte borders and the bright dots at the periphery of the hepatocytes appear to be larger than in the control[4,5,11]. In contrast to this, the lack of fluorescence increase in the ascites cells of phalloidin-poisoned animals (compare Figure 9*c* with 9*d*) and the finding by electron microscopy of some microfilamentous

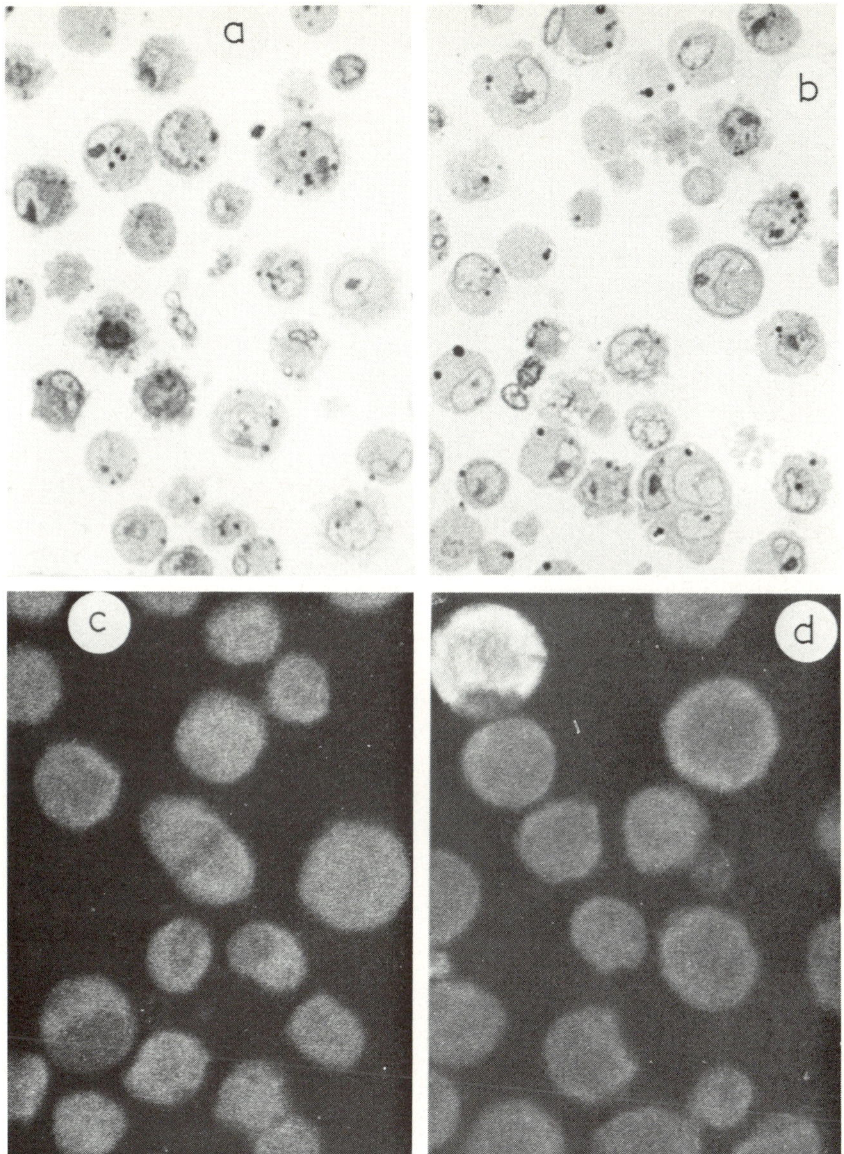

Figures 9a, b, c and d Micrographs of cells of AH 103 Yoshida ascites hepatoma from the same control (*a* and *c*) and poisoned animal (*b* and *d*) as in Figure 8. (*a*) and (*b*) Thick sections through pellets of ascites cells embedded in epon 812 stained with toluidine blue for light microscopy. Cells of poisoned sample (*b*) appear similar to those of the control (*a*). (× 1350). (*c*) and (*d*) Smear preparations challenged with antiactin antibodies (*cf.* Figure 8). No differences on degree of fluorescence can be detected between control (*c*) and poisoned sample (*d*). The strong fluorescence of a cell in (*d*) is most probably due to artifact (× 1600)

structures in the same cells of both untreated (Figure 10*a*) and poisoned animals (Figure 10*b*) strongly suggest that the ascites cells of Yoshida hepatoma are not penetrated by phalloidin.

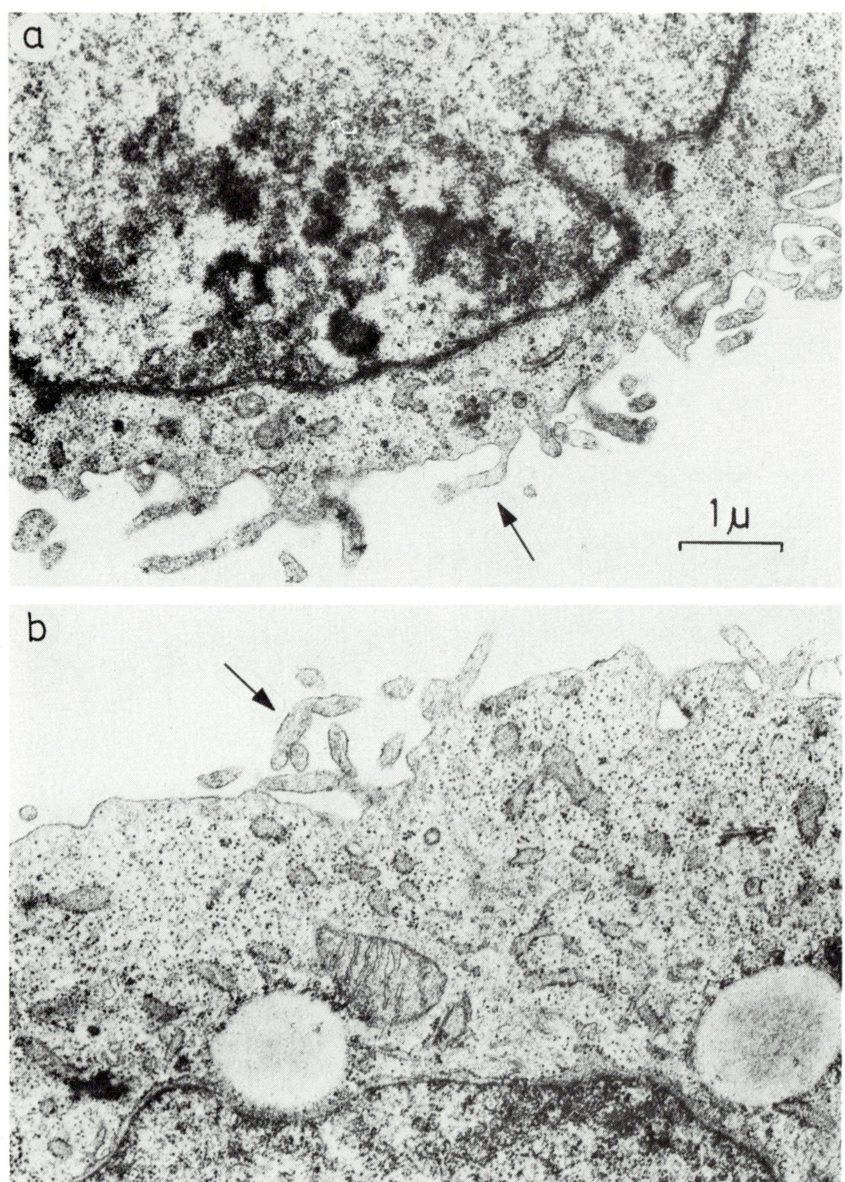

Figure 10 Electron micrographs of thin sections through the same pellets as in Figure 9*a* and *b*. Filamentous material appears to be contained in equal amount in the cytoplasmic evaginations (arrows) of both control (*a*) and phalloidin-poisoned sample (*b*) (× 13 500)

CONSEQUENCES OF ACUTE AND CHRONIC PHALLOIDIN POISONING

Adult experimental animals die within a few hours after parenteral poisoning due to the haemorrhagic shock which occurs concomitantly with the development of the haemorrhagic necrosis of the liver. No signs of pronounced liver damage, which could explain the rapid death as a consequence of an acute impairment of hepatic function, could be detected by chemical clinical investigation of the serum of acute poisoned animals[29]. However, positivity to the LP-X test[29], the recently discovered important tool to demonstrate or exclude cholestasis[30], was found in the serum of rats with serum phalloidin damage, which survived the acute poisoning, 24 and 48 hours respectively after phalloidin injection.

The presence of LP-X, the abnormal plasma lipoprotein characteristic of cholestasis[31], in phalloidin-poisoned animals can be related to the severe structural and functional damage to bile canaliculi and to the surrounding contractile apparatus (*cf.* Figure 5), and points out the importance of the hepatocyte skeleton for bile secretion.

Morphological and autoradiographic studies with [^{3}H]thymidine on adult rats treated with sublethal doses of phalloidin, and on very young animals which survive the intoxication, demonstrate that the phalloidin-induced alterations of the liver are reversible, with *restitutio ad integrum*, after removal of the necrotic tissue and reparative regeneration[26]. Moreover, no signs of fibrosis could be observed on late examination of poisoned animals. Also no increase of connective tissue could be found in the liver of up to 90-day-old rats which had been treated weekly with phalloidin beginning 3 days after birth. Some fibrosis with signs of initial cirrhosis was observed only in the liver of rats treated in the same way up to 110 days of age. However, neither did the latter rats show any definite biochemical signs of parenchymal or mesenchymal liver impairment[29].

Even if the role of phalloidin in spontaneous human intoxication is not yet clear[3], we should like to mention our finding of increased actin filaments in damaged hepatocytes of a 55-year-old woman who had eaten wild mushrooms[32].

CONCLUSION

Further studies are necessary to identify the location, chemical nature and function of the phalloidin-binding structure (receptor?) which makes the plasma membrane of normal hepatocytes, and most likely of transformed hepatocytes as long as they are in their original liver, like those of autochthon diethylnitrosamine-induced carcinoma, capable of interacting with the poison. The above results clearly indicate, however, the close relationship between plasma membrane and cytoskeleton, cell permeability and cell motility, and point out the importance of alteration of the hepatocyte plasma membrane for the onset of liver injury. Phalloidin still represents a very suitable tool for investigations in this field.

ACKNOWLEDGEMENTS

We thank Professor Th. Wieland for his generous donation of phalloidin, advice in part of the work and discussion of the results. Thanks are due to Professor U. Gröschel-Stewart, Zoologisches Institut der Technischen Hochschule, Darmstadt, and Professor H. P. Seelig, Institut für Immunologie und Serologie der Universität, Heidelberg, which donated the labelled antibodies against actomyosin of chicken stomach[37] and respectively antiactin prepared from patients with hepatitis B[38–40]. The rats with transplantable Morris hepatoma and with AH Yoshida ascites hepatoma were kindly provided by Professor D. Gericke, Fabwerke Hoechst AG, Frankfurt/M. Part of the experiments on rats with diethylnitrosamine-induced carcinoma were carried out in cooperation with Professor S. Ivankovic, Deutsche Krebsforschungszentrum, Heidelberg. His helpful discussion of the results is gratefully acknowledged. Thanks are due to Mrs. A. Völzke for her skilled photographic work.

References

1. Faulstich, H. and Weckauf-Bloching, M. (1974). Isolation and toxicity of two cytolytic glycoproteins from *Amanita phalloides* mushrooms. *Hoppe-Seyler's Z. Physiol. Chem.*, **355,** 1489
2. Wieland, Th. and Wieland, O. (1972). The toxic peptides of *Amanita* species. *Microbiol. Toxicol.* **8,** 249
3. Wieland, Th. (1976). Phallotoxine — Struktur und pharmakologische Wirkung. *Dtsch. Apotheker Z.*, **116,** 805
4. Agostini, B., Govindan, V. M. and Hofmann, W. (1975). Morphological changes induced by phalloidin in the rat liver. In: K. Keppler (ed.). *Pathogenesis and Mechanisms of Liver Cell Necrosis*, pp. 176–192. (Lancaster: MTP)
5. Agostini, B., Govindan, V. M., Hofmann, W. and Wieland, Th. (1975). Phalloidin-induced hyperplasia of actin filaments within rat hepatocytes. *Z. Naturforsch.*, **30c,** 793
6. Govindan, V. M., Faulstich, H., Wieland, Th., Agostini, B. and Hasselbach, W. (1972). *In vitro* effect of phalloidin on a plasma membrane preparation from rat liver. *Naturwissenschaften*, **59,** 521
7. Agostini, B. and Govindan, V. M. (1973). Elektronenmikroskopische Untersuchungen an der Zellmembran der Leber bei Phalloidin-Vergiftung. *Verh. Dtsch. Ges. Pathol.*, **57,** 436
8. Tuchweber, B., Kovacs, K., Khandekar, J. D. and Garg, B. D. (1973). Peliosis-like changes induced by phalloidin in the rat liver. *J. Med.*, **4,** 327
9. Govindan, V. M., Rohr, G., Wieland, Th. and Agostini, B. (1973). Binding of a phallotoxin to protein filaments of plasma membrane of liver cell. *Hoppe-Seyler's Z. Physiol. Chem.*, **354,** 1159
10. Lengsfeld, A., Löw, I., Wieland, Th., Dancker, P. and Hasselbach, W. (1974). Interaction of phalloidin with actin. *Proc. Natl. Acad. Sci. USA*, **71,** 2803

11. Gabbiani, G., Montesano, R., Tuchweber, B., Salas, M. and Orci, L. (1975). Phalloidin-induced hyperplasia of actin filaments in rat hepatocytes. *Lab. Invest.*, **33,** 562
12. Farquhar, M. G. and Palade, G. E. (1963). Junctional complexes in various epithelia. *J. Cell Biol.*, **17,** 375
13. Oda, M., Price, V. M., Fischer, M. M. and Phillips, M. J. (1974). Ultrastructure of bile canaliculi, with special reference to the surface coat and the pericanalicular web. *Lab. Invest.*, **31,** 314
14. French, S. W. and Davies, P. L. (1975). Ultrastructural localisation of actin filaments in rat hepatocytes. *Gastroenterology*, **68,** 765
15. Puchinger, H. and Wieland, Th. (1969). Suche nach einem Metaboliten bei Vergiftung mit Desmethylphalloin (DMP). *Eur. J. Biochem.*, **11,** 1
16. Agostini, B. and Hofmann, W. (1976). Studies by electron microscopy and immunofluorescence on phalloidin tolerance in newborn rats. In *Eleventh International Congress of International Academy of Pathology*. Abstracts Book, p. 36. (Washington, D.C., USA)
17. Agostini, B., Hofmann, W. and Wieland, Th. (1977). Phalloidin tolerance and hyperplasia of actin filaments in the hepatocytes of newborn rats. (In preparation)
18. Agostini, B., Wieland, Th., Ivankovic, S. and Hofmann, W. (1976). Phalloidin tolerance in rats with liver carcinoma induced by diethylnitrosamine. *Naturwissenschaften*, **63,** 438
19. Agostini, B., Wieland, Th., Ivankovic, S. and Hofmann, W. (1977). Different response to phalloidin intoxication of rats with response to autochthon and transplantable hepatoma. (In preparation)
20. Fiume, L. (1965). Mechanism of action of phalloidin. *Lancet*, **i,** 1284
21. Wieland, O. and Szabados, A. (1966). On the nature of phalloidin tolerance in newborn rats. In *Sixth International Congress of Clinical Chemistry*. Vol. VI, p. 59. (Basel, New York: S. Karger)
22. Szabados, A. (1971). Zum Phänomen der Phalloidintoleranz neugeborener Ratten — Ein Beitrag zur Pathologie der Knollenblätterpilzvergiftung. Dissertation (München)
23. Siess, E., Wieland, O. and Miller, F. (1970). Elektronenmikroskopische Untersuchungen zur Phalloidintoleranz neugeborener Ratten, Mäuse und Kaninchen. *Virchows Arch. Abt. B. Zellpathol.*, **6,** 151
24. Frimmer, M. and Schiscke, B. (1972). Decreased toxicity of phalloidin in partially hepatectomized rats. *Naunyn-Schmiedeberg's Arch. Pharmacol.*, **272,** 447
25. Tuchweber, B. Kovacs, K., Khandekar, J. D. and Garg, B. D. (1972). Prevention of phalloidin intoxication in rats by partial hepatectomy. *Arch. Toxikol.*, **29,** 311
26. Lesch, R., Meinhardt, K. and Agostini, B. (1977). Studies by light, electron microscopy and autoradiography on liver of newborn and adult rats following phalloidin poisoning with special respect to the proliferative response of the newborn rat liver. (In preparation)
27. Druckrey, H., Schmähl, D., Dischler, W. and Schildbach, A. (1962). Quantitative Analyse der experimentellen Krebserzeugung. *Naturwissenschaften*, **49,** 217
28. Barclay, M. and Terebus-Kekish, O. (1973). Enzyme activities and extrinsic proteins in plasma membranes from normal liver and Morris hepatoma 5123 tc. *J. Natl. Cancer Inst.*, **51,** 1709
29. Agostini, B. and Wieland, H. (1977). Correlative biochemical and morphological studies on acute and chronic phalloidin poisoning in the rat with special regard to the onset of cholestasis. (In preparation)
30. Seidel, D., Schmitt, E. A. and Alaupovic, P. (1970). Ein abnormes Low-Density-Lipoprotein bein Cholestase. II. Bedeutung in der Differentialdiagnose des Ikterus. *Dtsch. Med. Wochenschr.*, **95,** 1805
31. Seidel, D., Alaupovic, P. and Furman, R. H. (1969) A lipoprotein characterizing obstructive jaundice. A method for quantitative separation and identification of lipoproteins in jaundiced subjects. *J. Clin. Invest.*, **48,** 1211
32. Raute-Kreinsen, U., Hofmann, W. and Agostini, B. (1977). Damage to hepatocytes and involvement of actin filaments in mushroom poisoning. (In preparation)

33. Sabatini, D. D., Bensch, K. G. and Barnett, R. J. (1963). Cytochemistry and electron microscopy. The preservation of cellular ultrastructure and enzymatic activity by aldehyde fixation. *J. Cell Biol.*, **17,** 19
34. Millonig, G. (1961). Advantage of a phosphate buffer for OsO_4 solution in fixation. *J. Appl. Phys.*, **32,** 1637
35. Watson, M. L. (1958). Staining of tissue sections for electron microscopy with heavy metals. *J. Biophys. Biochem. Cytol.*, **4,** 475
36. Reynolds, E. S. (1973). The use of lead citrate at high pH as an electron-opaque stain in electron microscopy. *J. Cell Biol.*, **17,** 208
37. Gröschel-Stewart, U. (1971). Comparative studies of human smooth muscle myosin. *Biochim. Biophys. Acta*, **229,** 322
38. Johnson, G. D., Holborow, E. J. and Glynn, L. E. (1965). Antibody to smooth muscle in patients with liver disease. *Lancet*, **ii,** 878
39. Farrow, L. J., Holborow, E. J. and Brighton, W. D. (1971). Reaction of smooth muscle antibody with liver cells. *Nature* (*London*), **232,** 186
40. Gabbiani, G., Ryan, G. B., Lamelin, J. P., Vassalli, P., Majno, G., Bouvier, C. A., Cruchaud, A. and Lüschner, E. F. (1973). Human smooth muscle autoantibody. Its identification as antiactin antibody and a study of its bindings to 'non-muscular' cells. *Am. J. Pathol.*, **72,** 473

22
Mechanism of Phalloidin Intoxication I: Cell Membrane Alterations

M. FRIMMER and E. PETZINGER

Until 1967 it was assumed that phalloidin interacted primarily with the endoplasmic reticulum. Subsequent observations on whole liver and isolated hepatocytes indicated additional lesions (Table 1) and particularly an interaction of phalloidin with plasma membranes (Table 2). All these studies were facilitated by the development of methods for the preparation of isolated

Table 1 Effects of phalloidin poisoning on whole livers and isolated hepatocytes

	Whole liver (in situ *or perfused*)	*Isolated hepatocyte*
Swelling	Volume increase (100% and more)	No changes of the packed cell volume
Vacuolization	Parallels swelling and depends on the microcirculatory conditions	In presence of phalloidin no additional vacuoles
Protrusions		Rapid development of characteristic protrusions on the cell surface
Potassium loss	Depends on the circulatory conditions	Not as marked as in perfused livers
Release of lysosomal enzymes	Release starts immediately after the addition of phalloidin	Phalloidin does not cause any enzyme release beyond that of untreated controls
Specificity	No other tissue with primary sensitivity to phalloidin is known	Only intact hepatocytes develop protrusions; AS-30D hepatoma cells are insensitive to phalloidin
Antagonists	Only protective when given before phalloidin	All drugs protective *in vivo* prevent protrusions in isolated cells

Table 2 Evidence for the interaction of phalloidin with plasma membranes of liver cells

1967	Potassium loss after treatment of perfused livers with phalloidin	Frimmer *et al.*[9]
1972	Desmethyl-phalloin binds to isolated plasma membrane	Lutz *et al.*[10]
1972	Interaction of phalloidin with microfilamentous structures on the inside of plasma membranes	Govindan *et al.*[5]
1973	Trypsinization of isolated hepatocytes or of plasma membranes prevents the interactions of phalloidin with the membrane	Kroker and Hegner[8]
1974	Phalloidin produces characteristic protrusions on the surface of isolated hepatocytes	Frimmer *et al.*[3]
1974–1975	Phalloidin antagonists prevent the development of protrusions in isolated hepatocytes	Faulstich *et al.*[11]; Frimmer and Petzinger[4]
1976	Treatment of hepatocytes with trypsin, chymotrypsin, elastase, phospholipase A prevents the development of protrusions	Petzinger and Frimmer[12,13]

hepatocytes and purified plasma membranes. Phalloidin induces characteristic deformations on the surface of liver cells[1]. In the whole liver phalloidin causes invaginations of the membrane with subsequent excessive vacuolization of the parenchymal cells[2]. In isolated hepatocytes the deformation of the

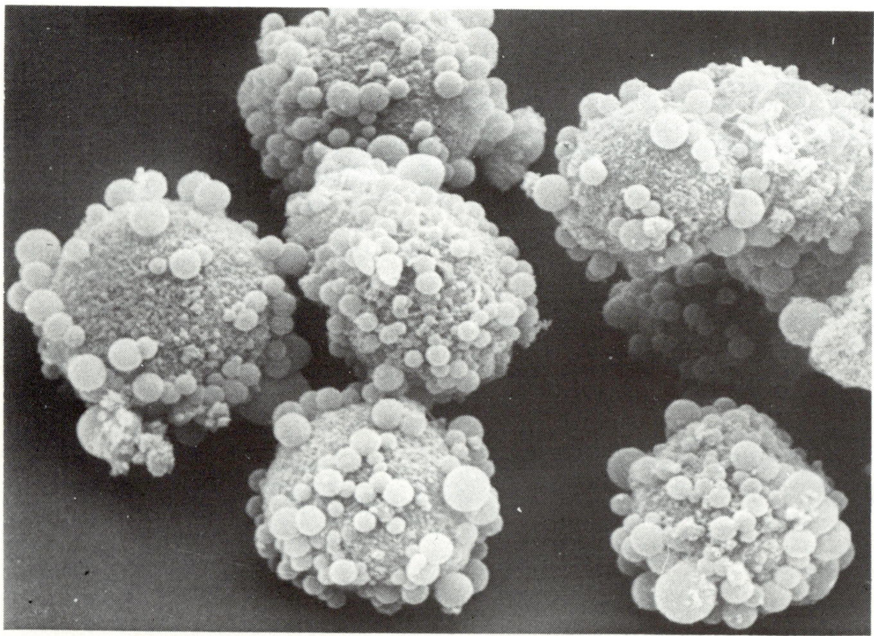

Figure 1 Isolated hepatocytes 20 min after the addition of 10 μg/ml of phalloidin. Scanning EM: Leitz AMR 1000 (× 1500)

surface goes in the opposite direction[1,3] (Figure 1). Counting the protrusions after treatment of hepatocytes with phalloidin is a useful quantification of the phalloidin response[4]. The above method replaced experiments with perfused livers for many problems.

The findings of Wieland's group[5–7] stimulated the concept that both protrusions and invaginations might be due to structural changes of microfilamentous actin-like structures on the inside of the membrane. This idea presumes that only membranes of liver cells recognize or transport phalloidin. However, molecular interactions of phalloidin with actin can neither explain the high specificity of phalloidin for intact liver cells nor the enrichment of the toxin in the whole liver. Several types of extrahepatic cells contain equal or larger amounts of microfilaments, but are insensitive to phalloidin. The complete insensitivity of AS-30D hepatoma cells to phalloidin is not due to a defective endowment with microfilamentous structures. It is more probable that these hepatoma cells cannot recognize phalloidin.

Some evidence for a component on the surface of intact liver cells which is indispensable for the action of phalloidin was recently found (Table 3). In 1973 Kroker and Hegner[8] showed that trypsinization of liver cells or of plasma membranes prevents the interaction of phalloidin with their binding sites. They assumed that trypsin solubilizes the binding sites for phalloidin on

Table 3 Influence on phalloidin sensitivity of pretreatment of isolated hepatocytes with different enzymes[12]

Neuraminidase: No influence

Proteases:

1. Preincubation with trypsin reduces the sensitivity of isolated hepatocytes to phalloidin
2. The degree of protection by trypsin depends on the concentration of trypsin and of phalloidin, but not on the duration of the preincubation
3. The protective effect of trypsin against phalloidin is reversible, when trypsin is removed prior to the addition of phalloidin
4. Chemical substitution of the active centre of trypsin abolishes its protective activity
5. Chymotrypsin and elastase are less active as protectors. Subtilisin, thrombin, collagenase and papain are ineffective at low concentrations

Phospholipases: Phospholipase A (bee venom) most effective

1. In presence of very low concentrations of phospholipase A hepatocytes become insensitive to phalloidin immediately
2. The protective effect depends on the concentration of the enzyme
3. When the enzyme is removed after a short period of preincubation, the protection of cells against phalloidin disappears
4. Preincubation of hepatocytes with phospholipase A for a longer period (30–60 min) abolishes their sensitivity to phalloidin even after removing the enzyme before the addition of phalloidin
5. Under these conditions (4), cells recover their sensitivity to phalloidin gradually during 30–60 min in the absence of phospholipase A

the surface of liver cells. The expectation that the 'solubilized receptor' could be isolated did not come true because:

(1) Trypsinization removes some glycoproteins from the cell surface, but not the complete structures needed for the response to phalloidin.

(2) Trypsin probably binds to the same area of the surface of liver cells as phalloidin does.

(3) The spontaneous disappearance of the protection of liver cells against phalloidin by washing off the trypsin indicates a process on the surface of liver cells is the first step of phalloidin response (Figure 2, Table 3).

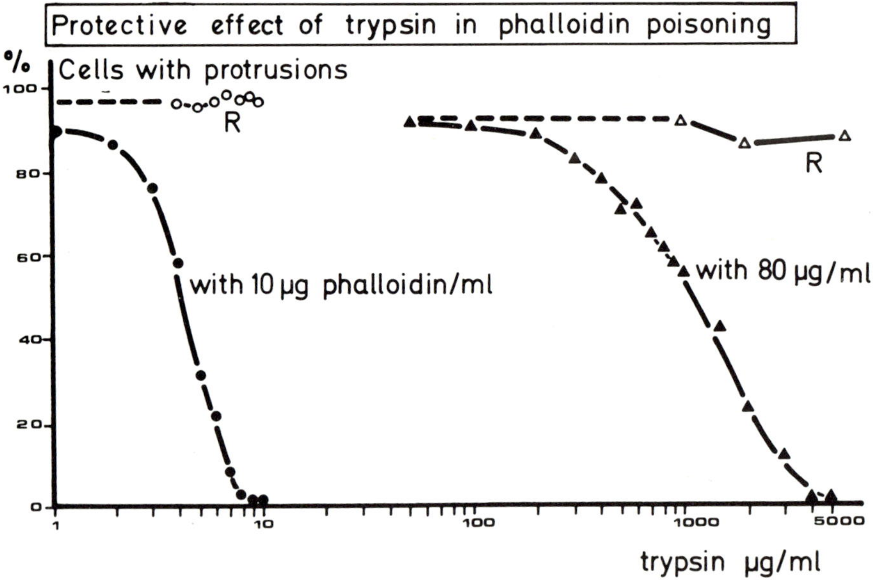

Figure 2 Influence of various concentrations of trypsin on the response of isolated liver cells to 10 and 80 μg phalloidin/ml. ● and ▲: The effect of phalloidin was tested in presence of trypsin; ○ and △: Incubation with phalloidin after preincubation with trypsin and after subsequent removal of the enzyme

(4) Experiments with phospholipase A show that lipid structures of the plasma membrane are also involved in the primary interaction of phalloidin with liver cells (Tables 3 and 4).

(5) The degradation of superficial phospholipids reduces the sensitivity of liver cells to phalloidin.

Apparently cells are able to restore their lipids by the acyltransferase system within 30–60 min (Table 4).

Table 4 Protective effect of phospholipase A (bee venom) in phalloidin poisoning of isolated hepatocytes

Preincubation of isolated hepatocytes with phospholipase		*Reincubation after removal of phospholipase*	*Ratio: Cells with protrusions/cells total after subsequent incubation with 10 μg phalloidin/ml*
Phospholipase A ng/ml	*Pretreatment min*	*min*	
50	<1	NR	0.87
50	60	NR	0.92
100	<1	NR	0.88
100	60	NR	0.73
100	120	NR	0.88
150	<1	NR	0.87
150	30	NR	0.30 (Partial protection)
150	60	NR	0.00 } (Total protection)
150	120	NR	0.00 }
150	<1	NR	0.87
150	30	NR	0.30 (Partial protection)
150	30	5	0.13
150	30	15	0.21
150	30	30	0.62
150	30	60	0.79 (Recovery of sensitivity to
150	30	120	0.80 phalloidin)

NR = no reincubation

CONCLUSIONS

Liver cells possess a specific mechanism for recognizing or taking up phalloidin. For this early reaction protein as well as lipid structures are needed. The existence of a superficial receptor is consistent with the action of phalloidin on microfilamentous structures. However, the link between phalloidin binding to actin and membrane deformation is hypothetical.

References

1. Frimmer, M. (1975). Phalloidin, a membrane-specific toxin. In: D. Keppler (ed.). *Pathogenesis and Mechanism of Liver Cell Necrosis*, pp. 163–173. (Lancaster: MTP)
2. Agostini, B., Govindan, V. M. and Hofmann, W. (1975). Morphological changes induced by phalloidin in the rat liver. In: D. Keppler (ed.). *Pathogenesis and Mechanism of Liver Cell Necrosis*, pp. 175–192. (Lancaster: MTP)
3. Frimmer, M., Kroker, R. and Porstendörfer, J. (1974). The mode of action of phalloidin: demonstration of rapid deformation of isolated hepatocytes by scanning electron microscopy. *Naunyn-Schmiedeberg's Arch. Pharmacol.*, **284**, 395

4. Frimmer, M. and Petzinger, E. (1975). Phalloidin-Antagonisten. 3. Mitteilung. *Arzneim. Forsch.* (*Drug Res.*), **25,** 1423
5. Govindan, V. M., Faulstich, H., Wieland, Th., Agostini, B. and Hasselbach, W. (1972). *In vitro* effect of phalloidin on a plasma membrane preparation from rat liver. *Naturwissenschaften,* **59,** 521
6. Govindan, V. M., Rohr, G., Wieland, Th. and Agostini, B. (1973). Binding of a phallotoxin to protein filaments of plasma membrane of liver cell. *Hoppe-Seyler's Z. Physiol. Chem.,* **354,** 1159
7. Lengsfeld, A. M., Löw, I., Wieland, Th., Danker, P. and Hasselbach, W. (1974). Interaction of phalloidin with actin. *Proc. Natl. Acad. Sci. USA*, **71,** 2803
8. Kroker, R. and Hegner, D. (1973). Solubilization of phalloidin-binding sites from rat hepatocytes and plasma membranes by trypsin. *Naunyn-Schmiedeberg's Arch. Pharmacol.,* **279,** 339
9. Frimmer, M., Gries, J., Hegner, D. and Schnorr, B. (1967). Studies of the mechanism of the action of phalloidin. Release of lysosomal enzymes and potassium. *Naunyn-Schmiedeberg's Arch. Pharmacol.,* **258,** 197
10. Lutz, F., Glossmann, H. and Frimmer, M. (1972). Binding of [^{3}H]desmethylphalloin to isolated plasma membranes from rat liver. *Naunyn-Schmiedeberg's Arch. Pharmacol.,* **273,** 341
11. Faulstich, H., Wieland, Th., Walli, A. and Birkmann, K. (1974). Antamanid protects hepatocytes from phalloidin destruction. *Hoppe-Seyler's Z. Physiol. Chem.,* **355,** 1162
12. Frimmer, M., Petzinger, E., Rufeger, U. and Veil, L. B. (In preparation)
13. Frimmer, M. (1976). The role of plasma membranes in hepatotoxic effects. Presented at the *Pharmacological Meeting,* September 14–17 (1976). *Naunyn-Schmiedeberg's Arch. Pharmacol.* (Suppl.) (In press)

DISCUSSION

Dr M. J. Phillips, Toronto (Canada): The protrusions of isolated hepatocytes described with phalloidin are not specific for phalloidin. They have been described also by Godman *et al.*[1] in a variety of epithelial cells treated with cytochalasin D. Moreover, they occur spontaneously in isolated hepatocytes and their formation can be prevented by cytochalasin B treatment[2]. Their enhanced formation with phalloidin is interesting and would suggest a role of actin filaments.

Dr Frimmer, Giessen (West Germany): Phalloidin is highly specific for hepatocytes. As yet, no other cell type was found to be sensitive to this toxin. In freshly prepared isolated hepatocytes the spontaneous occurrence of typical protrusions is very rare. We tested many other toxic agents on hepatocytes. No agent was found to produce identical or similar protrusions as seen after phalloidin. Some cytolytic agents (phospholipases, phallolysin, detergents) produce large protrusions quite different from those found after incubation with phalloidin. Other drugs lead to swelling of hepatocytes and cause disappearance of microvilli. Such changes often prevent a subsequent response to phalloidin.

Dr S. Erlinger, Clichy (France): Clinicians know that *Amanita phalloides* poisoning in man is followed by severe diarrhea and renal failure. Are there lesions of the intestinal cells and kidney tubular cells in the experimental animal?

Dr Frimmer, Giessen (West Germany): Intestinal injury after *Amanita phalloides* poisoning so far cannot be tested in laboratory animals, because amatoxins and phallotoxins are not, or hardly, absorbed after oral application in the common laboratory animal species. In our laboratory, excessive doses of whole mushrooms were given to rats orally without any effect. Therefore we do not know whether the gastrointestinal symptoms in man are caused by the toxic peptide itself or by other components of the mushroom. The renal failure is certainly caused by α-amanitin. In mice, injection of α-amanitin produces preferential damage of the kidney. Phalloidin injures liver tissue only.

References

1. Godman, G. C., Miranda, A. F., Deitch, A. D. and Tanenbaum, S. W. (1975). Action of cytochalasin D on cells of established lines. III. Zeiòsis and movements at the cell surface. *J. Cell Biol.*, **64,** 644
2. Phillips, M. S., Oda, M., Edwards, V. D., Greenberg, G. R. and Jeejeebhoy, N. (1974). Ultrastructural and functional studies of cultured hepatocytes. *Lab. Invest.*, **31,** 533

23
Mechanism of Phalloidin Intoxication II: Binding Studies

H. FAULSTICH, Th. WIELAND, H. SCHIMASSEK, A. K. WALLI and N. EHLER

The molecular mechanism of the phalloidin interaction with actin has been extensively studied in our laboratories. Actin of the rabbit muscle served as a model protein.

PHALLOIDIN INTERACTION WITH MUSCLE ACTIN

The outstanding property of actin is its ability to polymerize and depolymerize in an equilibrium reaction (Figure 1).

Filaments of actin (F-actin) can be obtained from monomers (G-actin) by addition of K^+ or Na^+ ions. By contrast, the filaments are depolymerized in a

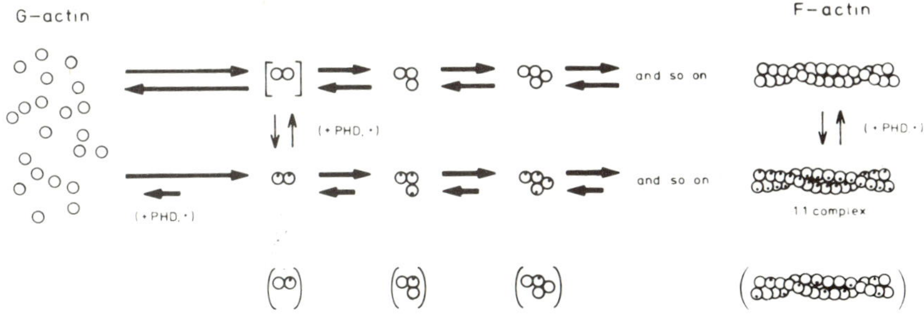

Figure 1 Interaction of phalloidin with actin. Possible mechanism of an accelerated polymerization and of the stabilization of aggregated forms of actin, even with ratios of toxin : actin < 1. (The polymerization scheme is modified from Wegner and Engel[4], by courtesy of *Biophys. Chem.*)

salt-free solution. Much evidence indicates that phalloidin binds exclusively to aggregated actin and not to the monomers. In the aggregates each actin unit can bind one toxin molecule with high affinity. In the resulting complex the actin units cannot depolymerize as long as toxin is bound to them. Phalloidin binding therefore stabilizes the actin filaments.

The stabilization of the filaments against KI-treatment was the first phallotoxin effect observed with actin and resulted in the discovery of this phalloidin-binding protein[1]. Today we know of a series of treatments, which nor-

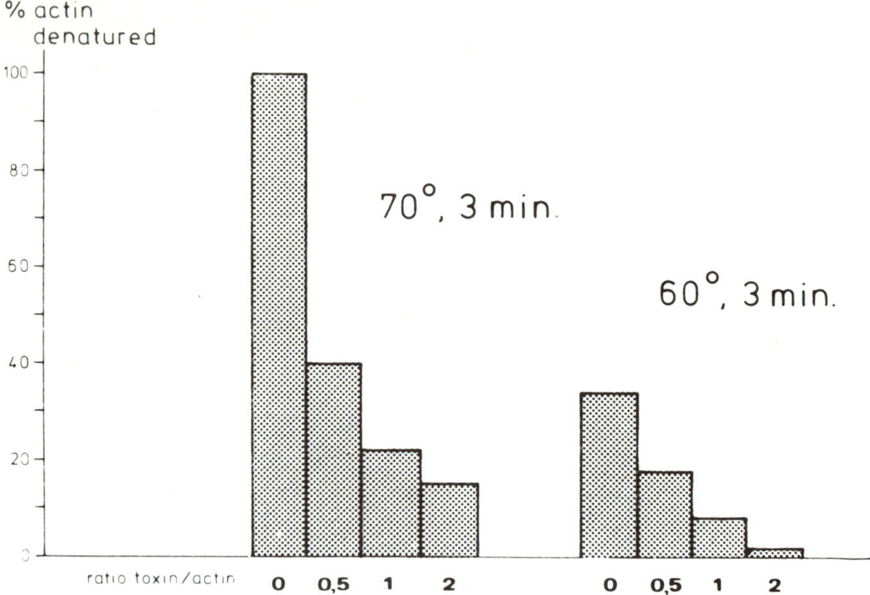

Figure 2 Protection from heat denaturation of F-actin (10^{-5} M, in KCl (0.1 M) Tris (1 mM) pH 7.4) by phalloidin ($0–2 \times 10^{-5}$ M). Protein was treated for 3 min at 70 °C or 60 °C, respectively

mally dissociate the filaments, but not if phalloidin is present. Among them are ultrasonication, incubation with DNAse I, or heating to temperatures where the protein is normally denatured[2]. Heat denaturation of actin and the effect of phalloidin on this process is presented in Figure 2 and illustrates the stabilizing capacity of the toxin.

A second effect of phalloidin (Figure 1), was observed by Dancker *et al.*[3] This is the acceleration of actin polymerization by phalloidin. The polymerization scheme, taken and modified from a publication of Wegner and Engel[4], assumes that there is a fast equilibrium of monomers and dimers, so fast in fact, that most of the dimers dissociate before a third monomer is attached to them.

Hence the low concentration of dimers sets the limits for the polymerization in that it determines the concentration of the trimers, which are regarded as the nuclei of polymerization. To explain the acceleration effect of phalloidin it may not only bind to filaments, but also to oligomers. Binding to the dimer, for example, would increase the concentrations of nuclei and hence accelerate the polymerization.

Even though the amounts of toxin are far below one equivalent, they still accelerate polymerization to quite a high extent. Evidently, there is a cooperativity in phalloidin action, which involves the polymerization process and also the filament stabilization.

The simplest explanation for the cooperativity in the acceleration of polymerization is that dimeric or trimeric aggregates are stabilized, if phalloidin is bound to only one of the actin units. Each actin unit has two actin binding sites, one in the filament axis, the other perpendicular to it. When phalloidin stabilizes both binding sites we would expect the maximum effect of phalloidin to occur at a ratio 1 : 3. Indeed, a similar ratio was obtained in most systems where phalloidin cooperativity was observed. The acceleration of polymerization is in fact also a stabilization effect, and may be regarded as a stabilization of the polymerization nuclei by the toxin.

This stabilizing effect was quantitated in equilibrium dialysis experiments[5] determining depolymerization of actin in dilution series. The depolymerization was measured by the release of ^{45}Ca and $[^{14}C]ADP$, which occurs from monomeric G-actin only and not from filamentous F-actin. For F-actin without toxin the half-value of depolymerization was 2.1×10^{-6} M. The presence of one equivalent of phalloidin diminished the half-value of depolymerization by 30-fold. This diminution was 90-fold with two equivalents of phalloidin. The findings are best understood by introducing the term of critical concentration $[G]_c = K_c^{-1}$ where $[G]_c$ is that concentration of monomeric actin, which must be exceeded to start the polymerization under appropriate conditions. $[G]_c$ for the normal equilibrium

$$\text{F-actin} \rightleftarrows \text{G-actin}$$

is 10^{-6} M; however, for the dissociation of the complex it is 3.6×10^{-8} M:

$$(\text{F-actin} \cdot \text{phalloidin}) \rightleftarrows \text{G-actin} + \text{phalloidin}$$

The same value was also determined for the dissociation of the labelled toxin. This means that at a concentration of 3.6×10^{-8} M the phalloidin–actin complex dissociates. However, F-actin evidently cannot exist in equilibrium at that concentration, and the filaments depolymerize concomitantly with the dissociation of phalloidin. The value of K_D for the demethylphalloin and muscle actin is in good agreement with the value measured for plasma membranes of the liver by Lutz and Frimmer[6]; they reported a value of 2.2×10^{-8} M, and the closeness of the values strongly suggests that actin of liver cells is the binding site of the toxin.

INTERACTION OF PHALLOIDIN WITH CELL ACTIN

Isolated hepatocytes

The binding capacity of isolated hepatocytes for the labelled toxin was compared with the binding capacity of the tissue homogenate. Up to a concentra-

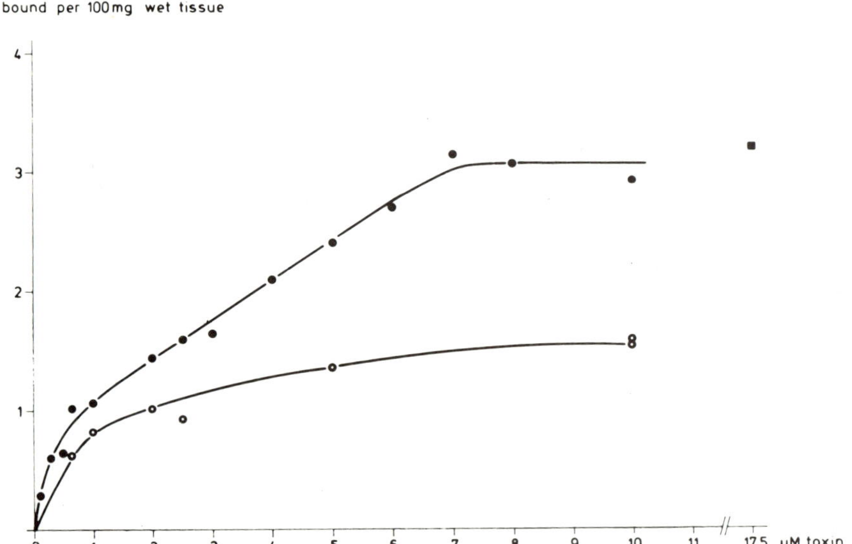

Figure 3 Binding of [^{3}H]demethylphalloin by isolated rat liver cells (●), and by their homogenate or ghosts (○). The homogenate was prepared in a Potter homogenizer and centrifuged 1 hour at 100 000 *g*. The ghosts were obtained by osmotic shock and centrifuged for 10 min at 600 *g*. Incubation time 30 min. For comparison, the binding of [^{3}H]demethylphalloidin by a perfused rat liver (■)

tion of 1 μM the uptake of the phalloidin by the hepatocyes and the homogenate runs parallel. At a concentration of 1.5 μg toxin/100 mg tissue the homogenate appears to be saturated. However, the hepatocytes continued to incorporate more toxin in a proportion linear to the toxin concentration. Under our experimental conditions, after incorporation of about 3 μg toxin/100 mg wet cell weight, no appreciable amounts of toxin were bound further. By comparing these values it appears that the cells take up twice as much toxin as the homogenate. Homogenized phalloidin-treated cells released an amount of phalloidin, equivalent to the above-noted difference. Thus it would appear that the binding capacities of the tissue homogenate and the hepatocyte homogenate are similar.

Isolated perfused livers gave similar results. In this system saturation was found at 3 μg/100 mg wet tissue; this amount was reduced by one-half after homogenizing the tissue. This means that rat liver cells either after isolation or during liver perfusion incorporated twice as much toxin as can be bound to the F-actin fraction of homogenate or ghosts. The tissue homogenates were centrifuged at 100 000 *g* for 1 hour. This procedure should spin down all filamentous actin. Compared to this the hepatocyte ghosts were spun down only at 600 *g* for a couple of minutes. This left F-actin, at least in muscle preparations, in the supernatant. For both these preparations, the binding capacities were similar, indicating that the F-actin-like binding capacity was associated with the particulate structure. Assuming that this specific binding capacity of the homogenate pellet is solely actin, we calculate an amount of about 1 mg of actin/g liver.

Isolated hepatocytes from rats pretreated with phalloidin (Gabbiani's hepatocytes)

Rats were injected with 500 μg/kg body weight with phalloidin daily according to the method of Gabbiani *et al.*[7] Their hepatocytes were isolated after 10–14 days of injection. According to Gabbiani *et al.* this treatment resulted in an enlargement of filamentous structures. The larger amount of actin present in these cells probably accounted for the higher saturation level of the homogenate (Figure 4). The homogenate of these cells bound nearly 100% more of the toxin than the homogenate of normal cells. But the hepatocytes from pretreated animals exhibited a diminished uptake of phalloidin as compared to normal hepatocytes. This is the first evidence that uptake of the toxin into cells and binding of the toxin to F-actin-like structures are two different events in phalloidin intoxication.

Cells other than hepatocytes

Behaviour similar to that of hepatocytes from phalloidin-pretreated rats was observed in transformed cells obtained from human lymphocyte cultures. After homogenization, these cells also had a high binding capacity for the toxin. This probably corresponded to a high amount of F-actin. In contrast, the intact lymphocytes were saturated with only 30% of the binding capacity of the homogenate.

Homogenates of HeLa cells cultures had lower binding capacity than hepatocyte homogenates. However, they behaved similarly to the hepatocytes,

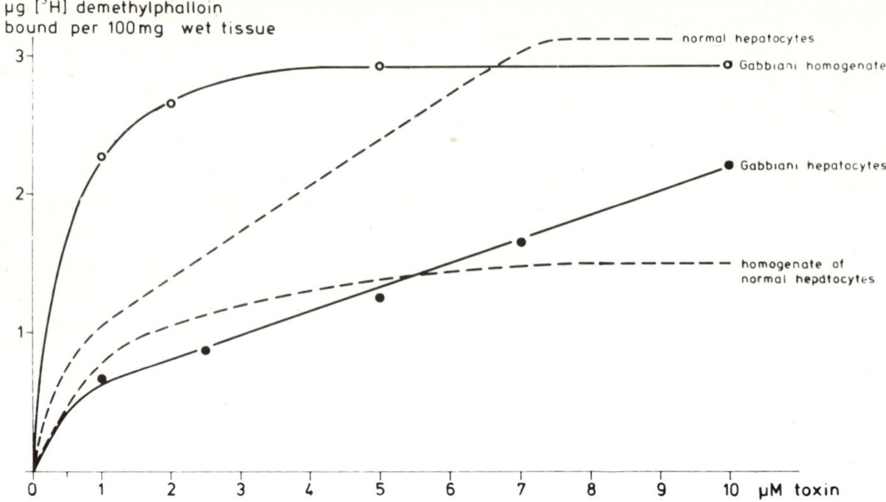

Figure 4 Binding of [^{3}H]demethylphalloin by liver cells isolated from rats, which were pretreated with sublethal doses of phalloidin (Gabbiani *et al.*[7]) (50 μg/day phalloidin for 10–14 days) (●). Binding by the homogenate of these cells, prepared as in Figure 1 (○). For comparison, the corresponding curves of isolated liver cells of untreated rats are shown. Incubation time 30 min

because they incorporated more of the toxin than they could bind in the homogenate. Thus two types of cells exist. One type has a high binding capacity but incorporates only about 30% of the binding capacity of the cell homogenate. The other type possesses lower binding capacity but incorporates more toxin than can be bound by the cell homogenate.

These results indicate that the actin content does not determine the incorporation of toxin into the cell.

PHALLOIDIN TOXICITY IN CELLS

The toxin incorporation has to be related to the poisonous effects on cells indicated by the protrusions of the membranes of hepatocytes (Figure 6). These protrusions were correlated to concentration of phalloidin in turn with the F-actin-like binding capacity of the homogenates of different cells. Cells with low actin-like binding capacity were more sensitive to the toxin since the protrusion appeared at low concentrations of phalloidin (Table 1). Isolated hepatocytes from phalloidin-pretreated rats, with a 3-fold higher binding capacity for the toxin in the cell homogenate than normal hepatocyte homogenate, had protrusions at a toxin concentration at least ten times higher than normal cells. Lymphocytes, with their high content of

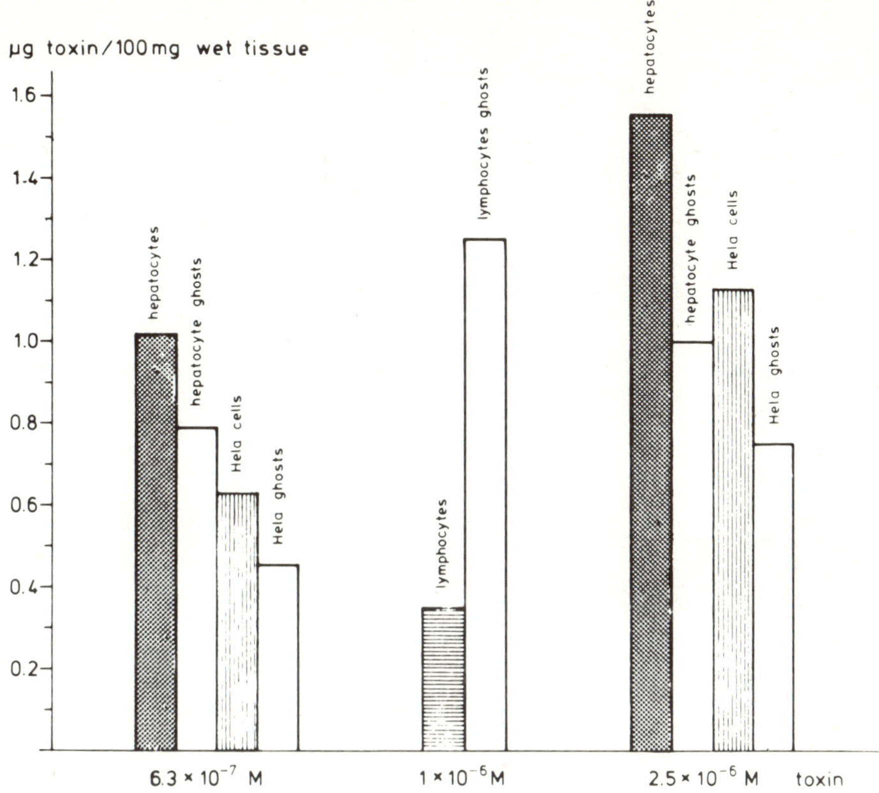

Figure 5 Binding of [^{3}H]demethylphalloin by HeLa S_3 cells, human transformed lymphocytes, and the ghosts of these cells at various toxin concentrations. Values for isolated rat liver cells are shown for comparison. Incubation time 30 min

actin, were not affected at all. They showed no changes in the membranes, nor inhibition of their growth in culture. Toxin concentrations $> 10^{-4}$ M

Table 1 Comparison of actin content and sensitivity against phalloidin of various cells

Cell type	*μg Toxin bound by 100 mg cells at 10^{-6} M toxin*	*μg Toxin bound by 100 mg homogenate at 10^{-6}M toxin (≡actin?)*	*Intoxication symptoms*
Hepatocytes	1	0.8	Protrusions at 0.25 μM toxin
Gabbiani's hepatocytes	0.6	2.3	Protrusions at 2 μM toxin
Lymphocytes	0.4	1.3	No protrusions No inhibition of growth up to 10^{-4} M

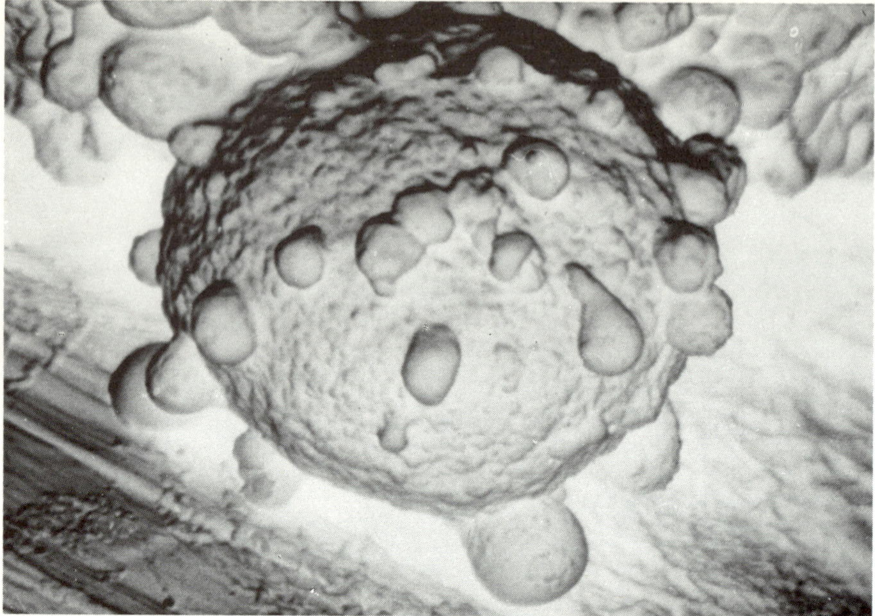

Figure 6 Isolated hepatocyte with typical protrusions of the plasma membrane after phalloidin treatment (2.5×10^{-6} M) for 15 min

only partially inhibited the growth. Thus large amounts of F-actin were associated with the incorporation of high amounts of toxin into the cells.

ACTION OF ANTAMANIDE

Antamanide is another type of cyclic peptide from *Amanita phalloides* mushrooms, which protects animals from lethal phalloidin intoxication. Antamanide also suppresses the development of the protrusions in isolated liver cells[8]. This protection was associated with a reduced incorporation of non-F-actin-bound toxin. Antamanide reduces the amount of toxin incorporated exactly to that amount which can be found by osmotically shocked cell ghosts (Figure 7).

CONCLUSIONS

No observation available explains how phalloidin, taken up in excess by cells, could be toxic. It is unlikely that unbound phalloidin causes lesions. More likely, the apparently unbound toxin is not free, but is complexed to a receptor still unknown. One possibility is the binding of phalloidin to G-actin, provided that G-actin itself is bound to other cell components. Such a possi-

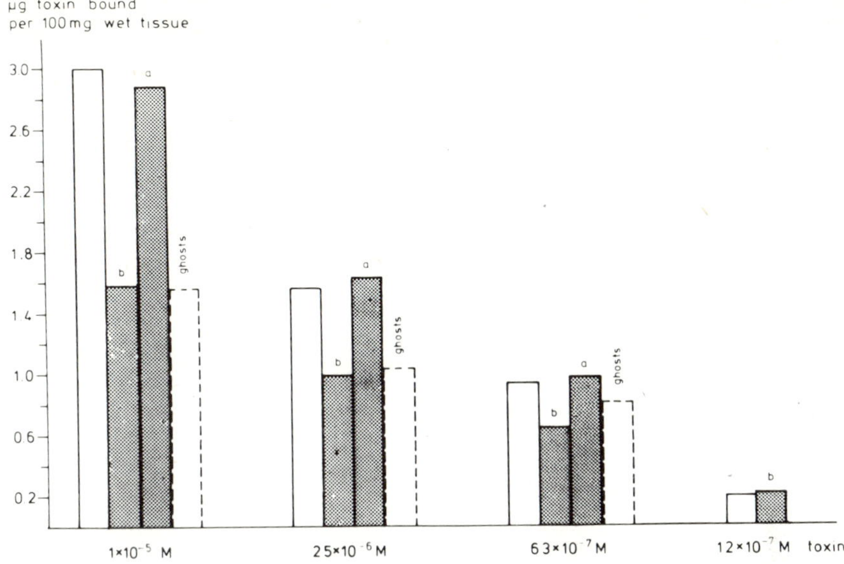

Figure 7 Inhibition of binding of [^{3}H]demethylphalloin by isolated rat liver cells with antamanide [10^{-5} M] at different toxin concentrations. Treatment with antamanide for 15 min before the toxin (b) and after the toxin treatment (a)

bility arises, according to Carlsson and co-workers[9], who isolated a protein of molecular weight 16 000, which binds to G-actin. This protein participates in the DNAse I inhibitor activity and evidently stabilizes the monomeric form of actin. If such a ternary complex of [protein 16 000 + G-actin + phalloidin] was present in our supernatants, our studies would not have detected it since we separated the particulate fraction together with filamentous actin only.

SUMMARY

Phalloidin binds to actin with a high affinity, but only to the aggregated part. Thus an equilibrium of monomeric actin and aggregated actin in the cell is disturbed, the significance of which is not yet understood. The only poisoning effects of phalloidin known are changes in the membranes of liver cells. These alterations include vacuoles and protrusions and resemble closely the features produced during endocytosis or exocytosis. Both processes depend greatly on contractile proteins, like F-actin. Hence, changes in the function of F-actin associated with plasma membranes can best explain the phalloidin effects to date. Moreover, the binding studies indicate that protrusions depend on the amount of phalloidin which is incorporated by cells, but which is not bound to F-actin filaments.

References

1. Govindan, V. M., Faulstich, H., Wieland, Th., Agostini, B. and Hasselbach, W. (1972). *In vitro* effect of phalloidin in a plasma membrane preparation from rat liver. *Naturwissenschaften*, **59**, 521
2. DeVries, J. X., Schäfer, A. J., Faulstich, H. and Wieland, Th. (1976). Protection of actin from heat denaturation by various phallotoxins. *Hoppe-Seyler's Z. Physiol. Chem.*, **357**, 1139
3. Dancker, P., Löw, I., Hasselbach, W. and Wieland, Th. (1975). Interaction of actin with phalloidin: polymerisation and stabilisation of F-actin. *Biochim. Biophys. Acta*, **400**, 407
4. Wegner, A. and Engel, J. (1975). Kinetics of the cooperative association of actin to actin filaments. *Biophys. Chem.*, **3**, 215
5. Faulstich, H., Schäfer, A. J. and Weckauf, M. (1977). The dissociation of the phalloidin actin complex. *Hoppe-Seyler's Z. Physiol. Chem.*, **358**, 181
6. Lutz, F., Glossmann, H. and Frimmer, M. (1972). Binding of [^{3}H]demethylphalloin to isolated plasma membranes from rat liver. *Naunyn-Schmiedeberg's Arch. Pharmacol.*, **273**, 341
7. Gabbiani, G., Montesano, R., Tuchweber, B., Salas, M. and Orci, L. (1975). Phalloidin-induced hyperplasia of actin filaments in rat hepatocytes. *Lab. Invest.*, **33**, 562
8. Faulstich, H., Wieland, Th., Walli, A. and Birkmann, K. (1974). Antamanide protects hepatocytes from phalloidin destruction. *Hoppe-Seyler's Z. Physiol. Chem.*, **355**, 1162
9. Carlsson, L., Nyström, L. E., Lindberg, U., Kannan, K. K., Cid-Dresdner, H., Lövgren, S. and Jörnvall, H. (1976). Crystallization of a non-muscle actin. *J. Mol. Biol.*, **105**, 353

DISCUSSION

Dr K. Decker, Freiburg (West Germany): Does biotransformation of phalloidin take place in the liver? How else would you explain the organ specificity and the accumulation of phalloidin in liver?

Dr Faulstich, Heidelberg (West Germany): More than 90% of phalloidin has been extracted unmetabolized from homogenized intoxicated rat livers (Puchinger and Wieland). Hence we no longer believe that phalloidin is toxified in the liver, e.g. by microsomal oxidation. The organ specificity and the accumulation of phalloidin in the liver is probably caused by the rapid uptake of the toxin. Saturation in perfused rat livers is achieved not later than 10 min after application of phalloidin.

24
Thick Microfilaments (Intermediate Filaments) and Chronic Alcohol Ingestion

S. W. FRENCH, J. S. SIM and M. G. CALDWELL

One of the hallmarks of the liver lesion seen in alcoholic patients is the Mallory body (MB), often designated as 'hyalin'. This body is located in the cytoplasm of hepatocytes and is composed of branching filaments which resemble thick microfilaments (intermediate filaments)[1]. Polyacrylamide gel electrophoresis of MB extracts failed to demonstrate a protein which had the same mobility as actin[2]. Recently, however, Nenci[3] reported that smooth muscle antibody (antiactin) bound MBs suggesting that an actin-like protein was present in these bodies.

In this report we have extended our studies of the MB proteins by comparing the antigenicity of human MBs with griseofulvin-induced MBs and phalloidin-induced microfilament hyperplasia in C_3H in mice using immunofluorescent techniques. Immunoabsorption of antiMB antibody by a variety of antigens was also tested. Immunodiffusion studies indicated that the antiMB antibody was specific for a protein in the solubilized MB protein fraction.

MATERIALS AND METHODS

Liver tissues

Five livers obtained at autopsy from patients who died of alcoholic hepatitis were utilized as a source of MBs for frozen section analysis and for the isolation of purified MB fractions.

MBs were experimentally induced in the livers of mice by feeding griseofulvin using the method of Denk *et al.*[4] Twenty male C_3H strain mice weighing 25–29 g were fed a semisynthetic complete standard diet[5] (Teklad Test Diet, Madison, Wisconsin) which contained 30% casein and 2.5% griseofulvin (Schering Corp., Ltd., Quebec, Canada). Five control rats were fed the basal diet without griseofulvin added. Five rats fed griseofulvin for 5 months were selected for study after MBs were demonstrated by light and electron-microscopy[6]. The MB-containing livers were stored at −18 °C until used for immunofluorescent studies.

Microfilament hyperplasia was induced in livers according to the method of Gabbiani *et al.*[7] Ten adult C_3H mice were maintained on purina chow *ad libitum.* Five were injected intraperitoneally daily with 1 mg/kg phalloidin in 0.9% saline (kindly donated by Dr Th. Wieland, Max-Planck Institute, Heidelberg, Germany). Five controls received saline alone. Two experimental mice, as well as two controls, were sacrificed at 19 and 40 days. The livers of these mice were used for light and electron microscopic examination. Frozen sections were made for immunofluorescent studies.

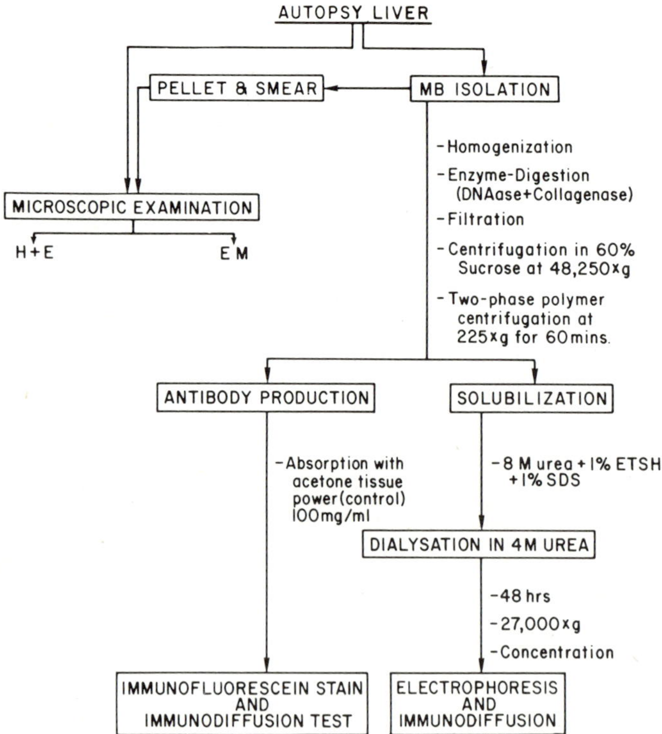

Figure 1 Flow sheet for the isolation of MBs for immunopathological studies and protein electrophoresis

Mallory body isolation and immunization

MBs were isolated by a modification of Okamura's method[8] using the two-phase polymer system. Prior to the two-phase polymer separation step (Figure 1) tissue homogenates were digested at 37 °C for 5–24 hours in 0.05 M Tris buffer pH 7.5 containing 0.1% DNAase (Sigma Chemical Co.) and 0.1% clostridia collagenase (Worthington Enzymes), 0.36 mM $CaCl_2$, 0.1 mM

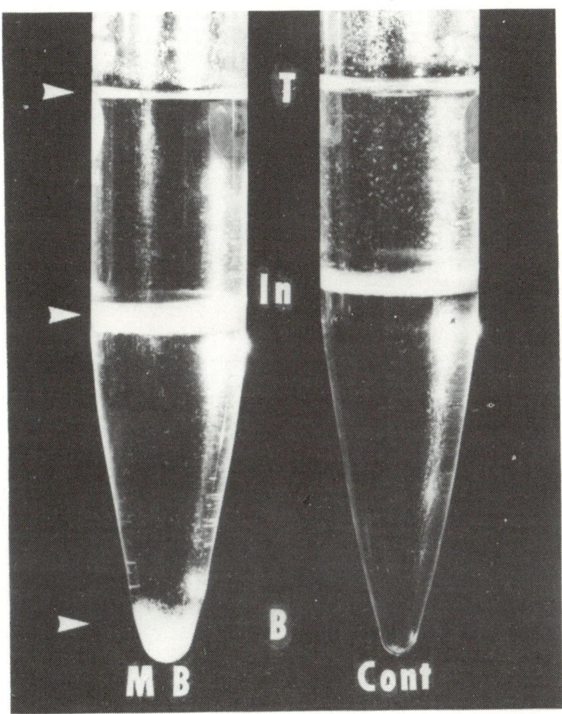

Figure 2 Two-phase polymer separation of alcoholic liver homogenate into a top layer (T), interphase (In) and a bottom (B) pellet. Note that the MB pellet is prominent whereas the control liver pellet is not visible

$MgSO_4$, 50 units of mycostatin and 0.25 mg/ml tetracycline. The enzyme pretreatment prevents aggregation and trapping of particles and removes contaminating collagen and nuclei. This increases the yield of MBs in the pellet (Figure 2) and improves the purity of the isolates. Purity was assessed by light and electron microscopy[6] (Figure 1).

The purest MB fraction prepared from 100 g of fresh liver was washed three times in distilled water by centrifugation. This fraction was then incubated in 5% deoxycholate at 37 °C for 24 hours[9] to remove any membranes, lipids or bile which may contaminate the fraction. The deoxycholate-treated MB fraction was then washed repeatedly in distilled water and lyophilized. The lyophilized MB fraction (8.8 mg) was dispersed in 0.5 ml saline and mixed with an equal volume of Freund's complete adjuvant (Cappel Laboratory). This MB antigen preparation was injected into the left thigh muscle of an adult male Wistar rat. The booster was injected 2 weeks later into the right thigh muscle; 2 weeks later the rat was bled by heart puncture. The serum was inactivated by 53 °C and stored at −18 °C prior to use.

Immunofluorescence

Cryostat sections of liver were cut at 2 μm. To remove soluble protein, duplicate unfixed sections were washed in PBS (0.14 M NaCl in 0.01 M phosphate buffer, pH 7.3) for 30 min at room temperature before staining. Two drops of antisera (1 : 20) (rat antiMB sera or human smooth muscle antibody) were layered on the specimens and allowed to react at room temperature for 30 min in a moist chamber. The smooth muscle antibody (SMA) was a gift from Dr H. J. Smith, Bioscience Laboratories. The antisera were washed off three times with PBS for 30 min using a magnetic stirrer. Two drops of fluorescein isothiocyanate (FITC) conjugated rabbit antirat globulins (Cappel Laboratory), diluted ten times in PBS, were flooded on the MB sections for 30 min. In the case of the SMA-treated sections, FITC-conjugated rabbit antihuman IgG (Cappel Laboratory) was used. The specimens were again washed for 30 min with PBS and mounted in 10% glycerol. The preparations were examined with a fluorescence microscope (Zeiss) equipped with a HBO 200 high pressure mercury vapour lamp and appropriate exciter and barrier combination. Serial cryostat sections were stained with haematoxylin and eosin to correlate areas of positive specific fluorescence with the same areas using light microscopy.

Using the same technique, MB smears were also tested to see if they would bind rabbit antineurofilament protein. The neurofilament antibody was obtained as a gift from Dr V. I. Kalnins[10], University of Toronto and Dr M. L. Shelanski, Harvard Medical School[11]. Preimmune serum from the immunized rabbit used by Kalnins was also tested as a control. The antiNF antibodies were used at a dilution of 1 : 20. Conjugated goat–antirabbit sera was 1 : 40.

Immunoabsorption

The immune specificity of the rat antiMB antibody for MBs and the SMA for microfilaments was tested by immunoabsorption. In the case of SMA, purified actin extracted from acetone-dried rabbit skeletal muscle (Worthington) was used to absorb actin-specific antibody. The purity of the actin used was established by polyacrylamide gel electrophoresis[12]. The SMA binding to smooth muscle of mouse intestine was also demonstrated. Absorbents for the MB-antibody included control human liver (CL), isolated MBs, actin and the salt-insoluble fraction of chicken gizzards smooth muscle. The latter fraction was used because it is rich in intermediate filaments (IF). The IF protein fraction was prepared according to the method of Cooke[13]. Acrylamide gel electrophoresis of this fraction showed that the fraction was contaminated with actin as well as unidentified proteins. Acetone-dried powders were prepared from each absorbent. Each absorbent powder was mixed with antisera (100 mg/ml) and was left overnight at 4 °C[14] except for actin (3 mg/ml) where the antiserum was absorbed three times as described by Nenci[3]. After centrifugation, the reactivated sera were tested by immunofluorescence and the results compared with those of the unabsorbed antisera.

Immunodiffusion studies

Double immunodiffusion[15] was performed in 1% agarose plates with barbital buffer pH 7.6, ionic strength 0.025. Antisera along with control rat sera were reacted for the presence of precipitating antibodies against solubilized MB antigen and control human liver pellet prepared by the two-phase polymer system (Figure 1). Antisera absorbed with acetone powder of human liver was similarly studied.

Aliquots (50 mg dry weight) of MB isolates were solubilized in 25 ml of 8 M urea, 1% β-mercaptoethanol in 0.01 M phosphate buffer (pH 7) containing one of the following surfactants: 1% sodium dodecyl sulphate (SDS), 5% deoxycholate (DOC), or 4% Triton X-100 (t). Solubilization was achieved by incubating at 37 °C for 16 hours. The dissolved Mallory body solutions were centrifuged at 100 000 *g* for 1 hour and the supernatants were dialysed for 24 hours against 0.01 M phosphate buffer pH 7.6 containing 4 M urea and 1% β-mercaptoethanol. The dialysates were centrifuged at 27 000 *g* for 20 min at 5 °C. The solubilities of the MBs in the three different detergents were examined by weighing the residues after lyophilization. Solubilities were determined by the difference between the initial sample weight and the final weight of the residue. The residues were below 1.5% of the original weight for all three detergents. Pellets of the residue were examined by light and electron microscopy and no MBs were found. Aliquots of the dialysed MB solution (1.2 mg protein/ml) were used as antigen for the immunodiffusion study. The antiserum was diluted 1 : 5 in PBS.

MB antigen was also tested against the antiNF sera of Kalnins and Shelanski using the technique of Yen *et al.*[11] The MB antigen, dissolved in 0.1% SDS, was diffused in 1% agar containing 0.1% SDS and 0.5% Triton X-100.

RESULTS

Griseofulvin-fed mice

MBs were found in liver cells of mice fed griseofulvin after 5 months on the diet. By light microscopy the intracytoplasmic eosinophilic hyalin was seen in swollen liver cells throughout the lobule. By electron microscopy the presence of MBs was confirmed (Figure 3). The hyalin was composed of branching filaments. All three types of MBs described by Yokoo *et al.*[16] in alcoholic hepatitis were seen in the mice livers. No MBs were seen in the five control mice. By 5 months of griseofulvin feeding the characteristic liver lesion was seen. The livers were enlarged compared to controls. The griseofulvin-fed mice livers averaged 20.8 g/100 g body weight and the controls averaged 6.5 g/100 g. Cholestasis, brown pigment deposits, bile duct and fibroblastic proliferation, and PMN infiltration were prominent microscopically.

Phalloidin-injected mice

Phalloidin treatment induced marked hyperplasia of the microfilaments around the bile canaliculi as visualized by electron microscopy (Figure 4) conforming to findings of Gabbiani *et al.*[7] Here the ectoplasm was markedly widened and microvilli reduced in number The ectoplasm around the bile canaliculi was focally thickened by microfilament hyperplasia. By light microscopy the pericanalicular thickening was most evident in the central zone of the lobule. The canalicular walls were thickened and appeared rigid. The lumens of the canaliculi were dilated and were slit-like or triangular in shape. Kupffer cells contained a large round sphere of microfilaments in their cytoplasm.

Immunofluorescent studies

The antiMB serum raised in the rat bound to human MBs as assessed by indirect immunofluorescence (Figure 5). Griseofulvin-fed mice livers containing MBs also bound MB antiserum using the same technique (Figure 6). The fluorescent appearance of human and mouse MBs was similar morphologically to the MBs by light microsopy. The MBs appeared as irregularly outlined cytoplasmic inclusions. The location of the fluorescent MBs consistently coincided with those seen in neighbouring tissue sections stained by haematoxylin and eosin. Other hepatic structures did not bind the MB antisera.

Immunoabsorption studies of the antiMB antibody showed that immunofluorescence disappeared only when the antisera was absorbed with isolated MBs or salt extracted smooth muscle homogenate. Acetone powders of control human liver or purified actin did not remove the antiMB antibody from the

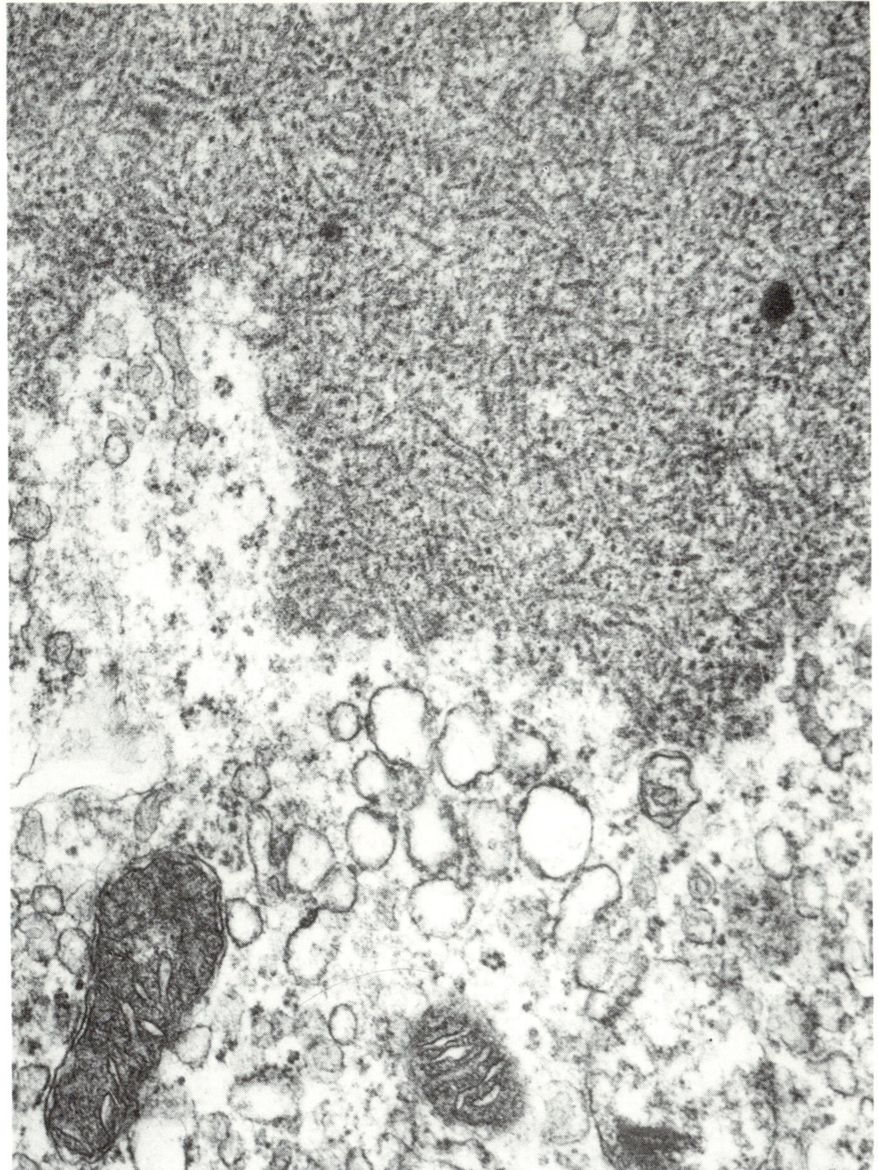

Figure 3 Griseofulvin-fed mouse liver showing a portion of an MB (above) in the cytoplasm of a hepatocyte. Uranyl acetate and lead citrate (× 53 000)

antisera (Table 1). The immunoabsorption results were consistent in human liver and mouse liver.

Sections from control human and mouse livers showed no fluorescent staining after incubation with the rat antiMB sera. Normal rat sera did not bind to

MBs or other cellular structures. The SMA did not bind to MBs of mouse or man. Nenci's[3] observation that MBs bind SMA was not confirmed. Our negative results may be due to the removal of actin during the PBS wash step prior to staining the liver slices with the SMA. Nenci[3] used acetone fixation which would prevent extraction of actin from the tissue. Evidence for the conclusion that PBS removed the actin from the tissue slices was the observation that the characteristic polygonal staining of the liver cells was absent

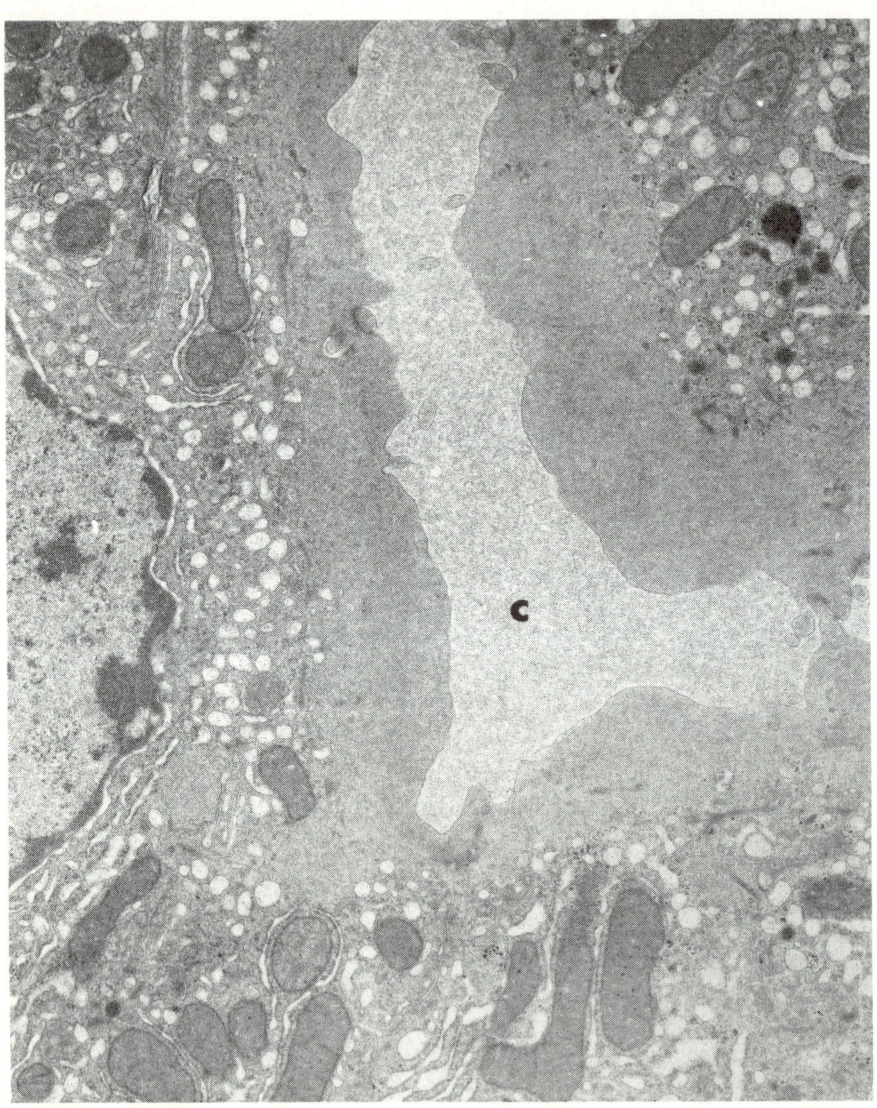

Figure 4 Phalloidin-injected mouse liver showing a portion of a bile canaliculus (c). The ectoplasm is expanded by microfilament hyperplasia. Uranyl acetate and lead citrate (× 18 400)

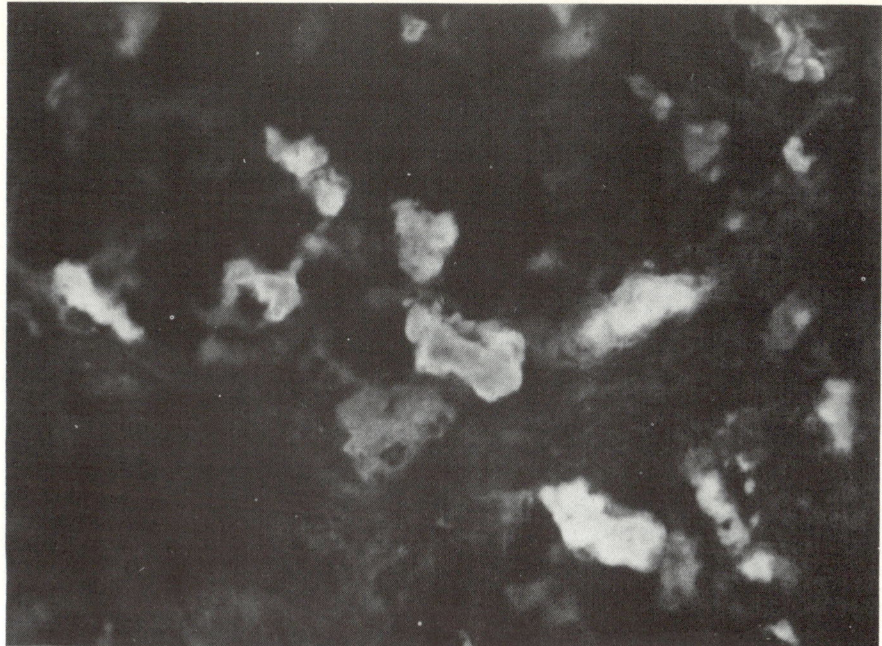

Figure 5 Frozen sections of an autopsy liver from a patient with alcoholic hepatitis. The MBs fluoresce apple-green by the indirect immunofluorescent technique using rat antiMB antibody (× 700)

Table 1 The effect of various absorbents (reacted with rat antiMB sera) on MB-specific antibody staining

*Absorbents**	*MBs in alcoholic livers*	*MBs in mice livers*
None	+	+
CL	+	+
MBs	–	–
IFs	–	–
Actin	+	+

* After absorption, the antisera supernatant was tested by immunofluorescent staining of duplicate tissue sections
+ Indicates that the MBs bound the antisera
– Indicates no staining was observed

from the liver slices. The phalloidin-induced microfilament hyperplasia was not removed by the PBS wash step because phalloidin treatment causes the actin filaments to resist extraction.

SMA sera bound the pericanalicular ectoplasmic zone of hyperplastic microfilaments in liver slices from phalloidin-treated mice. The binding was demonstrated by the indirect immunofluorescent technique (Figure 7). Fluorescent staining was completely prevented by immunoabsorption of the SMA sera with purified actin. Control human serum did not give a positive reaction. These results confirm the findings of Gabbiani *et al.*[7]

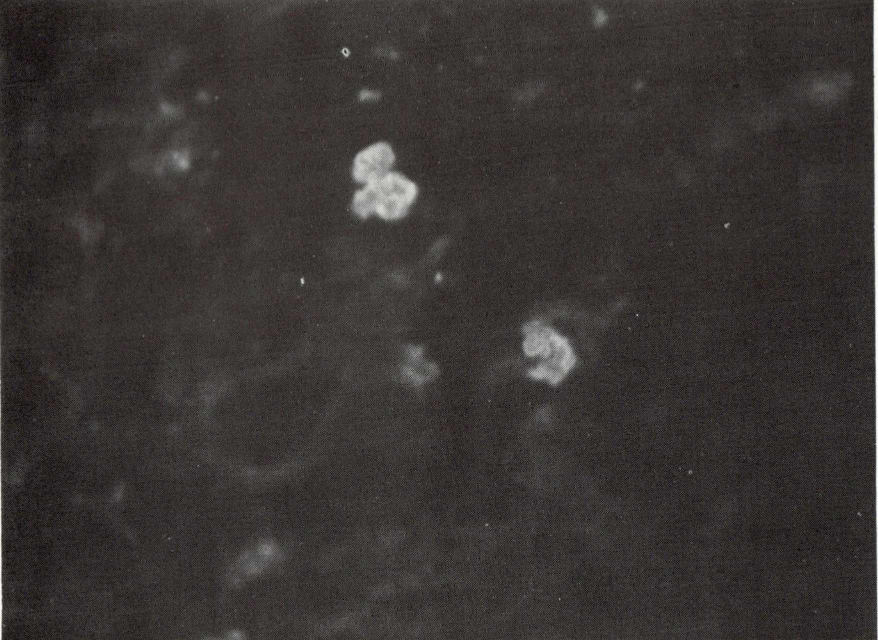

Figure 6 Frozen section of a griseofulvin-fed mouse liver. The antiMB antibody bound to the MB as indicated by the indirect immunofluorescence (× 700)

An unanticipated finding was the observation that antiMB sea bound the mouse hyperplasic microfilaments in the same way as did the SMA sera. This binding was completely prevented by absorption of the antiMB sera with purified actin. As already stated, actin did not remove the MB-binding antibody in the MB antisera. It is concluded that the MB antisera contained two antibodies, one of which was absorbed by actin, while the other was specific for MBs.

The presence of IF protein in MBs was looked for using antiNF antibody by the indirect immunofluorescent technique. Isolated MBs fixed to slides bound the antiNF antibody of Kalnins[10] and Shelanski[11]. The MBs stained with antiNF antibody appeared an apple-green colour with bright outlines. MBs also stained with control rabbit serum. The results are ambiguous because of the non-specific staining by control sera.

Double immunodiffusion studies

The MB antiserum was tested against the solubilized MB proteins extracted from the human MB isolates by double immunodiffusion. A well-defined precipitin line formed when the MB antiserum was reacted with the solubilized MB protein. This was true regardless of whether anionic or neutral detergents

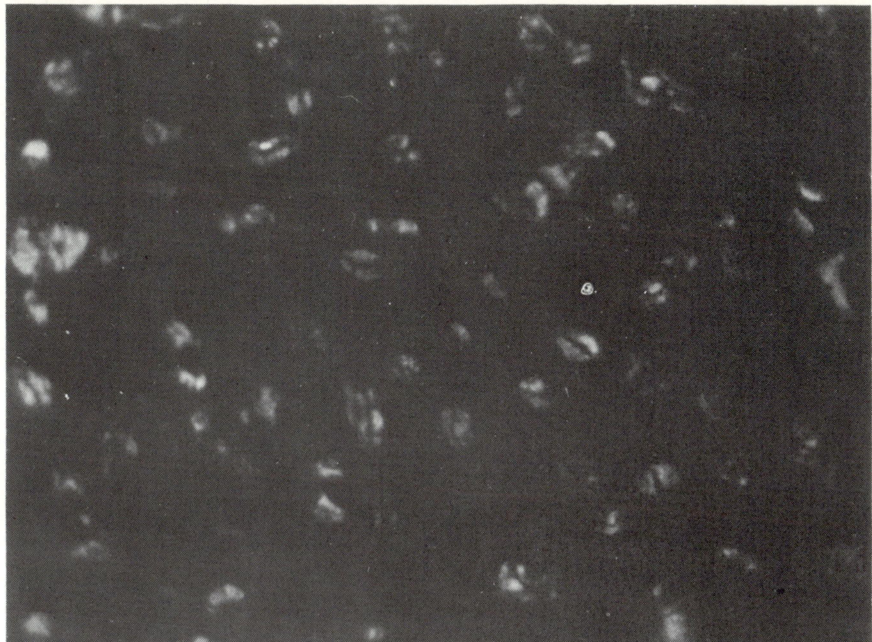

Figure 7 Frozen section of phalloidin-treated mouse liver. Note the pericanalicular staining of the microfilament hyperplasia zone flanking the central slit-like canalicular lumen. This distinctive staining pattern resembles that of a lipstick blot (× 400)

were used to solubilize the MB proteins (Figure 8). Extracts of control liver and the Cooke[13] smooth muscle extract which contained IF protein failed to form an immunoprecipitate with the MB antiserum. When the MB antiserum

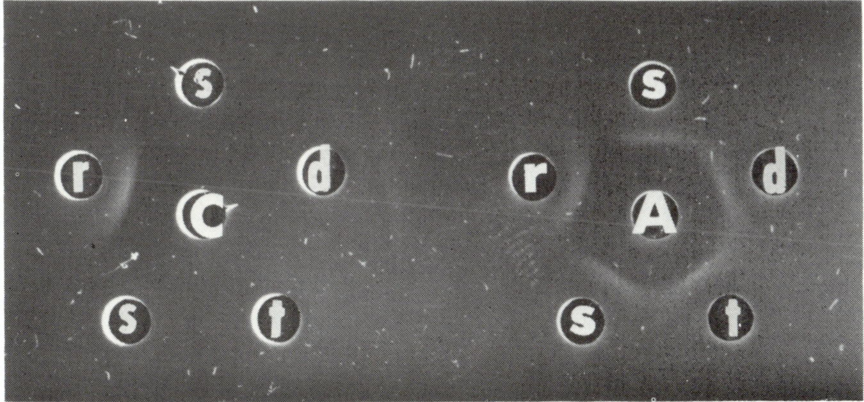

Figure 8 Agarose immunodiffusion of rat antiMB serum (A) and normal rat serum (C) against solubilized MB protein. The MB protein was solubilized by SDS (s), deoxycholate (d) and Triton X-100 (t). The detergents were replaced by 4 M urea by dialysis before immunodiffusion. Rabbit antirat IgG serum was used as a positive reference (r)

was replaced by normal rat serum no precipitin line was observed (Figure 8). The antirat IgG serum formed a precipitate with both the MB and control rat sera as a positive reference (Figure 8).

AntiMB serum absorbed with control human liver acetone powder formed an immunoprecipitate with the solubilized MB antigen (Figure 9). The antiserum did not react with control human liver. As in Figure 8, no reaction was seen between the control sera and the liver antigens. No precipitate formed when the antineurofilament rabbit antibody was tested against solubilized MBs.

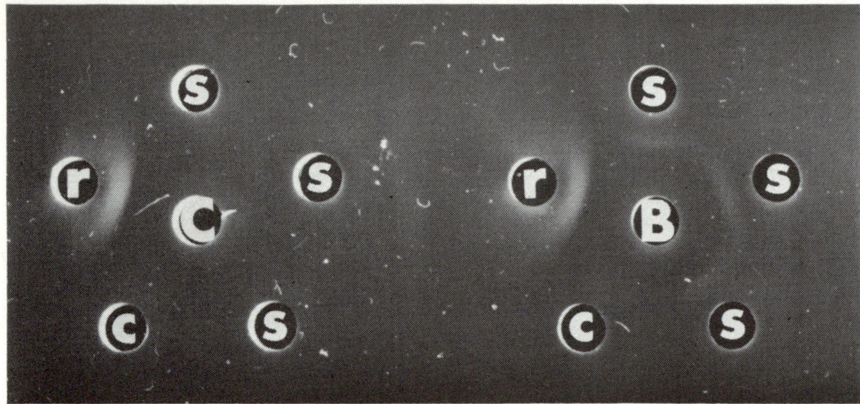

Figure 9 Agarose immunodiffusion of rat antiMB serum absorbed with control liver (B) and normal serum (C) against solubilized MB protein (S) and control liver protein (c)

DISCUSSION

The results reported here suggest that MBs contain IF protein. The significance of this observation resides in the fact that it furnishes a clue to the pathogenesis of MB formation. It is known, for instance, that intermediate filaments accumulate in a variety of cell types as a consequence of treatment with the antitubulin drugs colchicine and colcemid[17,18]. The accumulation of IFs was first observed in skeletal muscle cells cultured from chick embryos treated with colchicine or colcemid.[19]. The accumulation of IFs in muscle cells was further exaggerated by pretreatment with cytochalasin B[20]. De Brabander *et al.*[21] showed that colchicine, vinblastine or vincristine induced microtubule dissolution and an increase in IFs. The later phase of IF accumulation was inhibited by the protein synthesis inhibitor cycloheximide. The disappearance of the accumulated IFs following the removal of colchicine was gradual. Thus, if MBs are composed of IFs, the formation of MBs could result from the antitubulin action of alcohol[22] or griseofulvin[23]. Denk[24] has observed that although the antitubulin colchicine does not induce MBs in mice treated with colchicine alone, it does induce MB formation in mice which have developed MBs from prior griseofulvin treatment followed by a disappearance of the MBs after griseofulvin withdrawal. Since the experimental MB in the griseofulvin-fed mice stained with the rat antihuman MB sera, it is reasonable to postulate that the pathogenesis of the MBs in the two species was similar.

Prior studies have demonstrated that actin-containing microfilaments are located deep to the plasma membrane[25–28] in liver cells. Both thick (100 Å) and thin (50 Å) microfilaments occur here[28]. The thick microfilaments do not bind heavy meromyosin indicating that they are not actin filaments[26]. It is possible, therefore, that the normal liver cell contains intermediate filaments which are similar to IFs found in other cells.

It is important to emphasize the presence of intermediate filaments in cells of many types and are found in many species (chicken, mouse, man). Ishikawa *et al.*[19] noted IFs in fibroblasts, chondrocytes and myoblasts. They noted that the cells containing IFs did not bind antibody to myosin or actin. Yen *et al.*[11] characterized a similar intermediate filament in brain tissue (neurofilament (NF) and astroglial filament). They showed that antisera to NFs cross-reacted with glial filaments[10,11]. AntiNF did not cross-react with actin[11], tubulin[10,11] or tropomyosin[10]. AntiNF also binds IFs in endothelial cells[29].

The fact that the antiNF antibody did not cross-react with MBs as reported in the present paper is a point against the hypothesis that MBs are composed of IF protein.

SUMMARY

Evidence supporting the hypothesis that Mallory bodies (alcoholic hyalin) are composed of intermittent filaments was reviewed. Since Mallory bodies (MB) can be experimentally produced in mice fed griseofulvin it was further postulated that MB formation is the result of an antitubulin effect. To further strengthen the evidence it was necessary to try and determine if MBs from alcoholic patients were composed of the same protein as the griseofulvin-induced MBs. An antibody to human MBs was raised in a rat. The specificity of the antibody raised was demonstrated by immunofluorescent, immunodiffusion and immunoabsorption techniques. The MB antisera contained two antibodies one of which bound phalloidin-induced microfilament hyperplasia (actin) and was absorbed out by actin and the other of which was specific for MBs. The antisera was found to bind to both the human and experimental MBs suggesting that MBs from both species are the same. The results lend support to the concept that MB filament accumulation in liver cells is a result of antitubulin action.

DISCUSSION

Dr H. Denk, Wien (Austria): Mallory bodies (MB) were induced in all mice by continuous griseofulvin (GF) feeding (2.5% GF incorporated into the diet) for 3 months. After a GF-free period of 1 month and disappearance of MB, the hepatocytes responded almost immediately (within 4 days) with MB formation to a GF challenge ('primary' effect of GF pretreatment). MB could be induced by colchicine in the GF-pretreated ('primed') mouse liver. Since intermediate-sized filaments increase in cells in response to antitubulin agents, this observation supports the hypothesis that MB are related to intermediate filaments.

References

1. French, S. W. and Davies, P. L. (1975). The Mallory body in the pathogenesis of alcoholic liver disease. In: J. M. Khanna, Y. Israel and H. Kalant (eds.). *Alcoholic Liver Pathology*, pp. 113–143. (Toronto: Addition Research Foundation of Ontario)
2. Rubin, E. and Lieber, C. (1975). Relation of alcoholic liver injury in cirrhosis. *Clin. Gastroenterol.*, **4,** 247
3. Nenci, I. (1975). Identification of actin-like protein in alcoholic hyalin by immunofluorescence. *Lab. Invest.*, **32,** 247
4. Denk, H., Gshnait, F. and Wolff, K. (1975). Hepatocellular hyalin (Mallory bodies) in long-term griseofulvin-treated mice: a new experimental model for the study of hyalin formation. *Lab. Invest.*, **32,** 773
5. French, S. W. (1966). Effect of chronic ethanol ingestion on liver enzyme changes induced by thiamine, riboflavin, pyridoxine or choline deficiency. *J. Nutr.*, **88,** 291
6. Wiggers, K. D., French, S. W., French, B. A. and Carr, B. N. (1973). The ultrastructure of Mallory body filaments. *Lab. Invest.*, **29,** 652
7. Gabbiani, G., Montesano, R., Tuchweber, B., Salas, M. and Orci, L. (1975). Phalloidin-induced hyperplasia of actin filaments in rat hepatocytes. *Lab. Invest.*, **33,** 562
8. Okamura, K., Harwood, T. R. and Yokoo, H. (1975). Isolation and electrophoretic study on Mallory bodies from the livers of alcoholic cirrhosis. *Lab. Invest.*, **33,** 193
9. French, S. W., Ihrig, T. J. and Norum, M. L. (1972). A method of isolation of Mallory bodies in a purified fraction. *Lab. Invest.*, **26,** 240
10. Jorgensen, A. O., Subrahmanyan, L., Turnbull, C. and Kalnins, V. I. (1976). Localization of the neurofilament protein in neuroblastoma cells by immunofluorescent staining. *Proc. Natl. Acad. Sci. USA*. (In press)
11. Yen, S. H., Dahl, D., Schachrer, M. and Shelanski, M. L. (1976). Biochemistry of the filaments of brain. *Proc. Natl. Acad. Sci. USA*, **73,** 529
12. Weber, K. and Osborn, M. (1969). The reliability of molecular weight determinations by dodecyl sulfate–polyacrylamide gel electrophoresis. *J. Biol. Chem.*, **244,** 4406
13. Cooke, P. (1976). A filamentous cytoskeleton in vertebrate smooth muscle fibers. *J. Cell Biol.*, **68,** 539
14. Kyriaris, A. P. and Wissler, R. W. (1972). Demonstration and purification of a specific antigen in Morris hepatoma. *Lab. Invest.*, **26,** 178
15. Ouchterlony, O. (1958). Diffusion in gels. Methods for immunological analysis. *Prog. Allergy*, **5,** 1
16. Yokoo, H., Minick, O. T., Batti, F. and Kent, G. (1972). Morphologic variants of alcoholic hyalin. *Am. J. Pathol.*, **69,** 25

17. Daniels, M. P. (1973). Fine structural changes in neurons and nerve fibers associated with colchicine inhibition of nerve fiber formation *in vitro*. *J. Cell Biol.*, **58,** 463
18. Blose, S. H. and Chacko, S. (1976). Rings of intermediate (100 Å) filament bundles in the perinuclear region of vascular endothelial cells. *J. Cell Biol.*, **70,** 459
19. Ishikawa, H., Bischoff, R. and Holtzer, H. (1968). Mitoses and intermediate-sized filaments in developing skeletal muscle. *J. Cell Biol.*, **38,** 538
20. Holtzer, H., Croop, J., Dienstman, S., Ishikawa, H. and Somlyo, A. P. (1975). *Proc. Natl. Acad. Sci. USA*, **72,** 513
21. De Brabander, M., Aerts, F., Van de Veire, R. and Borders, M. (1975). Evidence against interconversion of microtubules and filaments. *Nature* (*London*), **253,** 119
22. French, S. W., Sim, J. S. Franks, K. E., Burbige, E. J., Denton, T. and Caldwell, M. G. (1976). Alcoholic hepatitis, In: M. M. Fisher and J. G. Rankin (eds.). *Proceedings of the Symposium on Alcohol and the Liver*. (New York: Plenum Press). (In press)
23. Weber, K., Wehland, J. and Herzof, W. (1976). Griseofulvin interacts with microtubules both *in vivo* and *in vitro*. *J. Mol. Biol.*, **102,** 817
24. Denk, H. (1976). (Personal communications)
25. Agostini, B., Govindan, V. M., Hofman, W. and Wieland, Th. (1975). Phalloidin-induced proliferation of actin filaments within rat hepatocytes. *Z. Naturforsch.*, **30,** 793
26. French, S. W. and Davies, P. L. (1975). Ultrastructural localization of actin-like filaments in rat hepatocytes. *Gastroenterology*, **68,** 765
27. Holborow, E. J., Trenchev, P. S., Dorling, J. and Webb, J. (1975). Demonstration of smooth muscle contractile protein antigens in liver and epithelial cells. *Ann. N.Y. Acad. Sci.*, **254,** 489
28. Oda, M., Price, V. M., Fisher, M. M. and Phillips, M. J. (1974). Ultrastructure of bile canaliculi with special reference to the surface coat and the pericanalicular web. *Lab. Invest.*, **31,** 314
29. Blose, S. H., Shelanski, M. L. and Chacko, S. (1976). Localization of bovine neurofilament antibody to intermediate filaments in endothelial cells. *J. Cell Biol.*, **70,** 94a

25

Effect of Ethanol on Lipoprotein and Protein Export from the Liver and its Relationship to Progressive Alcoholic Liver Injury

C. S. LIEBER, E. BARAONA,* S. A. BOROWSKY† and M. A. LEO‡

One of the most conspicuous manifestations of alcohol abuse is hepatomegaly. This enlargement of the liver has been generally attributed to massive fat accumulation. It must be pointed out, however, that experimental hepatomegaly already occurs at an early stage when fat infiltration is only moderate and accounts for only about half the increase in liver weight[1]. Therefore, we considered whether, in addition to fat, some other factors could play a role in the development of hepatomegaly. This investigation[2] led to the finding that ethanol consumption results in protein accumulation in the liver in association with a decrease in microtubular mass which may contribute to disturbances in the secretion of both protein and lipoprotein.

EFFECT OF ETHANOL ON LIVER SIZE, PROTEIN AND FAT

To determine whether there are cell constituents other than fat which contribute to ethanol-induced hepatomegaly, male rat littermates were pair-fed

* Recipient of USPHS Career Development Award AA 70409.
† Research Associate, Medical Research Service, Veterans Administration.
‡ Recipient of a fellowship from the Add. Did. Scient., University of Bari (Italy).

liquid diet as described[3] for 4 to 8 weeks. The diet contained 18% of calories as protein, 35% as fat, 11% as carbohydrate and 36% as either ethanol or additional carbohydrate. Ninety minutes after an intragastric administration of their respective diets (6 ml/100 g body weight), the rats, weighing 180 to 360 g, were killed under ether anaesthesia by exsanguination from the aorta. The livers were excised and their weights (wet and dry), volume, total lipid[2], protein[4], and DNA[5] contents were measured.

Livers from ethanol-fed rats increased 30% both in volume and in wet weight (4.28 ± 0.13 g per 100 g of body weight compared to 3.34 ± 0.06 in controls; means ± SEM of 24 pairs; $p < 0.001$ by the paired Student's t test). The specific gravity of the liver was unchanged (1.009 ± 0.010 g/cm^3

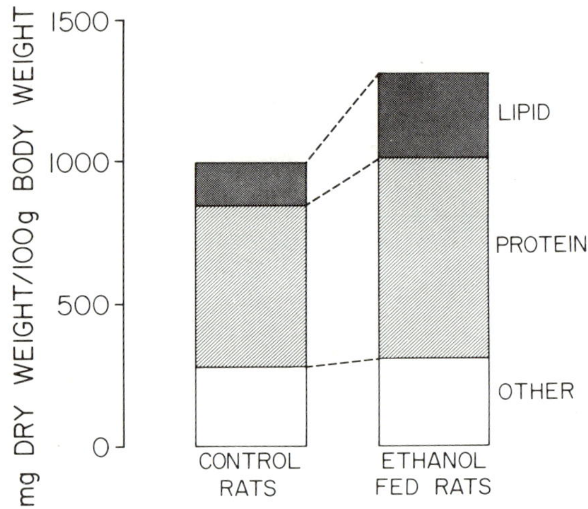

Figure 1 Effect of ethanol feeding (3–4 weeks) on hepatic dry weight, lipid and protein contents in the rat. These three differences are significant ($p < 0.01$)[2]

compared to 1.016 ± 0.017 in controls) despite the doubling of hepatic lipids in the rats that were fed ethanol (Figure 1). The increase in liver fat (151 ± 29 mg per 100 g body weight) accounted for only half of the increase in liver dry weight (304 ± 40 mg per 100 g body weight). In addition to the increase in liver lipid, there was a concomitant increase in liver protein (Figure 1). The increase in protein (132 ± 41 mg per 100 g of body weight) accounted for almost all of the other half of the increased hepatic dry weight. Hepatic protein concentration did not change, indicating that intracellular water also increased in proportion to the increase in protein. Liver wet weight/dry weight ratios were similar in both groups of animals.

In contrast to the increase in protein, DNA content of the liver remained unchanged (10.8 ± 0.5 mg per 100 g body weight compared to 10.3 ± 0.4 in controls), suggesting that hepatomegaly was due to an increase in cell size rather than to an increase in cell number. This was verified in histological sections of the liver. Hepatocytes occupied a significantly larger area (602 ± 43 μm^2 per cell) in livers from rats that were fed ethanol than in those from pair-fed controls (442 ± 36; 8 pairs; $p < 0.01$), even in zones where there was little visible fat.

To determine the subcellular distribution of the increase in liver protein, mitochondria, microsomes and cytosol were isolated. Recovery of each fraction was assessed by appropriate markers[6–8]. Protein increased in all three fractions, but the total increase in mitochondrial and microsomal proteins accounted for less than half of the total increase in liver protein (Figure 2). The

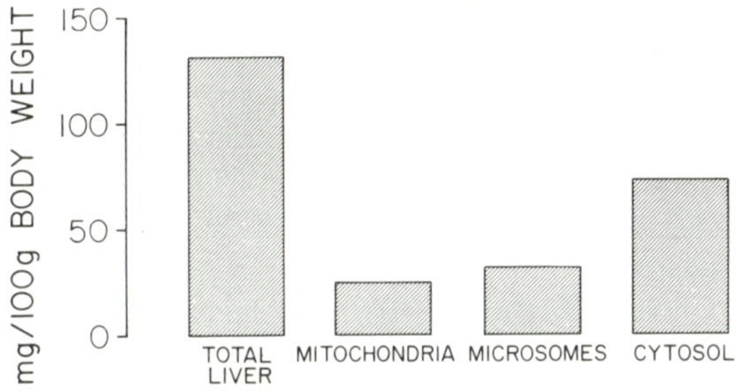

Figure 2 Distribution among subcellular fractions of the increase in liver protein induced by ethanol feeding. The increase in mitochondrial protein is not significant. All other increases are significant ($p < 0.01$)[2]

major protein increase occurred in the 100 000 *g* supernatant, or cytosolic fraction. Thus, in addition to the known increase in organelle protein[9], there was an even greater increase in soluble protein. Moreover, the increase in protein was associated with a change in its composition characterized by increased concentration of proteins normally exported into the plasma. The concentration of export proteins (such as albumin and transferrin) was significantly increased in the deoxycholate extracts of the livers from ethanol-fed rats compared to those from pair-fed controls. Since intrahepatic albumin concentration varies widely according to the method used[10,11], we measured albumin by isotopic dilution of labelled serum albumin in the immunoprecipitates of the liver extracts[10] in one report[2] and by radial immunodiffusion[12] in another study[13]. Though the absolute values for intrahepatic albumin differ in both methods, a similar accumulation of albumin in alcohol-treated rats was observed. Contamination with extracellular albumin remaining after perfusion of the liver was assessed with labelled rat albumin and found to be similar per g of liver in both types of rats; thus subtraction of contaminating extracellular albumin did not affect the difference between alcohol-treated and control animals.

Contrasting with the increased hepatic concentration of export proteins, total liver protein concentration did not change. The concentration of soluble constituent proteins (such as ferritin, organic anion-binding proteins[14] and alcohol dehydrogenase[15]) have been reported either to be unchanged or decreased in rats fed ethanol chronically. Because of the hepatomegaly in the ethanol-fed rats, the differences in concentration of export proteins reflect an even larger increase in their total hepatic content.

Increased concentration of export proteins in the liver could result either from increased synthesis or from decreased export into the plasma. Acute ethanol administration to animals has been reported to inhibit rather than to increase hepatic synthesis of both constituent[16–18] and export[19–22] proteins *in vivo* and *in vitro*. Other investigators have reported that chronic ethanol administration increases protein synthesis in ribosomes[16] and microsomes[23], whereas acute administration produces opposite effects. In keeping with these observations, we found enhanced protein synthesis after chronic ethanol feeding. Though the incorporation of [^{14}C]leucine into total liver protein was unchanged, the label was markedly diluted in the leucyl-tRNA pool. This was due to the increased concentrations of leucine both in serum and liver of ethanol-fed rats[13]. At 10 min, when there was not yet secretion of newly labelled albumin into plasma and the labelling of liver albumin was unaffected by changes in secretion, leucine incorporation into hepatic immunoreactive albumin was unchanged. Hepatic albumin synthesis appeared to be increased if one considers the isotopic dilution in the leucyl-tRNA of ethanol-fed rats. However, there is no evidence to suggest that albumin was synthesized at a faster rate than other liver proteins of which the concentration did not change. Furthermore, if increased albumin synthesis were the primary cause of the accumulation, one would expect this to be associated with a parallel increased output into the plasma. However, 30 min after leucine injection, when secretion of newly labelled albumin was taking place at maximal rates, ethanol feeding produced opposite changes in liver and serum albumin labelling (Figure 3): [^{14}C]leucine incorporation into liver albumin was significantly increased whereas incorporation into serum albumin was significantly decreased. Thus, the triad of increased albumin concentration in the liver, delayed transport of newly labelled albumin into the serum and retention of albumin label in the liver, indicates that ethanol feeding interferes with the export of albumin, leading to accumulation in the liver.

Export proteins are known to be produced by ribosomes attached to the membranes of the rough endoplasmic reticulum[24,25] and are released into the microsomal cisternae. Small amounts of these proteins are also found in the liver cytosol even after removal of blood by perfusion[10,11], but this is generally attributed to contamination with serum proteins remaining in the vascular or interstitial spaces. It is therefore intriguing that, after alcohol

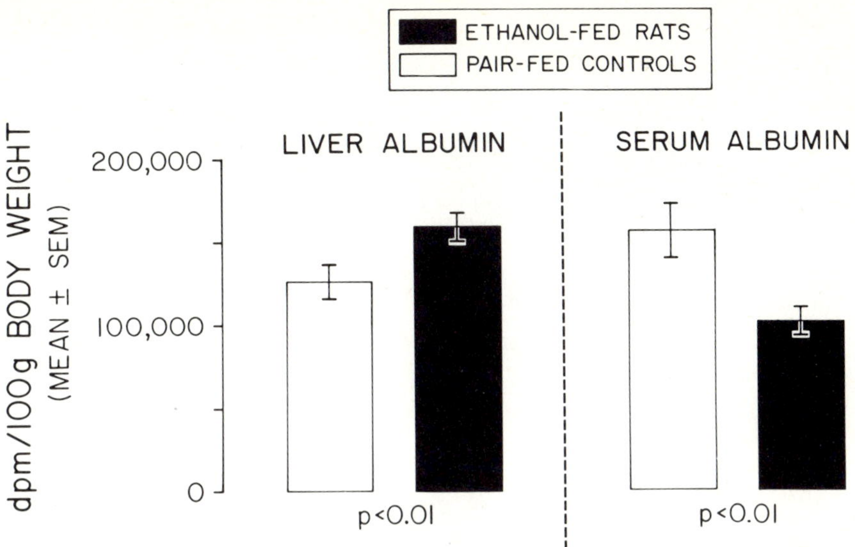

Figure 3 Labelling of liver albumin (left) and serum albumin (right) 30 min after the injection of [^{14}C]leucine in ethanol-fed rats and controls. Ethanol feeding for 3–4 weeks resulted in a delayed appearance of labelled albumin in the plasma, associated with the retention in the liver

feeding, albumin and transferrin concentration increased, not only in the microsomal fractions but also in the cytosol, despite subtraction of the extracellular protein contaminating the liver extracts[13]. The possibility that microsomes from ethanol-fed rats could leak out more of their content during centrifugation is not likely since 10 min after the administration of labelled leucine newly labelled albumin remained in the microsomal fraction, with only minimal labelling of the cytosolic albumin. This observation suggests that cytosolic albumin originates in microsomes and is not merely formed by ribosomes detached from the endoplasmic reticulum. The latter possibility had been raised because after chronic ethanol consumption, degranulation occurs in the rough endoplasmic reticulum which is decreased on electron microscopy[26–29], and also by chemical fractionation[9]. Since leakage into the cytosol or local synthesis appear unlikely, one must then wonder whether the cytosol becomes a storage site for retained export proteins. The retention of these proteins may contribute to the total increase in liver protein found in ethanol-fed rats. However, the magnitude of this contribution was difficult to assess, since only two proteins of the export class have been measured[13], the intrahepatic concentration of export protein does not parallel that of the serum, and retained proteins may undergo a change in their immunological characteristics preventing full recognition by antibodies against serum proteins. The reported increased urea excretion in ethanol-fed rats[30] also suggests that protein catabolism may have been enhanced. The alteration produced by ethanol feeding on the export of proteins from the liver could take place at many possible sites of this complex process. Since it appears to affect more than a single protein of the export class, a common site in this pathway should be looked for. One such possible site could be the microtubular system. This system has been postulated to play a role in the export of proteins in the liver as well as in other organs[31,32]. The evidence is based mainly on the effects of colchicine and other related alkaloids which bind the microtubular protein (tubulin), interfere with the assembly of microtubules[33] and inhibit the export of proteins from the liver[32,34–37]. From the difference in colchicine binding before and after microtubule depolymerization, it was calculated that ethanol feeding reduces the amount of polymerized tubulin (Figure 4) as well as that of total tubulin. Microtubules decreased not merely in concentration, but also

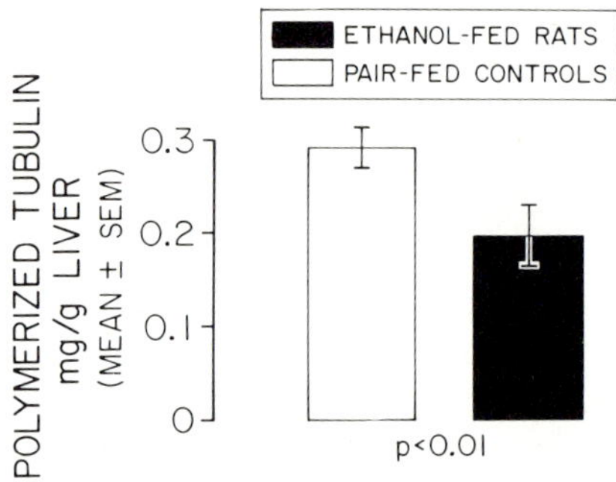

Figure 4 Liver microtubules in control rats and littermates pair-fed an isocaloric adequate ethanol contained diet for 3–4 weeks: there was a significant decrease in microtubules after chronic ethanol consumption

when expressed per total liver; thus, this decrease cannot be accounted for by dilution. The mechanism of this hitherto unrecognized hepatic effect of ethanol remains to be elucidated. In any event, the alteration in microtubules induced by ethanol feeding may account, at least in part, for the delayed export of serum protein from the liver. This alteration in export leads to intrahepatic accumulation of protein, which contributes to the increase in liver protein after ethanol feeding. At early stages of alcoholic liver damage, the delay in export is not sufficient to change serum protein pool or total output of albumin into the serum. Increased synthesis of nascent albumin in the liver may reflect an homeostatic compensation to maintain plasma levels. In the course of advancing alcoholic liver damage, decreased concentration of serum protein could result from either progressive deterioration of the secretory mechanism and/or concomitant failure in protein synthesis, as well as from increased catabolism.

EFFECT OF ETHANOL ON LIPOPROTEINS

As has been illustrated by Stein and Stein at this symposium, and previously[36,37], interference with microtubular functions in the liver (after colchicine administration) results in a defect of protein export, including decreased lipoprotein secretion. Accordingly, in view of the decreased microtubular mass after chronic ethanol consumption discussed already one would expect ethanol feeding to result in impaired lipoprotein secretion. Actually, the opposite can be found in the initial stages of chronic alcohol consumption, which are commonly associated with hyperlipaemia.

The alcohol-induced hyperlipaemia can occur in the fasting state[38]; it is markedly exaggerated, however, when alcohol is given with a fat-containing diet[39–41]. This alcohol effect does not result solely from caloric overload, since no comparable hyperlipaemia was produced by isocaloric amounts of either carbohydrate or lipids[42]. Incorporation of intragastrically administered [^{3}H]palmitate and intravenously injected [^{14}C]lysine into serum VLDL is significantly increased by alcohol administration[43]. These as well as other studies suggested that the ethanol-induced hyperlipaemia results from enhanced lipoprotein production. This contrasts with choline[44–46] deficiency which produces the opposite effect.

Fatty acids are esterified[47] and lipoproteins are formed[48] in the endoplasmic reticulum. Moreover, ethanol consumption enhances the activity of hepatic microsomal L-α-glycerophosphate acyltransferase[49], as well as that of other acyltransferases, depending upon the dietary conditions[50]. The mechanisms of the alterations of these microsomal functions produced by ethanol has not been clarified. It could be linked indirectly to the fact that ethanol can be oxidized at this key metabolic site[15]. Possibly as an adaptive response, chronic ethanol consumption results in increased activity of this microsomal ethanol oxidizing system (MEOS)[15], with an associated rise in cytochrome P_{450}[51] and a proliferation of the membranes of the smooth endoplasmic reticulum[26–29]. This, in turn, is accompanied by an increase in activity of a variety of microsomal enzymes[51], including those involved in drug and steroid metabolism, and also those which participate in glycerolipid synthesis[49]. Moreover, ethanol feeding was also found to enhance the activity of glycosyltransferase[52] which participates in the incorporation of carbohydrate into lipoprotein in the Golgi apparatus.

Thus, contrasting with the delay in protein secretion, there is an apparent increase in lipoprotein release. Actually, the dissociation between proteins and lipoproteins may be more apparent than real. Indeed, a commonly held view is that one determinant of the degree of lipoprotein secretion is the magnitude of the fat load presented to the liver. Indeed, ethanol could act by enhancing the availability of fatty acids, which, in turn, can induce hepatic synthesis of lipoproteins[53].

Thus, in the presence of alcohol, which replaces fat as the liver fuel[54] and also promotes lipogenesis[55], the availability of fat in the liver is markedly increased and it is our working hypothesis that even in the presence of a relative defect in lipoprotein secretion, hyperlipaemia may occur if the fat load is inordinately great. There are at least two arguments in favour of this relative deficiency in lipoprotein secretion. First, there is fat accumulation in the liver despite the hyperlipaemia which illustrates a relative insufficiency of the hyperlipaemic response. Moreover, with time and continuous alcohol intake, the relative deficiency in lipoprotein response becomes an absolute one, as shown in Figure 5. In volunteers, chronic ethanol administration resulted in initial hyperlipaemia[56]. However, blood lipid content declined after 2 weeks (Figure 5) implying that lipoprotein output fell with progressing alcoholic liver disease. This concept is supported by the results of Marzo *et al.*[57] who correlated serum

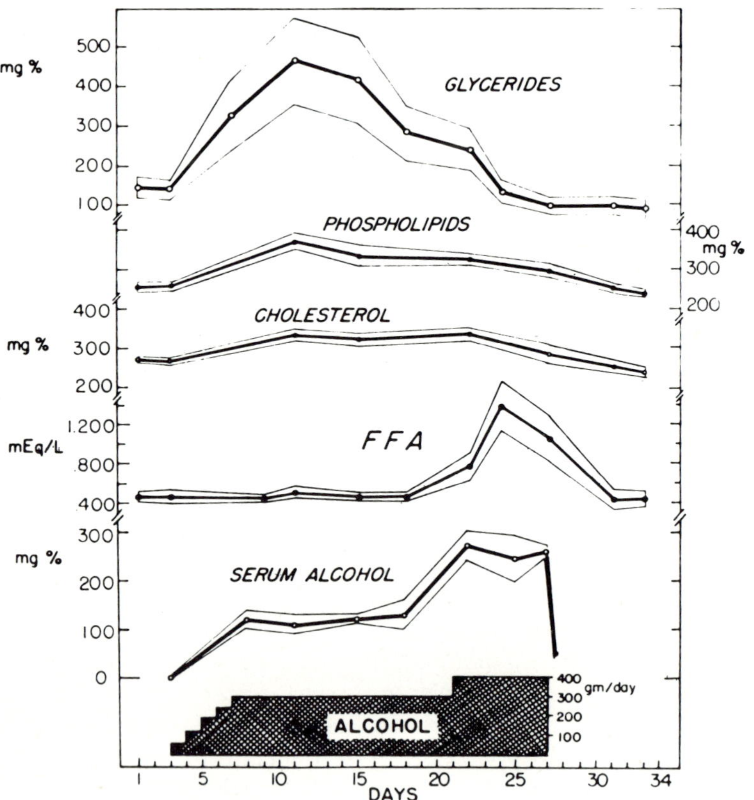

Figure 5 Effect of prolonged alcohol intake on serum lipids in seven chronic alcoholic individuals (average results ± SE of the mean). Contrasting with the initial increase in serum glycerides, after about 10 days, there was a progressive decrease, despite continuation of the same ethanol dose. An increase of the ethanol dose resulted in even further reduction of serum glycerides, despite the concomitant rise of serum FFA[54]

lipids with histologic stage in 90 alcoholics. Peak serum lipid values were found during the stage of fatty metamorphosis. During the succeeding stages of steatosis and hepatitis, a progressive decrease in serum lipids occurred. The decrease was predominant in the triglyceride and cholesterol fractions. The changes of circulating lipids in the fasting state reflect corresponding alterations in lipoproteins. In well-established cirrhosis circulating lipoproteins are generally low[58]. In addition, serum alpha lipoproteins are absent in some cirrhotic patients[59], which illustrates another lipoprotein abnormality. By contrast, patients with alcoholic fatty liver have elevated fasting pre-β-lipoproteins[60].

The progression of liver injury to alcoholic hepatitis in primates is associated

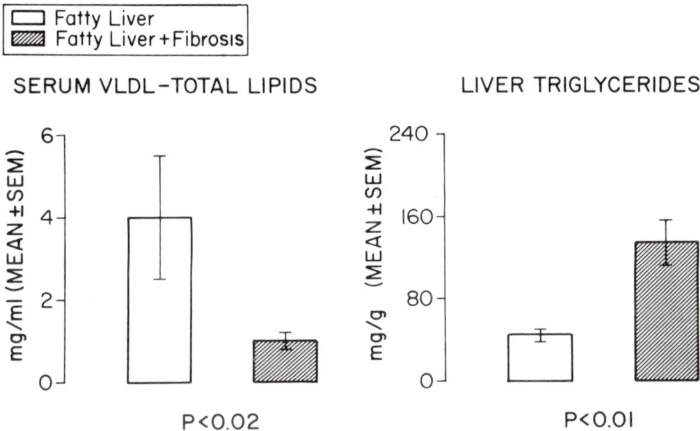

Figure 6 Hepatic triglycerides and serum VLDL in baboons fed ethanol chronically. Whereas the stage of alcoholic fatty liver was associated with a marked increase in VLDL and moderate steatosis, when lesions more severe than fatty liver developed, a significant reduction in plasma VLDL ensued, associated with a striking enhancement of triglyceride accumulation in the liver

with an enhancement of the steatosis[61]. Our preliminary observations also indicate that serum lipoproteins decrease with advancing liver damage[60], suggesting the disappearance of the compensatory role that alcoholic hyperlipaemia exerts on the development of fatty liver.

Indeed, as shown in Figure 6, the increased accumulation of lipids in the liver was associated with a parallel decrease in the VLDL level in the blood of baboons. Similarly, in patients given a dietary fat load, those with fatty liver had a striking increase in their pre-β-lipoprotein or VLDL response in the blood, whereas those with cirrhosis had virtually no pre-β-lipoprotein response[60]. The decreased capacity of the liver to respond to a lipid load with VLDL secretion may actually serve as a sensitive index for the transition of the fatty liver to more advanced stages of alcoholic liver injury.

The disappearance of the hyperlipaemia during progression of alcoholic liver damage indicates, at least in part, an alteration in the ability of the liver to secrete lipoproteins. This decompensation, in turn, might promote further accumulation of fat in the liver. Of course, other mechanisms which could lower serum lipids in cirrhotic patients are not excluded. Fat malabsorption in liver cirrhosis is common[62], but patients with such defects were excluded from our study[61]. Similarly, a possible role for malnutrition was excluded in the baboons which were given ethanol as part of an adequate diet[60]. Malabsorption was not present under these conditions (Lindenbaum and Lieber, unpublished observation). Furthermore clinical protein malnutrition is associated with characteristic plasma amino acid abnormalities, including a depression of branched chain amino acids[63] whereas the alcohol-induced liver injury in the well-fed baboon was associated with opposite amino acid changes[64]. Branched chain amino acids and alpha amino-*n*-butyric acid were found to be increased 2 to 3-fold and 7-fold respectively in the plasma of baboons fed alcohol as 50% of total calories, whereas these amino acids are all depressed when measured in protein deficiency.

It should also be pointed out that the effect of ethanol is dose-dependent. *In vitro*, high ethanol concentration may decrease hepatic lipoprotein release[65]. By contrast, when livers are perfused with ethanol in lower concentrations no inhibition of lipoprotein secretion is found[66]. Similarly, contrasting with the hyperlipaemia which is commonly associated with the administration of moderate to large amounts of ethanol[38,43,56], an extremely high dose has been reported to decrease serum triglyceride[67], very low density lipoproteins[68], high density lipoproteins[69] and the incorporation of glucosamine into the carbohydrate moiety of serum lipoproteins[70] in the rat.

Thus, in addition to the relative impairment of lipoprotein secretion, associated with the initial phase of ethanol abuse, lipoprotein secretion may be reduced because of progressive liver deterioration or because of the high level of ethanol used. When such an absolute decrease in lipoprotein secretion occurs, hepatic fat accumulation will be strikingly enhanced, as illustrated in the case of the baboon in Figure 6. This, in conjunction with retention of soluble protein and the associated increase in intracellular water, may be the basis for the 'ballooning' of the hepatocyte that is commonly observed in alcoholic liver disease. Ballooning, in turn, is frequently associated with hepatic necrosis and inflammation. It is therefore tempting to speculate that one of the mechanisms whereby the alcoholic fatty liver progresses to the hepatitis stage is the impairment of hepatocytic functions secondary to the distension accompanying the ballooning, with associated necrosis and inflammation.

In conclusion, chronic ethanol consumption results in hepatic protein retention, possibly secondary to decreased protein export associated with reduced microtubules. This is accompanied by a relative decrease in lipoprotein secretion which, after prolonged ethanol intake or after high doses, progresses to an absolute reduction in lipoprotein output. The combined accumulation of fat and protein (with associated water retention) results in increased size of the hepatocyte which, when exaggerated, leads to ballooning, a possible key lesion in the transition phase of fatty liver to alcoholic hepatitis.

SUMMARY

Alcohol-induced liver injury is characterized by a swelling of the hepatocyte which, in the most severe cases, leads to 'ballooning'. This increase in cell size (and the associated hepatomegaly) has been attributed to the conspicuous fat accumulation. However, we found that in addition that there is an equally important retention of proteins. Fat accumulates because of an imbalance between the capacity of the hepatocyte to secrete lipoproteins and the intracellular availability of lipids which is very much increased, since ethanol displaces fat as a fuel for the liver. Although in absolute amounts lipoprotein secretion may be enhanced at the early phase of alcoholic abuse, relative to the availability of fat, it is insufficient. Lipoprotein accumulation in the Golgi apparatus may reflect this relative deficiency in lipoprotein secretion. In addition, there is an absolute impairment of the secretion of other proteins resulting in hepatic accumulation of proteins, primarily those of an export type such as albumin and transferrin. One can speculate that the decrease in hepatic microtubules we found in alcohol-fed animals may be responsible for the reduction in protein secretion. In the early stages, protein synthesis is not impaired. With high doses of ethanol or prolonged intake, there is reduction of protein synthesis with a dissociation of ribosomes from the rough endoplasmic reticulum. By contrast, there is proliferation of the smooth membranes of the endoplasmic reticulum; there is an increase of total cytochrome P_{450} with enhanced activity, not only of various microsomal drug metabolizing and steroid hydroxylating enzymes, but also of enzymes involved in lipoprotein assembly, such as glycerophosphate acyltransferase. This, in turn, may facilitate the increased lipoprotein response to the excess lipid load. However, with continuing alcohol abuse, the lipoprotein response fades; the relative lipoprotein deficiency becomes an absolute one. Whereas patients with alcoholic fatty liver have an enhanced plasma VLDL or pre-β-lipoprotein response to a fat meal, patients with cirrhosis have virtually no such response. In the baboon, the decline in circulating VLDL coincided with a progression of fatty liver to more severe stages of alcoholic liver disease, with a further aggravation of the steatosis. Accumulation of fat and protein then results in such severe swelling or 'ballooning' of the hepatocytes that it may well contribute to the development of cell necrosis, a characteristic feature of alcoholic hepatitis.

Acknowledgements

Original studies reported in this paper were supported by the Medical Research Service of the Veterans Administration and US Public Health Service Grants AA-00224 and AM-12511.

References

1. Lieber, C. S., Jones, D. P. and DeCarli, L. M. (1965). Effects of prolonged ethanol intake: production of fatty liver despite adequate diets. *J. Clin. Invest.*, **44,** 1009
2. Baraona, E., Leo, M. A., Borowsky, S. A. and Lieber, C. S. (1975). Alcoholic hepatomegaly: accumulation of protein in the liver. *Science*, **190,** 794
3. DeCarli, L. M. and Lieber, C. S. (1967). Fatty liver in the rat after prolonged intake of ethanol with a nutritionally adequate new liquid diet. *J. Nutr.*, **91,** 331
4. Lowry, O. H., Rosebrough, N. J., Farr, A. and Randall, R. J. (1951). Protein measurement with the folin phenol reagent. *J. Biol. Chem.*, **193,** 265
5. Martin, R. F., Donohue, D. C. and Finch, L. R. (1972). New analytical procedure for the estimation of DNA with *p*-nitrophenylhydrazine. *Anal. Biochem.*, **47,** 562
6. Cederbaum, A. I., Lieber, C. S., Toth, A., Beattie, D. S. and Rubin, E. (1973). Effects of ethanol and fat on the transport of reducing equivalents into rat liver mitochondria. *J. Biol. Chem.*, **248,** 4977
7. Greim, H. (1970). Syntheseteigerung und Abbauhemmung bei der Vermehrung der Mikrosomalen Cytochrome P-450 und *b*-5 durch Phenobarbital. *Naunyn-Schmiederberg's Arch. Pharmacol.*, **266,** 261
8. Tottmar, S. O. C., Pettersson, H. and Kiessling, K.-H. (1973). The subcellular distribution and properties of aldehyde dehydrogenases in rat liver. *Biochem. J.*, **135,** 577
9. Ishii, H., Joly, J.-G. and Lieber, C. S. (1973). Effect of ethanol on the amount and enzyme activities of hepatic rough and smooth microsomal membranes. *Biochim. Biophys. Acta*, **291,** 411
10. Gordon, A. H. and Humphrey, J. H. (1961). Measurement of intracellular albumin in rat liver. *Biochem. J.*, **78,** 551
11. Peters, T., Jr. (1962). The biosynthesis of rat serum albumin. II. Intracellular phenomena in the secretion of newly formed albumin. *J. Biol. Chem.*, **237,** 1186
12. Mancini, G. A., Carbonara, O. and Heremans, J. F. (1965). Immunochemical quantitation of antigens by single radial immunodiffusion. *Immunochemistry*, **2,** 235
13. Baraona, E., Leo, M. A., Borowsky, S. A. and Lieber, C. S. (1977). Pathogenesis of alcohol-induced accumulation of protein in the liver. *Clin. Invest.* (In press)
14. Reyes, H. A., Levi, J., Gatmaitan, Z. and Arias, I. M. (1971). Studies of Y and Z, two hepatic cytoplasmic organic anion-binding proteins: Effects of drugs, chemical, hormones, and cholestasis. *J. Clin. Invest.*, **50,** 2242
15. Lieber, C. S. and DeCarli, L. M. (1970). Hepatic microsomal ethanol-oxidizing system: *in vitro* characteristics and adaptive properties *in vivo*. *J. Biol. Chem.*, **245,** 2505
16. Kuriyama, K., Sze, P. Y. and Rauscher, G. E. (1971). Effects of acute and chronic ethanol administration on ribosomal protein synthesis in mouse brain and liver. *Life Sci.*, **10,** 181
17. Perin, A., Scalabrino, G., Sessa, A. and Arnaboldi, A. (1974). *In vitro* inhibition of protein synthesis in rat liver as a consequence of ethanol metabolism. *Biochim. Biophys. Acta*, **366,** 101
18. Perin, A. and Sessa, A. (1975). *In vitro* effects of ethanol and acetaldehyde on tissue protein synthesis. In: *Satellite Symp., 6th Int. Congress Pharmacology, Vol. XXIII*, p. 105. (Helsinki: The Finnish Foundation for Alcohol Studies)

19. Rothschild, M. A., Oratz, M., Mongelli, J. and Schreiber, S. S. (1971). Alcohol-induced depression of albumin synthesis: reversal by tryptophan. *J. Clin. Invest.*, **50,** 1812
20. Rothschild, M. A., Oratz, M. and Schreiber, S. S. (1974). Alcohol, amino acids and albumin synthesis. *Gastroenterology*, **67,** 1200
21. Jeejeebhoy, K. N., Phillips, M. J., Bruce-Robertson, A., Ho, J. and Sodtke, U. (1972). The acute effect of ethanol on albumin, fibrinogen and transferrin synthesis in the rat. *Biochem. J.*, **126,** 1111
22. Kirsch, R. E., Frith, L. O., Stead, R. H. and Saunders, S. J. (1973). Effect of alcohol on albumin synthesis by the isolated perfused rat liver. *Am. J. Clin. Nutr.*, **26,** 1191
23. Renis, M., Giovine, A. and Bertolino, A. (1975). Protein synthesis in mitochondrial and microsomal fractions from rat brain and liver after acute and chronic ethanol administration. *Life Sci.*, **16,** 1447
24. Hicks, S. J., Drysdale, J. W. and Munro, H. N. (1969). Preferential synthesis of ferritin and albumin by different populations of liver polysomes. *Science*, **164,** 584
25. Redman, C. M. (1969). Biosynthesis of serum proteins and ferritin by free and attached ribosomes of rat liver. *J. Biol. Chem.*, **244,** 4308
26. Lane, B. P. and Lieber, C. S. (1966). Ultrastructural alterations in human hepatocytes following ingestion of ethanol with adequate diets. *Am. J. Pathol.*, **49,** 593
27. Iseri, O. A., Lieber, C. S. and Gottlieb, L. S. (1966). The ultrastructure of fatty liver induced by prolonged ethanol ingestion. *Am. J. Pathol.*, **48,** 535
28. Lieber, C. S. and Rubin, E. (1968). Alcoholic fatty liver in man on a high protein and low fat diet. *Am. J. Med.*, **44,** 200
29. Rubin, E. and Lieber, C. S. (1967). Early fine structural changes in the human liver induced by alcohol. *Gastroenterology*, **52,** 1
30. Rodrigo, C., Antezana, C. and Baraona, E. (1971). Fat and nitrogen balances in rats with alcohol-induced fatty liver. *J. Nutr.*, **101,** 1307
31. Rice, R. H. and Means, G. E. (1971). Radioactive labeling of proteins *in vitro*. *J. Biol. Chem.*, **246,** 831
32. Redman, C. M., Banerjee, D., Howell, K. and Palade, G. E. (1975). Colchicine inhibition of plasma protein release from rat hepatocytes. *J. Cell Biol.*, **66,** 42
33. Olmsted, J. B. and Borisy, G. G. (1973). Microtubules. *Ann. Rev. Biochem.*, **42,** 507
34. Le Marchand, Y., Singh, U. A., Assimacopoulos-Jeannet, F., Orci, L., Rouiller, C. and Jeanrenaud, B. (1973). A role for the microtubular system in the release of very low density lipoproteins by perfused mouse livers. *J. Biol. Chem.*, **248,** 6862
35. Le Marchand, Y., Patzelt, C., Assimacopoulos-Jeannet, F., Loten, E. G. and Jeanrenaud, B. (1974). Evidence for a role of the microtubular system in the secretion of newly synthesized albumin and other proteins by the liver. *J. Clin. Invest.*, **53,** 1512
36. Stein, O. and Stein, Y. (1973). Colchicine-induced inhibition of very low density lipoprotein release by rat liver *in vivo*. *Biochim. Biophys. Acta*, **306,** 142
37. Stein, O., Sanger, L. and Stein, Y. (1974). Colchicine-induced inhibition of lipoprotein and protein secretion into the serum and lack of interference with secretion of biliary phospholipids and cholesterol by rat liver *in vivo*. *J. Cell Biol.*, **62,** 90
38. Jones, D. P., Losowsky, M. S., Davidson, C. S. and Lieber, C. S. (1963). Effects of ethanol on plasma lipids in man. *J. Lab. Clin. Med.*, **62,** 675
39. Brewster, A. C., Lankford, H. G., Schwartz, M. G. and Sullivan, J. F. (1966). Ethanol and alimentary lipemia. *Am. J. Clin. Nutr.*, **19,** 255
40. Barboriak, J. J. and Meade, R. C. (1968). Enhancement of alimentary lipemia by preprandial alcohol. *Am. J. Med. Sci.*, **255,** 245
41. Wilson, D. E., Schreibman, P. H., Brewster, A. C. and Arky, R. A. (1970). The enhancement of alimentary lipemia by ethanol in man. *J. Lab. Clin. Med.*, **75,** 264
42. Losowsky, M. S., Jones, D. P., Davidson, C. S. and Lieber, C. S. (1963). Studies of alcoholic hyperlipemia and its mechanism. *Am. J. Med.*, **35,** 794

43. Baraona, E. and Lieber, C. S. (1970). Effects of chronic ethanol feeding on serum lipoprotein metabolism in the rat. *J. Clin. Invest.*, **49,** 769
44. Oler, A. and Lombardi, B. (1970). Further studies on a defect in the intracellular transport and secretion of proteins by the liver of choline-deficient rats. *J. Biol. Chem.*, **245,** 1282
45. Flores, H., Pak, N., Maccioni, A. and Monckeberg, F. (1970). Lipid transport in kwashiorkor. *Br. J. Nutr.*, **24,** 1005
46. Seakins, A. and Waterlow, J. C. (1972). Effect of a low-protein diet on the incorporation of amino acids into rat serum lipoproteins. *Biochem. J.*, **129,** 793
47. Stein, Y. and Shapiro, B. (1958). Glyceride synthesis by microsome fractions of rat liver. *Biochim. Biophys. Acta*, **30,** 271
48. Jones, A. L., Ruderman, N. B. and Herrera, M. G. (1967). Electron microscopic and biochemical study of lipoprotein synthesis in the isolated perfused rat liver. *J. Lipid Res.*, **8,** 429
49. Joly, J.-G., Feinman, L., Ishii, H. and Lieber, C. S. (1973). Effect of chronic ethanol feeding on hepatic microsomal glycerophosphate acyltransferase activity. *J. Lipid Res.*, **14,** 337
50. Mendenhall, C. L., Bradford, R. H. and Furman, R. H. (1969). Effect of ethanol on glycerolipid metabolism in rat liver. *Biochim. Biophys. Acta*, **187,** 501
51. Joly, J.-G., Ishii, H., Teschke, R., Hasumura, Y. and Lieber, C. S. (1973). Effect of chronic ethanol feeding on the activities and submicrosomal distribution of reduced nicotinamide adenine dinucleotide phosphate (NADPH)-cytochrome P-450 reductase and the demethylases for aminopyrine and ethylmorphine. *Biochem. Pharmacol.*, **22,** 1532
52. Gang, H., Lieber, C. S. and Rubin, E. (1973). Ethanol increases glycosyl transferase activity in the hepatic Golgi apparatus. *Nature New Biology*, **243,** 123
53. Alcindor, L. G., Infante, R., Soler-Argilaga, C. Raisonnier, A. Polonvski, J. and Caroli, J. (1970). Induction of the hepatic synthesis of β-lipoproteins by high concentrations of fatty acids. Effect of actinomycin D. *Biochem. Biophys. Acta*, **210,** 483
54. Lieber, C. S., Lefevre, A., Spritz, N., Feinman, L. and DeCarli, L. M. (1967). Difference in hepatic metabolism of long and medium-chain fatty acids: the role of fatty acid chain length in the production of the alcoholic fatty liver. *J. Clin. Invest.*, **46,** 1451
55. Lieber, C. S. (1974). Effects of ethanol upon lipid metabolism. *Lipids,* **9,** 103
56. Lieber, C. S., Jones, D. P., Mendelson, J. and DeCarli, L. M. (1963). Fatty liver, hyperlipemia and hyperuricemia produced by prolonged alcohol consumption, despite adequate dietary intake. *Trans. Ass. Am. Physicians*, **76,** 289
57. Marzo, A., Chirardi, P., Sardini, B., Prandini, D. and Albertini, A. (1970). Serum lipids and total fatty acids in chronic alcoholic liver disease at different stages of cell damage. *Heft. Unfallheilk*, **48,** 949
58. Cachera, R., Lamotte, M. and Lamotte-Barrillon, S. (1950). Biologique et histologique des steatoses du foie chez les alcooliques. *Sem. Hôp., Paris*, **26,** 3497
59. Papadopoulos, N. M. and Charles, M. A. (1970). Serum lipoprotein patterns in liver disease. *Proc. Soc. Exp. Biol. Med.*, **134,** 797
60. Borowsky, S. A., Perlow, W., Baraona, E. and Lieber, C. S. (1976). Disappearance of alcoholic hyperlipemia as a sign of advancing liver damage. *Gastroenterology,* **70,** A-120
61. Lieber, C. S. and DeCarli, L. M. (1974). An experimental model of alcohol feeding and liver injury in the baboon. *J. Med. Primatol.,* **3,** 153
62. Baraona, E., Orrego, H., Fernandez, O., Amenebar, F. *et al.* (1962). Absorptive function of the small intestine in liver cirrhosis. *Am. J. Dig. Dis.,* **7,** 318
63. Holt, L. E., Jr., Snyderman, S. E., Norton, P. M. and Roitman, E. (1968). The plasma aminogram as affected by protein intake. In: J. H. Leathem (ed.). *Protein Nutrition and Free Amino Acids Patterns*, p. 32. (New Brunswick, N.J.: Rutgers University Press)
64. Shaw, S. and Lieber, C. S. (1975). Plasma amino acids in alcoholic liver injury: contrast with protein malnutrition. *Clin. Res.,* **23,** 459A
65. Schapiro, R. H., Drummey, G. D., Shimizu, Y. and Isselbacher, K. J. (1964). Studies on the pathogenesis of the ethanol-induced fatty liver. II. Effect of ethanol on palmitate-1-C^{14} metabolism by the isolated perfused rat liver. *J. Clin. Invest.*, **43,** 1338

66. Gordon, E. R. (1972). Effect of an intoxicating dose of ethanol on lipid metabolism in an isolated, perfused rat liver. *Biochem. Pharmacol.*, **21,** 2991
67. Dajani, R. M. and Kouyoumjian, C. (1967). A probable direct role of ethanol in the pathogenesis of fat infiltration in the rat. *J. Nutr.*, **91,** 535
68. Madsen, N. P. (1969). Reduced serum very low-density lipoprotein levels after acute ethanol administration. *Biochem. Pharmacol.*, **18,** 261
69. Koga, S. and Hirayama, C. (1968). Disturbed release of lipoprotein from ethanol induced fatty liver. *Experientia*, **24,** 438
70. Mookerjea, S. and Chow, A. (1969). Impairment of glycoprotein synthesis in acute ethanol intoxication in rats. *Biochim. Biophys. Acta*, **184,** 83

DISCUSSION

Dr H. Greim, Neuherberg-Munich (West Germany): What is the mechanism of the altered composition of the endoplasmic reticulum including an alcohol-specific cytochrome P_{450}? Is it a newly synthesized haemoprotein or the original P_{450} changed in substrate specificity due to altered lipoprotein composition of the endoplasmic reticulum?

Dr Lieber, New York (USA): Chronic alcohol consumption results in striking changes of the endoplasmic reticulum, including proliferation and alteration in composition. However, the substrate specificity for ethanol of cytochrome P_{450} is not changed solely because of altered lipoprotein composition. The altered electrophoretic mobility and the increased capacity to promote ethanol oxidation of the 'ethanol-specific' cytochrome persisted with the purified form (separated from the bulk of the lipoproteins of the microsomal membranes) upon reconstitution of the system with a synthetic phospholipid. The 'ethanol-specific' form is actually one of the many forms of P_{450} normally present, but its amount is considerably increased after chronic ethanol feeding.

Dr M. U. Dianzani, Torino (Italy): With Drs Gabriel and Bonelli, the involvement of the microtubular system was investigated in the early phases of fat accumulation in the liver of CCl_4-poisoned rats. It was assumed that aldehydes, or some other substances likely formed by lipid peroxidation of unsaturated fatty acids, may damage tubulin. Several aldehydes, including some presumably produced from liver lipids, blocked the binding of labelled colchicine to the supernatant tubulin from rat liver, starting from an *in vitro* concentration of 1 mM. It is doubtful whether these concentrations are reached in specific key sites of the cell, but we cannot exclude activity *in vivo* of lower concentrations. Acetaldehyde, the major metabolite of ethanol, was not tested, but it may behave similarly and it might reach a 1 mM concentration in the cytoplasm of ethanol-poisoned animals.

Dr Lieber, New York (USA): That acetaldehyde generated from ethanol may interfere with the action of tubulin is possible and we are currently testing this with encouraging results.

26
Microfilaments and Cholestasis: Selected Electron Microscopic and Cytochemical Observations

M. J. PHILLIPS, M. ODA, I. YOUSEF, M. M. FISHER, K. N. JEEJEEBHOY and K. FUNATSU

The mechanism of intrahepatic cholestasis is not well understood but a number of theories have been proposed[1–3], and it is likely that the mechanism is diverse. There is increasing evidence that an abnormality in bile canalicular (BC) membranes may be primary in certain forms of cholestasis[4–9], but canalicular changes are found in virtually all forms of cholestasis so that this problem has not been adequately resolved. Recently, a number of differences have been identified in BC membranes as compared to BC-free liver plasma membranes[10–14]. These reports provide more evidence that the canalicular membranes are specialized and that chemical and structural changes in these membranes might be the primary cause of certain types of intrahepatic cholestasis. The present report deals in particular with BC membrane and microfilament changes in the intrahepatic cholestasis caused by cytochalasin B.

EXPERIMENTAL CHOLESTASIS INDUCED BY CYTOCHALASIN B

Cytochemical markers for three components of bile canaliculi were used in this study: ruthenium red (RR) which stains the glycoprotein-rich surface coat

(RRSC) of canalicular and plasma membranes, Mg^{2+}-activated ATPase (Mg^{2+}-ATPase) as a marker for integral (canalicular) membrane proteins, and uranyl acetate (ua) *en bloc* and RR stains for the canalicular-associated microfilaments.

The effects of cytochalasin B (CB) on hepatocytes *in vivo* were examined as follows: the liver was infused via a mesenteric vein catheter with 50, 100 and 150 μg/ml CB and 0.2% dimethyl sulphoxide (DMSO) at a speed of 0.15

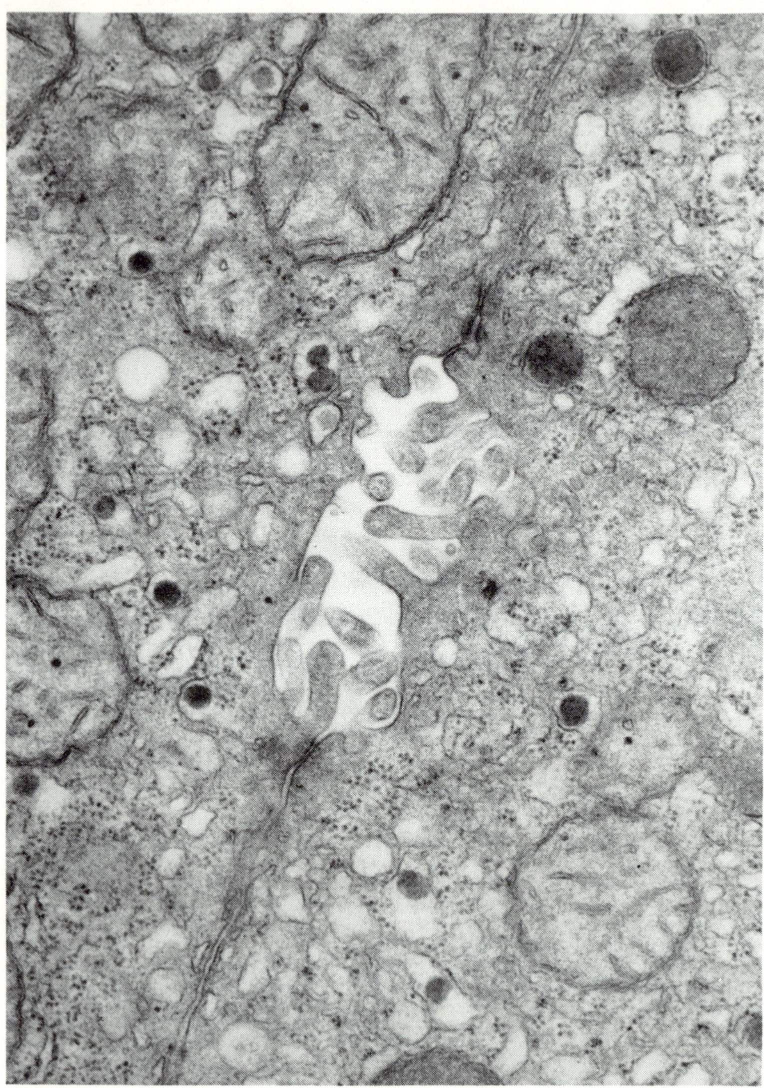

Figure 1 Rat liver from control group infused with 0.2% DMSO. Note normal hepatocellular structure. In the centre of the figure a normal bile canaliculus is shown. Lead citrate (× 47 150)

ml/min for up to 3 hours. In controls, only 0.2% DMSO was infused and normal bile flow and hepatic ultrastructure were observed (Figure 1). In all experiments, prior to infusion, the common bile duct was cannulated to measure bile flow. As reported in previous studies of similar design[25–28], a reduction in bile flow and canalicular changes with dilation and loss of microvilli (Figure 2) were consistently found. The severity of the cholestasis varied both

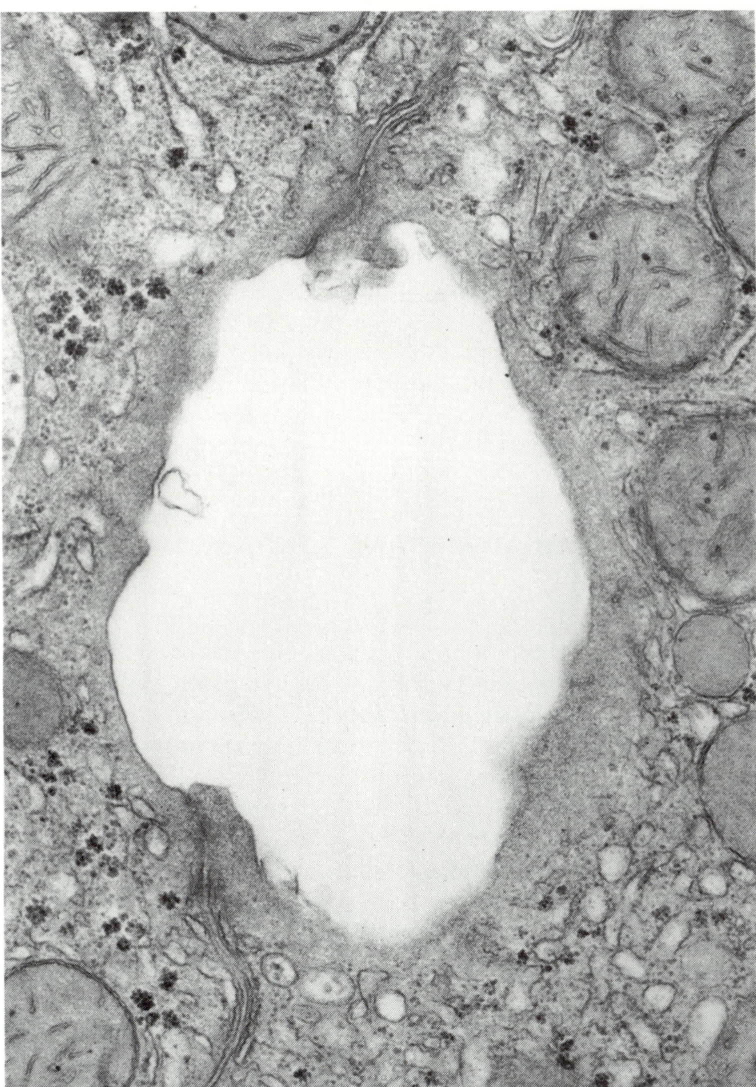

Figure 2 Rat liver from CB-infused animal. Note the dilated bile canaliculus, the absence of microvilli, the prominent ectoplasmic granular zone and the intact tight junctions. Other cell components are normal. Lead citrate (× 40 750)

functionally and structurally with the dosage used. In ruthenium red (RR) preparations stained by the method of Oda *et al.*[21,26], the RRSC on the outer (luminal) surface of the BC membranes was absent. In the pericanalicular region of ua *en bloc* stained preparations, a thick layer of granular material and a few short filamentous fragments replaced the pericanalicular network of microfilaments observed in the controls; Mg^{2+}-activated ATPase (Mg^{2+}-ATPase) activity was also assessed in these preparations. The electron cytochemical procedure for Mg^{2+}-activated adenosine triphosphatase (Mg^{2+}-ATPase) was a modification of the Wachstein–Meisel method[29]. A strong reaction was found on the outer left (luminal) leaflet of all controls, but was absent on the CB-treated BC membranes.

In normal rat liver, canalicular preparations isolated by the method of Song[30] and by its modifications[10] yield intact bile canaliculi and a rich network of canalicular membrane-associated thick (100 Å) and thin (50 Å) microfilaments (Figure 3). We have demonstrated actin in the thin filament component of this network by heavy meromyosin binding in isolated bile canalicular fractions[22]. Since this network might be involved in regulation of bile canalicular function and in cholestasis, we examined *in vitro* canalicular fractions isolated from CB-induced cholestatic liver. They were found *in vitro* devoid of their microfilamentous network[26] (Figure 4). The RRSC and Mg^{2+}-ATPase activity were consistently found in controls but were always absent from the CB-treated animals.

THE CANALICULAR MEMBRANE AND ASSOCIATED FILAMENTS

Actin filaments have been found in many non-muscle cells[15–19] including hepatocytes. They are present throughout the cytoplasm but are particularly concentrated near the plasma membranes[20]. It has been suggested that they participate in the movement of vesicles and other components within the cell[20]. There is a notable concentration of microfilaments in the bile canalicular region of hepatocytes. Some of these pericanalicular filaments contain actin as shown by their appropriate size[21], by their heavy meromyosin binding[20,22] and by immunofluorescence and immunoelectron microscopy[23,24]. We have suggested that these pericanalicular microfilaments might provide a contractile force that serves to facilitate the generation of bile secretory pressure and bile flow in the canalicular system[21,22]. This hypothesis was subsequently supported by experiments with cytochalasin B in which pericanalicular microfilament alterations and cholestasis were observed[25].

The precise mechanism of action of CB in molecular terms is not known, and the reports to date are conflicting[15,31–40]. However, there is extensive evidence that CB has an effect on the structure and function of actin-containing microfilaments in many cell types[31,38–48]. In a recent study using [^{3}H]CB, the

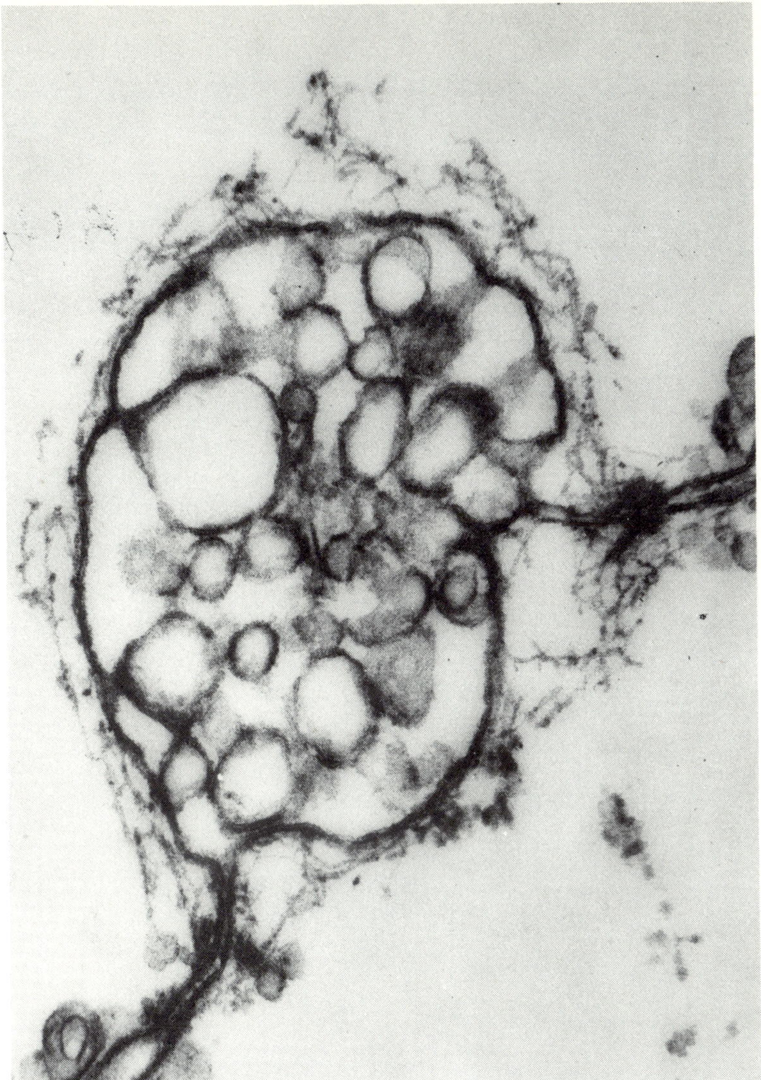

Figure 3 Bile canaliculus isolated from control group infused with 0.2% DMSO. Note the network of pericanalicular microfilaments. The surface coat on the outer (luminal) aspect of the bile canalicular membranes is seen on the microvilli. Ruthenium red (× 62 700)

major binding site of CB was the plasma membrane[49] and there is increasing evidence from electron microscopic and biochemical studies that actin-like microfilaments are associated with isolated plasma membranes[21,50–54].

The combined alterations in the canalicular surface coat and the pericanalicular microfilaments raise the possibility of an action of CB on the

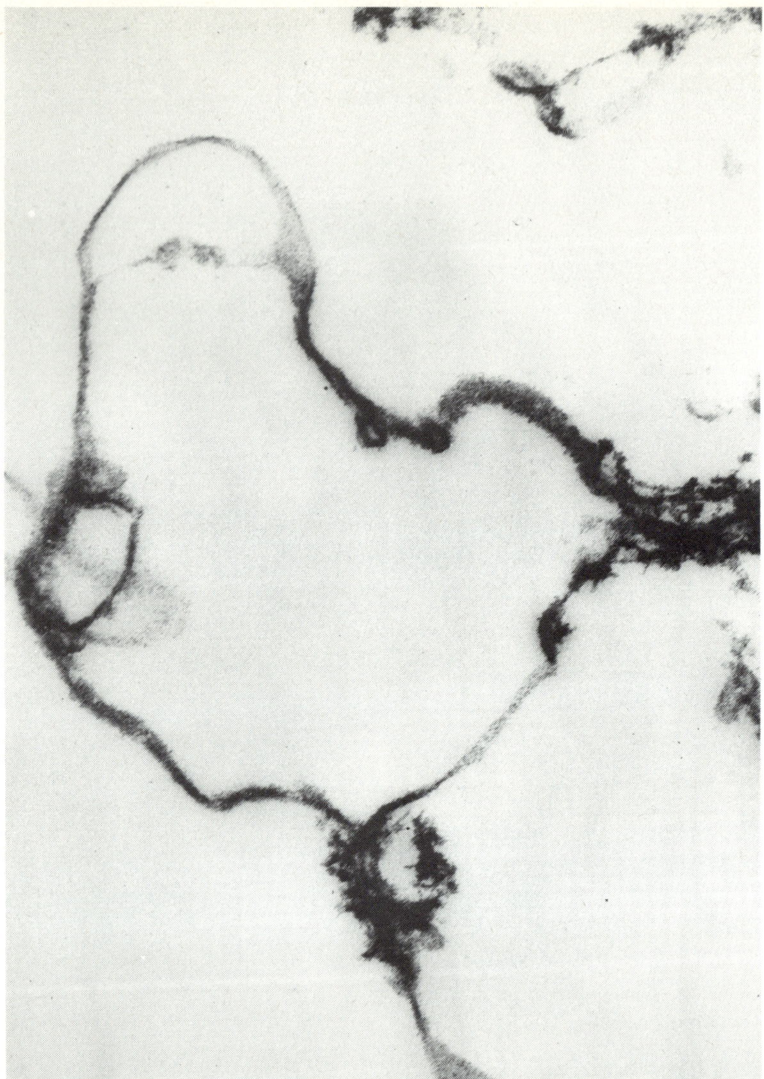

Figure 4 Bile canaliculus from CB-infused liver. Note the naked appearance of the canalicular membranes. Microvilli, microfilaments and the RRSC are all absent. Ruthenium red stain (× 82 500)

bile canalicular membrane itself with resultant changes in these membrane-associated components. This possibility is especially interesting in view of the recent proposed[55] transmembrane cytoskeletal control; cytoskeletal elements (microfilaments and microtubules) may be linked in some way to integral membrane components on the outer surface of the membrane. Perhaps CB affects integral membrane proteins in the BC membrane with dissociation of the RRSC and the microfilaments. Alternately, a direct action of CB on microfilaments with their dissociation from the BC membrane may itself lead to the other membrane alterations noted. There is further evidence from experiments reported elsewhere[25–28] that CB-induced cholestasis is the result of canalicular injury. When CB is added to culture media containing isolated hepatocytes *in vitro*, morphological and cytochemical changes of cholestasis are found in these hepatocytes including bile canalicular dilation with reduction or loss of microvilli, selective loss of the canalicular RRSC and the formation of a thick granular ectoplasmic zone replacing the pericanalicular filaments. In another study, using bile canalicular membranes *in vitro*, a direct action of CB on the canalicular membrane has also been shown. Incubation of BC membranes with CB *in vitro* results in dissociation of the microfilaments, loss of the RRSC and also inactivation of Mg^{2+}-ATPase activity, changes identical to those found *in vivo*. Hence, it is likely that CB-induced cholestasis results from bile canalicular membrane alterations with dissociation of microfilaments.

Acknowledgements

This research was supported by grants-in-aid of the Medical Research Council of Canada.

SUMMARY

To examine the possible role of bile canaliculi and their associated microfilaments in the pathogenesis of cholestasis, studies were carried out using cytochalasin B (CB)-infused rat liver *in vivo*, and using bile canaliculi isolated from normal and CB-infused rat livers. Using dosages of CB that produce cholestasis, structural changes in bile canaliculi, in the pericanalicular microfilaments, and in canalicular surface coat material were noted. When the bile canaliculi were isolated from CB-treated animals, it was found that they lacked the surface coat material and Mg^{2+}-ATPase activity as anticipated but their BC membranes were dissociated from their normal investment of attached microfilaments. These and other findings suggest that CB-induced cholestasis results from an action on bile canalicular membranes and their associated microfilaments.

References

1. Popper, H. and Schaffner, F. (1970). Pathophysiology of cholestasis. *Human Pathol.*, **1**, 1
2. Schaffner, F. and Popper, H. (1969). Cholestasis is the result of hypoactive hypertrophic smooth endoplasmic reticulum in the hepatocyte. *Lancet*, **ii**, 355

3. Simon, F. R. and Arias, I. M. (1973). Alteration of bile canalicular enzymes in cholestasis. A possible cause of bile secretory failure. *J. Clin. Invest.*, **52,** 765
4. Desmet, V. J., Krstulovic, B. and Van Damme, B. (1968). Histochemical study of rat liver in alphanaphthylisothiocyanate (ANIT) induced cholestasis. *Am. J. Pathol.*, **52,** 401
5. Goldfischer, S., Arias, I. M., Essner, E. and Novikoff, A. B. (1962). Cytochemical and electron microscopic studies of rat liver with reduced capacity to transport conjugated bilirubin. *J. Exp. Med.*, **115,** 467
6. Miyai, K., Price, V. M. and Fisher, M. M. (1971). Bile acid metabolism in mammals. Ultrastructural studies on the intrahepatic cholestasis induced by lithocholic and chenodeoxycholic acids in the rat. *Lab. Invest.*, **24,** 292
7. Miyai, K., Mayr, W. M. and Richardson, A. L. (1975). Acute cholestasis induced by lithocholic acid in the rat. A freeze-fracture replica and thin section study. *Lab. Invest.*, **32,** 527
8. Schaffner, F., Popper, H. and Perez, V. (1960). Changes in bile canaliculi produced by norethandrolone: electron microscopic study of human and rat liver. *J. Lab. Clin. Med.*, **56,** 623
9. Schaffner, F. and Javitt, N. B. (1966). Morphologic changes in hamster liver during intrahepatic cholestasis induced by taurolithocholate. *Lab. Invest.*, **15,** 1783
10. Fisher, M. M., Bloxam, D. L., Oda, M., Phillips, M. J. and Yousef, I. M. (1975). Characterization of rat liver cell plasma membranes. *Proc. Soc. Exp. Biol. Med.*, **150,** 177
11. Gregory, D. H., Vlahcevic, Z. R., Schatzki, P. and Swell, L. (1975). Mechanism of secretion of biliary lipids. I. Role of bile canalicular and microsomal membranes in synthesis and transport of biliary lecithin and cholesterol. *J. Clin. Invest.*, **55,** 105
12. Boyer, J. L. and Reno, D. (1975). Properties of Na^+–K^+-activated ATPase and rat liver plasma membranes enriched with bile canaliculi. *Biochim. Biophys. Acta*, **401,** 59
13. Yousef, I. M., Bloxam, D. L., Phillips, M. J. and Fisher, M. M. (1975). Liver cell plasma membrane lipids and the origin of biliary phospholipid. *Can. J. Biochem.*, **53,** 989
14. Ronchi, G. and Desmet, V. J. (1973). Histochemical study of gamma-glutamyltranspeptidase (GGT) in experimental intrahepatic and extrahepatic cholestasis. *Beitr. Pathol.*, **150,** 316
15. Wessells, N. K., Spooner, B. S., Ash, J. F., Bradley, M. D., Luduena, M. A. and Taylor, E. L. (1971). Microfilaments in cellular and developmental processes. *Science,* **171,** 135
16. Spooner, B. S., Yamada, K. M. and Wessells, N. J. (1971). Microfilaments and cell locomotion. *J. Cell Biol.*, **49,** 595
17. Schroeder, T. E. (1969). The role of 'contractile ring' filaments in dividing Arbacia eggs. *Biol. Bull.*, **137,** 413
18. Tilney, L. G. and Cardell, R. R. (1970). Factors controlling the reassembly of the microvillous border of the small intestine of salamander. *J. Cell Biol.*, **47,** 408
19. Allison, A. C., Davies, P. and dePetris, S. (1971). Role of contractile microfilaments in macrophage movement and endocytosis. *Nature New Biology,* **232,** 153
20. French, S. W. and Davies, P. L. (1971). Ultrastructural localization of actin-like filaments in rat hepatocytes. *Gastroenterology,* **68,** 765
21. Oda, M., Price, V. M., Fisher, M. M. and Phillips, M. J. (1974). Ultrastructure of bile canaliculi with special reference to the surface coat and the pericanalicular web. *Lab. Invest.*, **31,** 314
22. Phillips, M. J., Oda, M., Mak, E. and Fisher, M. M. (1975). Bile canalicular structure and function. In: C. A. Goresky and M. M. Fisher (eds.). *Jaundice*, pp. 367–382. (New York: Plenum Press)
23. Gabbiani, G., Montesano, R., Tuchweber, B., Salas, M. and Orci, L. (1975). Phalloidin-induced hyperplasia of actin filaments in rat hepatocytes. *Lab. Invest.*, **33,** 562
24. Holborow, E. J., Trenchev, P. S., Dorling, J. and Webb, J. (1975). Demonstration of smooth muscle contractile protein antigens in liver and epithelial cells. *Ann. N.Y. Acad. Sci.*, **254,** 489

25. Phillips, M. J., Oda, M., Mak, E., Fisher, M. M. and Jeejeebhoy, K. N. (1975). Microfilament dysfunction as a possible cause of intrahepatic cholestasis. *Gastroenterology*, **69,** 48
26. Oda, M., Funatsu, K., Yousef, I., Fisher, M. M., Jeejeebhoy, K. N. and Phillips, M. J. (1977). The mechanism of cytochalasin B-induced intrahepatic cholestasis. (Submitted for publication)
27. Phillips, M. J., Oda, M., Yousef, I., Funatsu, K. and Fisher, M. M. (1977). Direct action of cytochalasin B on bile canalicular membranes. (Submitted for publication)
28. Oda, M., Yousef, I., Funatsu, K., Fisher, M. M. and Phillips, M. J. (1975). Studies on isolated bile canaliculi. Selected effects of hydrolytic enzymes on the surface coat, on microfilaments and on membrane-bound enzymes. (Submitted for publication)
29. Oda, M. and Phillips, M. J. (1975). Electron microscopic cytochemical characterization of bile canaliculi and bile ducts *in vitro*. *Virchows Arch. B. Cell Pathol.*, **18,** 109
30. Song, C. S., Rubin, W., Rifkind, A. B. and Kappas, A. (1969). Plasma membranes of rat liver. Isolation and enzymatic characterization of a fraction rich in bile canaliculi. *J. Cell Biol.*, **41,** 124
31. Axline, S. G. and Reaven, E. P. (1974). Inhibition of phagocytosis and plasma membrane mobility of the cultivated macrophage by cytochalasin B. Role of subplasmalemmal microfilaments. *J. Cell Biol.*, **62,** 647
32. Baudhuin, H., Stock, C., Vincent, D. and Grenier, J. F. (1975). Microfilamentous system and secretion of enzyme in the exocrine pancreas. *J. Cell Biol.*, **66,** 151
33. Bluemink, J. C. (1971). Cytokinesis and cytochalasin-induced furrow regression in the first cleavage zygote of *Xenopus laevis*. *Z. Zellforsch.*, **121,** 102
34. Goldman, R. D. (1972). The effects of cytochalasin B on the microfilaments of the baby hamster kidney (BHK-21) cells. *J. Cell Biol.*, **52,** 246
35. Forer, A., Emmerson, J. and Behnke, O. (1972). Cytochalasin B: does it affect actin-like filaments? *Science*, **175,** 774
36. Mizel, S. B. and Wilson, L. (1972). Inhibition of the transport of several hexoses in mammalian cells by cytochalasin B. *J. Biol. Chem.*, **247,** 4102
37. Schroeder, T. E. The contractile ring. I. Fine structure of dividing mammalian (HeLa) cells and effects of cytochalasin B. *Z. Zellforsch.*, **109,** 431
38. Spudich, J. A. (1973). Effects of cytochalasin B on actin filaments. *Cold Spring Harbor Symp., Quant. Biol.*, **37,** 585
39. Taylor, E. L. and Wessells, N. K. (1973). Cytochalasin B: alterations in salivary gland morphogenesis not due to glucose depletion. *Dev. Biol.*, **31,** 421
40. Yamada, K. M. and Wessells, N. K. (1973). Cytochalasin B: effects on membrane ruffling, growth cone and microscopic activity and microfilament structure. Not due to altered glucose transport. *Dev. Biol.*, **31,** 413
41. Lin, S., Santi, D. W. and Spudich, J. A. (1974). Biochemical studies on the mode of action of cytochalasin B. *J. Biol. Chem.*, **249,** 2269
42. Manasek, J. F., Burnside, B. and Stroman, J. (1972). The sensitivity of developing cardiac myofibrils to cytochalasin B. *Proc. Natl. Acad. Sci. USA*, **69,** 308
43. Orci, L., Gabbay, K. H. and Malaisse, W. J. (1972). Pancreatic beta-cell web: its possible role in insulin secretion. *Science*, **175,** 1128
44. Boyle Kay, M. M. and Fudenberg, H. H. (1973). Inhibition and reversal of platelet activation by cytochalasin B or colcemid. *Nature* (*London*), **244,** 288
45. Shepro, D., Belmarich, J. A., Robblee, L. and Chao, F. C. (1970). Antimobility effect of cytochalasin B observed in mammalian clot retraction. *J. Cell Biol.*, **47,** 544
46. Spooner, B. S. and Wessells, N. K. (1970). Effects of cytochalasin B upon microfilaments involved in morphogenesis of salivary epithelium. *Proc. Natl. Acad. Sci. USA*, **66,** 360
47. Wrenn, J. T. and Wessells, N. K. (1970). Cytochalasin B: effects upon microfilaments involved in morphogenesis of estrogen-induced glands of oviduct. *Proc. Natl. Acad. Sci. USA*, **66,** 904

48. Yamada, K. M., Spooner, B. S. and Wessells, N. K. (1970). Axon growth: roles of microfilaments and microtubules. *Proc. Natl. Acad. Sci. USA*, **66**, 1206
49. Mayhew, E., Poste, G., Cowden, M., Tolson, N. and Maslow, D. (1974). Cellular binding of [^{3}H]cytochalasin B. *J. Cell Physiol.*, **84**, 373
50. Grunstein, E., Rich, A., Weihing, R. R. (1975). Actin associated with membranes from 3Y3 mouse fibroblast and HeLa cells. *J. Cell Biol.*, **64**, 223
51. Mooseker, M. S. and Tilney, L. G. (1975). Organization of actin filament–membrane complex. Filament polarity and membrane attachment in the microvilli of intestinal epithelial cells. *J. Cell Biol.*, **67**, 725
52. Pollard, T. and Korn, E. D. (1973). Electron microscopic identification of actin associated with isolated amoeba plasma membranes. *J. Biol. Chem.*, **248**, 448
53. Rohlich, P. (1975). Membrane-associated actin filaments in the cortical cytoplasm of the rat mast cell. *Exp. Cell Res.*, **93**, 293
54. Wickus, G., Gruenstein, E., Robbins, P. W. and Rich, A. (1975). Decrease in membrane-associated actin of fibroblasts after transformation by Rous sarcoma virus. *Proc. Natl. Acad. Sci. USA*, **72**, 746
55. Nicolson, G. L. (1975). Transmembrane control of the receptors on normal and tumour cells. I. Cytoplasmic influence over cell surface components. *Biochim. Biophys. Acta*, **457**, 57

DISCUSSION

Dr G. Paumgartner, Berne (Switzerland): (1) What is the extent of the cholestasis seen morphologically after cytochalasin B? (2) Do the cores of microvilli contain microfilaments?

Dr Phillips, Toronto (Canada): (1) With 50 μg/ml cytochalasin B the canalicular dilatation observed by light microscopy is localized to the periportal region, but with 100 or 150 μg/ml and continued infusion it becomes more severe and panlobular. (2) Microfilaments are seen in the cores of canalicular microvilli but less than expected. We may lose filaments during the isolation procedure.

27

Mechanisms of Cholestasis — Taurolithocholate Alters Canalicular Membrane Composition, Structure and Permeability

J. L. BOYER, T. J. LAYDEN and Z. HRUBAN

Lithocholate (3-hydroxycholanoic acid) regularly produces cholestasis when infused intravenously at sufficient doses into animals or isolated perfused liver preparations[1–9]. A variety of other hepatotoxic effects have also been described depending on dose, duration of administration and the animal species[10,11]. Although this secondary bile acid is excreted in bile, its hepatotoxicity is normally limited in man by sulphation at the 3-hydroxyl position. Sulphation also diminishes the intestinal absorption of this bile acid which accounts for its high concentration in faeces. However, lithocholate is continuously produced in the intestine by bacterial 7-alpha dehydroxylation of chenodesoxycholic acid and undergoes an enterohepatic circulation where it is conjugated in the liver with taurine and glycine. The majority is also sulphated, yet small amounts of unsulphated taurolithocholate or glycolithocholate are normally excreted in bile[12–14]. Interest in the model of lithocholate cholestasis is derived from the possibility that lithocholate may accumulate in the liver under pathological conditions in man and result in hepatic injury and cholestasis. Even though only small amounts of lithocholate are normally present in human bile, elevated levels of lithocholate have been found in plasma and other body fluids in some, but not all progressive cholestatic disorders of infancy and childhood[15–19]. The growing use of chenodesoxycholic acid to dissolve gallstones has also contributed to the interest in mechanisms of lithocholate toxicity. Recently several studies have helped to clarify how this monohydroxy bile acid may impair biliary secretion. Their results suggest that lithocholate produces cholestasis by modifying the structure and permeability of the membrane of the bile canaliculus.

EXPERIMENTAL OBSERVATIONS

Effects on bile secretion

Immediately following an infusion of 0.6 μm sodium taurocholate (TLC) per min in a 300 g male Sprague–Dawley rat (Table 1), bile flow diminishes progressively as previously described[2–9]. Bile flow is also transiently inhibited following administration of a single bolus of lithocholate, indicating that lithocholate cholestasis is reversible[7]. It is uncertain which fraction of bile is inhibited, although work with the isolated perfused hamster liver suggests that lithocholate may block bile acid independent canalicular secretion[6], which is thought to be largely dependent on sodium transport[20,21]. However, TLC

Table 1 **Bile secretion after 60 min of taurolithocholate (% of preinfusion values)**

I	Taurolithocholate (8)	40 ± 29%
II	Taurolithocholate and taurocholate (11)	138 ± 17%
III	Taurolithocholate and dehydrocholate (11)	83 ± 27%

Bile secretion expressed as the percent of a 30 min preinfusion collection period, 60 min following infusion of tauro-lithocholate (0.2 μm/min per 100 g rat) – Group I: taurocholate and taurolithocholate (0.23 μm/min per 100 g rat) – Group II: sodium dehydrocholate (0.23 μm/min per 100 g rat) and taurolithocholate – Group III. The numbers in parentheses represent the number of experiments in each group while the values are given as the mean ± SD

Table 2

	Mg^{2+}-ATPase	*Na^+–K^+-ATPase*	*5′-Nucleotidase*
Controls (8)	80.7 ± 14.3	15.7 ± 3.0	60.4 ± 9.5
10% Albumin (8)	83.0 ± 9.3	17.1 ± 2.3	72.7 ± 19.3
Taurolithocholate (6)	81.0 ± 17.0	15.6 ± 4.1	57.1 ± 12.7

Enzyme activities in canalicular membranes isolated from rat liver 60 min following infusion of taurolithocholate (0.2 μm per min per 100 g rat) in 10% albumin–normal saline. Control membranes were obtained from animals infused with 10% albumin–normal saline alone. Membranes were isolated and enzymes assayed as previously described except that the membranes were washed in 1 mM EDTA (21). Numbers in parentheses are the experiments in each group. The data is represented as the mean ± SD expressed as μm P_i mg^{-1} protein h^{-1}

has no effect on Na^+–K^+-ATPase activity in bile canalicular plasma membranes of the rat isolated immediately following cholestatic infusions of TLC (Table 2), thus making it unlikely that TLC produces cholestasis by inhibiting the putative sodium transport-dependent fraction of canalicular secretion.

Morphological effects

Morphological studies in the hamster and rat indicate that TLC has a primary effect on membrane structure[5,7–9]. The earliest morphological changes are limited to the membrane of the bile canaliculus and occur simultaneously with the impairment of bile secretion. Later, bile ducts proliferate if lithocholate is fed chronically to rabbits or mice[10,11]. These observations suggest that the initial toxic effects of TLC are at the level of the bile canaliculus, and the associated components within the surrounding pericanalicular ectoplasm. Distinctive morphological changes, which are unique of this particular model of cholestasis, have been observed by both electron and scanning electron microscopy studies in the rat[5,8,9]. Although non-specific changes occur which are seen in most forms of cholestasis, such as widening of the canalicular lumen, loss of microvilli and thickening of the pericanalicular ectoplasm[22], the ectoplasm increases substantially following TLC administration, and is transformed in focal areas into unusual ectoplasmic lamellae which protrude into the lumen of the bile canaliculus (Figure 1). In some areas, the limiting membrane of the canaliculus is disrupted and intracellular organelles and membranous structures enter the lumen. This intrabiliary particulate cytoplasmic material can be seen at higher magnification in Figure 2 and is reminiscent of morphological descriptions of canalicular bile in patients with a progressive cholestatic disease of infancy (Byler's disease) which has been termed 'Byler's bile'[15]. In some regions, the lumen of the canaliculus is filled with lamellar projections (Figure 3), which consist primarily of thickened pericanalicular ectoplasm composed of abundant filamentous structures which are believed to be microfilaments[23]. An increase in density of these microfilaments can also be seen at their site of attachment to the junctional complex (Figures 1, 2 and 3), suggesting that lithocholate may impair the normal function of these actin-containing filaments.

Although morphometric analysis of intracellular organelles has not yet been made following administration of lithocholate, previous qualitative studies suggest that lithocholate also results in an increase in secondary lysosomes, dilatation of the Golgi, disruption of the rough endoplasmic reticulum and proliferation of smooth endoplasmic reticulum[4]. However, the most prominent findings in our own studies have been localized to the bile canaliculus and surrounding structures[9].

These disturbances in canalicular membrane structure are most evident on scanning electron microscopy which provides a representative view of the

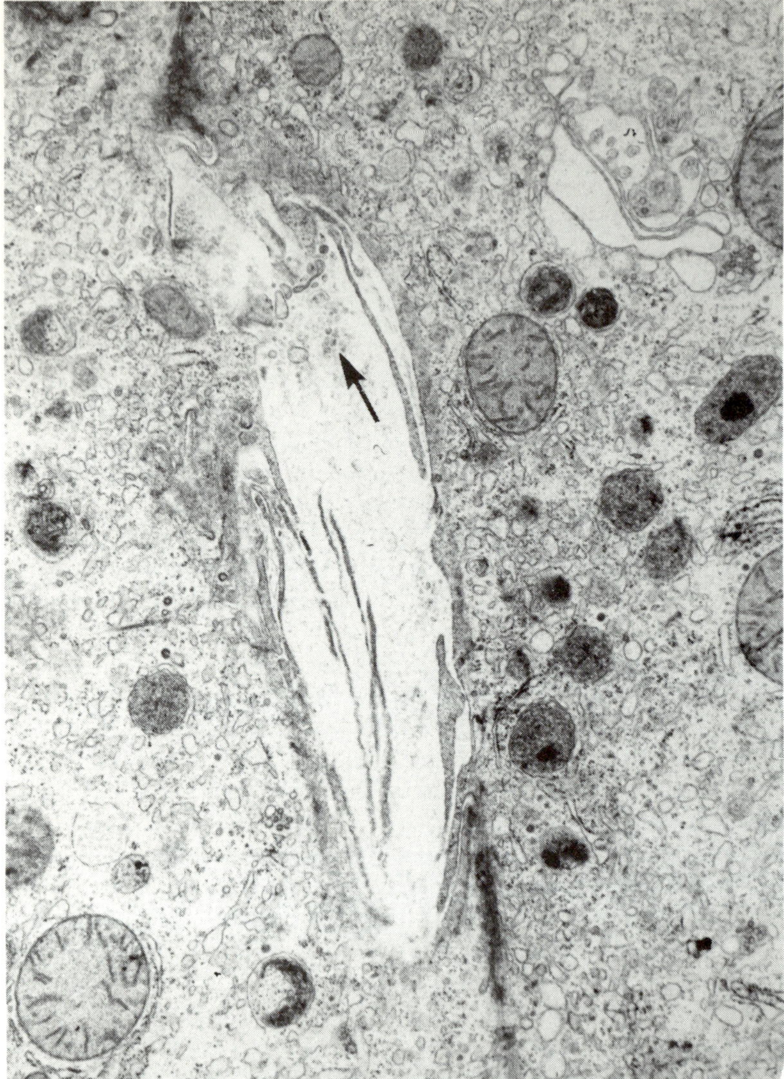

Figure 1 Taurolithocholate infusions (0.2 μm per min per 100 g rat body weight) transform the canalicular membrane into bizarre lamellae which are seen in the lumen of the canaliculus. Particulate cytoplasmic material can also be seen within the canaliculus where the limiting membrane has been disrupted (arrow). 10% Osmium tetroxide fixation — uranyl acetate and lead citrate staining ($\times$ 15 600)

distribution of abnormalities within the hepatic lobule. In addition to the focally distributed tongue-like protrusions of portions of the membrane of the canaliculus (Figure 4), intracellular diverticuli can be seen which penetrate into the interior of the cell. While irregular dilatations of the canalicular

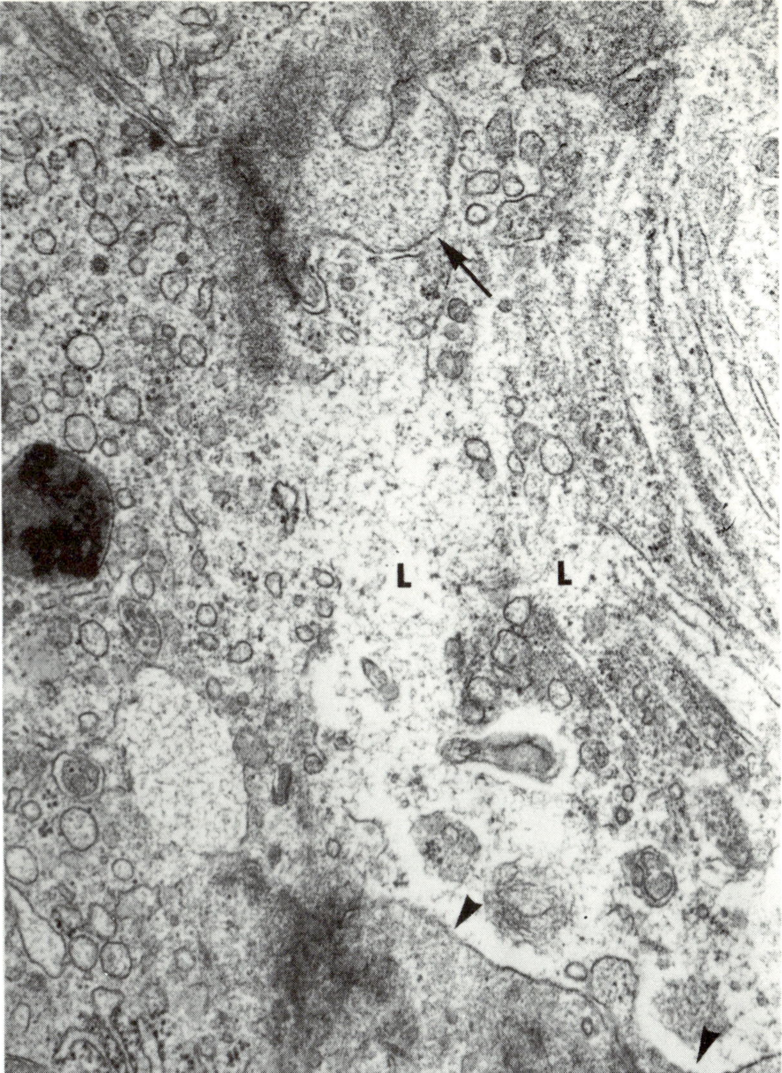

Figure 2 Taurolithocholate disrupts the canalicular membrane and induces bleb formation (arrow). Cytoplasmic material and organelles fill the canalicular lumen (L) and are reminiscent of 'Byler's bile' (reference [14]). Arrowheads mark the remaining portion of the membrane of the canaliculus. Osmium tetroxide fixation, uranyl acetate and lead citrate staining (× 33 600)

lumen are observed throughout the lobule, the most severe disruptions of the canalicular membrane are located in the blind-ended or terminal portions of canaliculi (found throughout the entire lobule) where the lumen of the canaliculus is often widened abnormally to more than half the diameter of the hepatocyte surface (Figure 5). Microvilli are blunted in these areas and largely obliterated

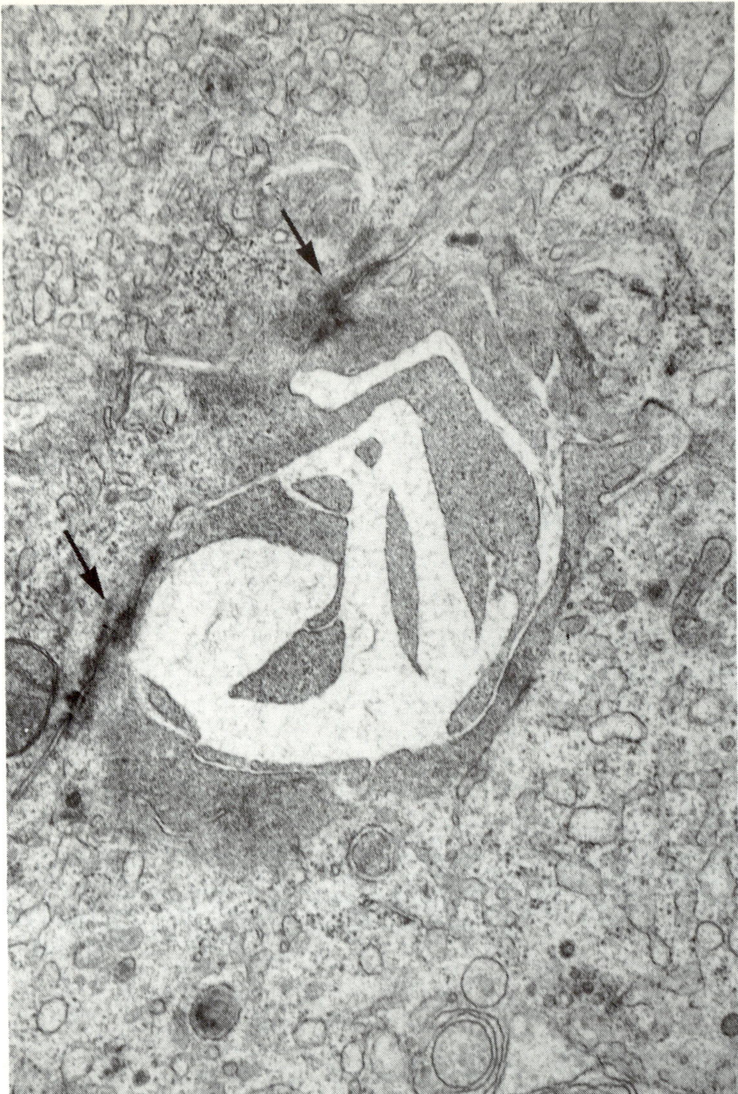

Figure 3 Taurolithocholate increases the pericanalicular ectoplasm which consists of microfilaments, and results in protrusions of these filaments into the lumen of the canaliculus. Microvilli are completely lost. Microfilaments in the area of the junctional complex are also increased in density (arrows). Osmium tetroxide fixation — uranyl acetate and lead citrate staining (Reproduced from *Gastroenterology*, **69,** 724[9]) (× 31 000)

while others are transformed into blebs and irregular tongue-like projections. Although these blebs and lamellae of the canaliculus are observed in all zones of the lobule, they occur primarily in hepatocytes which are located at the 'headwaters' of the canalicular stream. Since taurocholate may be taken up

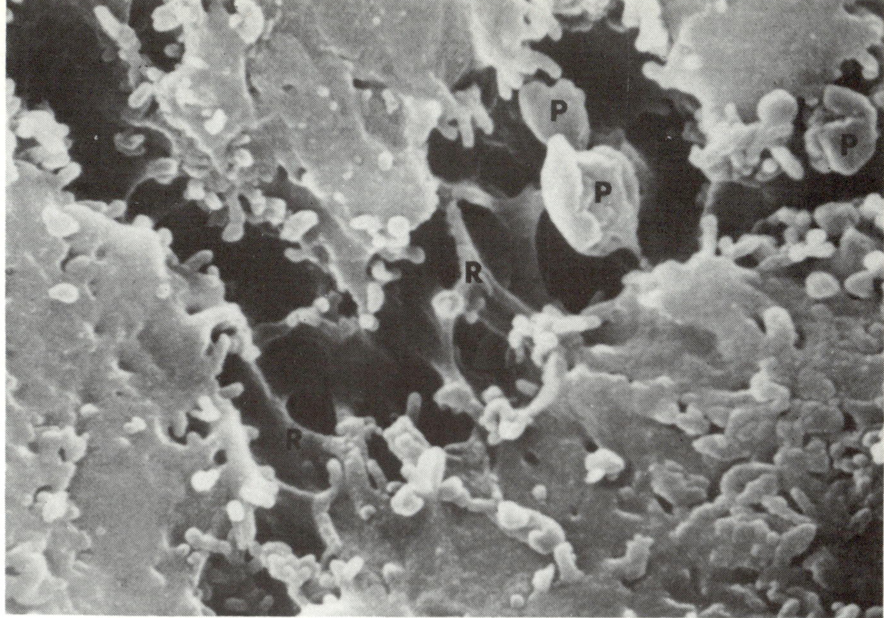

Figure 4 Scanning electron micrograph of a dilated bifurcated canaliculus from rat liver following infusion of taurolithocholate. Note the loss of microvilli, the intracanalicular ridges (R) and lamellar projections (P). (Reproduced from *Gastroenterology*, **69**, 724[9])

more efficiently than TLC by cells which are closer to the periportal sinusoid (near the portal tracts), the lesions in the terminal canaliculi may be localized to areas in the lobule where the ratio of lithocholate to taurocholate is particularly high. It is not entirely clear that these lesions are important to the pathogenesis of the cholestasis; they probably represent a toxic manifestation of lithocholate which is not essential to the initiation of cholestasis since they can be prevented by infusing the sulphated form of lithocholate (Figure 6), even though a mild degree of cholestasis persists[9].

Protection from lithocholate toxicity

Sulphation at the 3-hydroxyl position increases the aqueous solubility of lithocholate. However, it is not known whether this chemical modification diminishes toxicity by enabling a greater fraction to be excreted in the bile or urine, or by attenuating bile acid–membrane physical–chemical interactions. Studies in man indicate that sulphated derivatives of lithocholate are actually excreted more slowly in bile than the non-sulphated form, yet only traces appear in the urine[12–14], so that it seems possible that sulphation modifies membrane interactions.

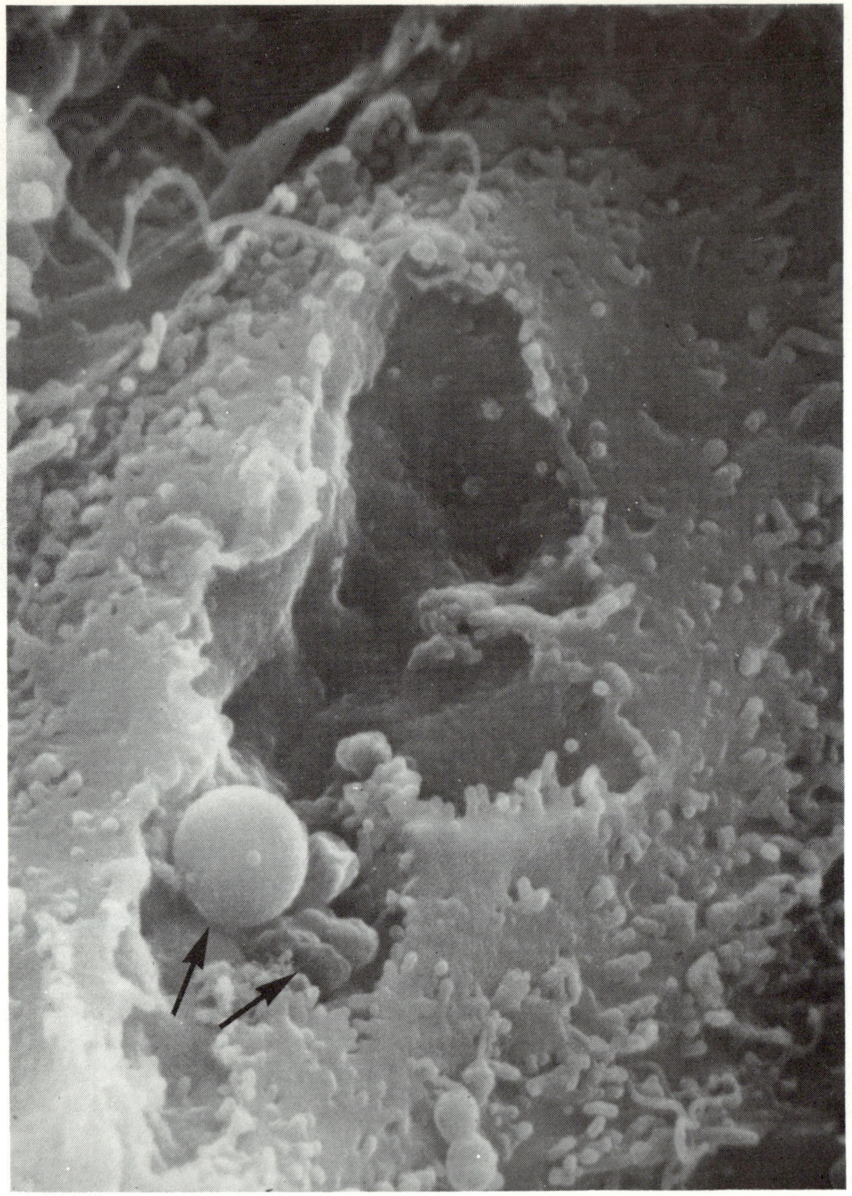

Figure 5 Scanning electron micrograph of terminal portions of a grossly distorted bile canaliculus. The lumen is widened to over half the diameter of the cell surface. Blunted microvilli line the margins near the tight junctions, but are rarely observed along the cell surface in the lumen. Blebs of distorted membrane can also be seen (arrows). (Reproduced from *Gastroenterology*, **69,** 724[9]) (× 5000)

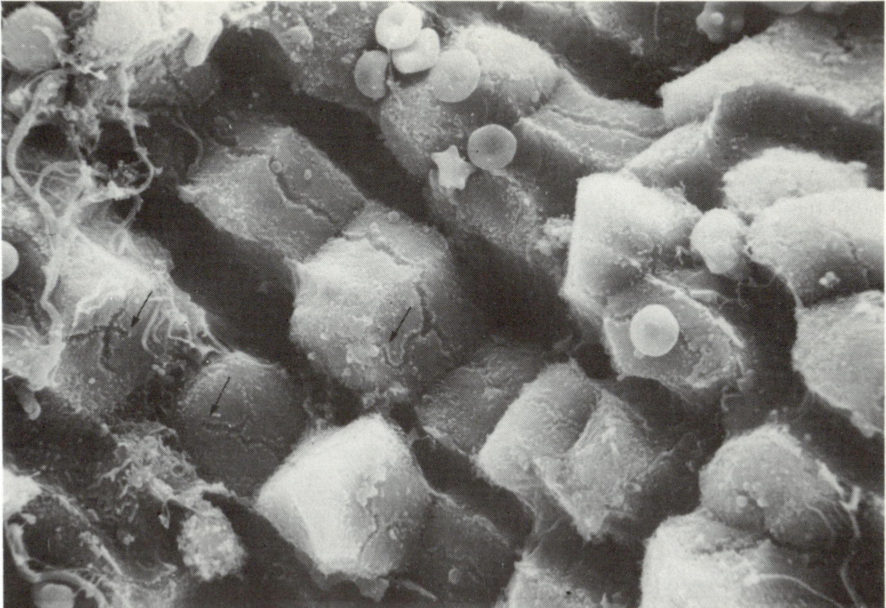

Figure 6 Scanning electron micrograph of the hepatic lobule following infusion of 0.2 μm lithocholate sulphate per 100 g rat body weight. Note the normal appearance of the terminal portions of bile canaliculi (arrows) as well as the presence of microvilli. The appearance is similar to controls ($\times$ 2000)

Complete prevention of TLC cholestasis occurs with simultaneous infusions of equimolar amounts of taurocholate[3,9]. Infusions of taurocholate not only totally pre-empt the cholestatic effects of TLC, but actually stimulate bile secretion (Table 1). The magnitude of this choleresis is equivalent to the effects on bile flow when taurocholate is infused alone. Some insight into the mechanism of the bile acid reversal of TLC toxicity can be gained by comparing the results of this experiment with an identical study which substitutes a non-micelle-forming potent choleretic (sodium dehydrocholate), for the equimolar infusion of sodium taurocholate. Although dehydrocholate normally stimulates bile secretion to a greater extent than taurocholate in the rat, dehydrocholate infusions only partially prevented the cholestatic effects of lithocholate (Table 1). Thus, the protective effects of taurocholate may be related to its detergent properties, and the solubilization of lecithin and lithocholate into mixed micelles during the process of biliary secretion[24]. Alternatively, taurocholate might wash out lithocholate precipitation from the canaliculus[3]. This possibility seems less likely as a primary event since dehydrocholate is not as protective as taurocholate.

Membrane composition changes

Several interesting compositional changes in the canalicular membrane have been described which accompany the physiological and ultrastructural derangements produced by taurolithocholate. In a preliminary communication, Yousef *et al.* report a 7-fold increase in the cholesterol content of canalicular membranes isolated from rats 15 min after infusions of 0.1 to 2.0 μm of lithocholate per min per 100 g body weight[25]. Increased concentrations of lithocholate in the membrane were also observed. Phospholipid content did not change, and there were no increases in lithocholate or lipids in plasma membranes isolated from the sinusoidal pole of the hepatocyte[25]. These provocative preliminary findings support the concept that lithocholate has a direct effect on the macromolecular organization of the canalicular membrane, a view which is supported further by the freeze-fracture replica studies of Miyai *et al.*[8] (Figures 7 and 8). Intramembranous granules can be identified which are evenly dispersed within the bilayer of the canalicular microvilli (Figure 7) and presumably represent integral proteins which may be related to bile secretory transport function. Following lithocholate infusions, the granules are lost from the lipid bilayer in the areas of the membrane which are distorted and transformed into lamellar projections (Figure 8).

Alterations in membrane permeability

An analysis of these diverse effects of TLC on canalicular membrane structure leads us to conclude that cholestasis must be mediated by a direct effect of TLC on the membrane or its associated components. How then might cholestasis occur? Studies of the effects of TLC on the permeability of the canalicular membrane carried out in our laboratory, provide one possible explanation. We examined the biliary clearance of two inert non-transported solutes: [^{3}H]inulin, which is largely excluded from bile because of its large molecular size, and [^{14}C]erythritol, which is smaller and freely diffusible in canalicular water and therefore used as a marker of canalicular bile secretion[24]. The clearance of the smaller inert solute [^{14}C]erythritol (124 daltons) declines in parallel as bile flow diminishes (Figure 9, right panel), suggesting that canalicular bile production is predominantly affected[26]. However, the clearance of [^{3}H]inulin does not decline, but remains at or slightly above control levels during the entire period of the progressive reduction in biliary secretion after correction for biliary dead space and changes in transit time[27].

When one corrects for the increase in diffusion of inulin into bile, which occurs at low rates of biliary secretion, the magnitude of the lithocholate-induced increase in the [^{3}H]inulin bile–plasma ratios (Figure 9, left panel), can only be explained by an increase in inulin permeability (for more complete development of this argument, see [24]). The site of change in permeability of the mem-

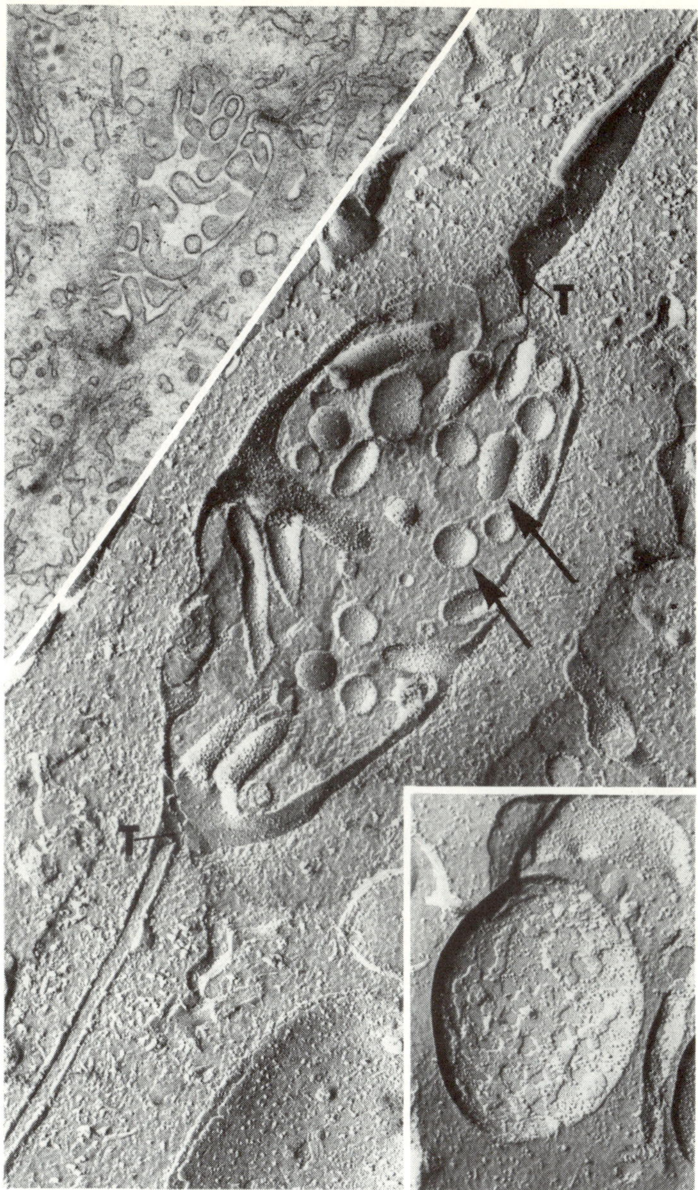

Figure 7 Freeze-fracture replica of a normal rat liver. Bile canaliculus is illustrated in the central figure (magnification × 49 650), and demonstrates intramembranous granules on the convex and concave fracture surfaces of microvilli (arrows). The tight junctions are designated by 'T'. An electron micrograph of a normal bile canaliculus of a rat is placed for comparison in the upper inset (original magnification × 20 000) and a freeze-fracture replica of a mitochondrion is in the lower right-hand inset (original magnification × 23 000). These figures were kindly provided by Dr K. Miyai[8]

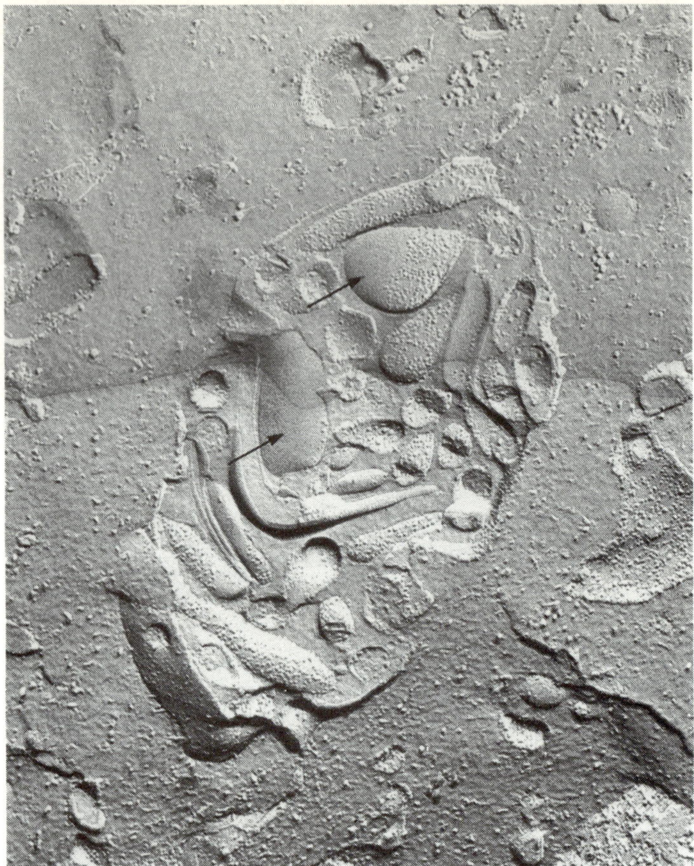

Figure 8 Freeze-fracture replica of a bile canaliculus following a 3-hour infusion of lithocholic acid (0.2 μm/min per 100 g rat), reproduced with the permission of Dr K. Miyai[8]. Some microvilli have become elongated and flattened and demonstrate a variable loss of the intramembranous granules (arrows) (× 55 200)

brane is speculative, but on the basis of morphological findings, could involve the intercellular space and junctional complex, or more likely the membrane of the canaliculus. However, increased entry of inulin across bile duct epithelia cannot be entirely excluded.

This increase in [^{3}H]inulin permeability can be entirely prevented by simultaneous infusions of TLC but only partially reversed by equimolar infusions of DHC, as noted previously for TLC-induced changes in bile flow. The reversal of TLC-induced changes in inulin permeability are also associated with a simultaneous improvement in the scanning electron microscopic abnormalities in the canalicular membrane, such that taurocholate but not dehydrocholate prevented the severe damage observed in the terminal, blind-ended bile

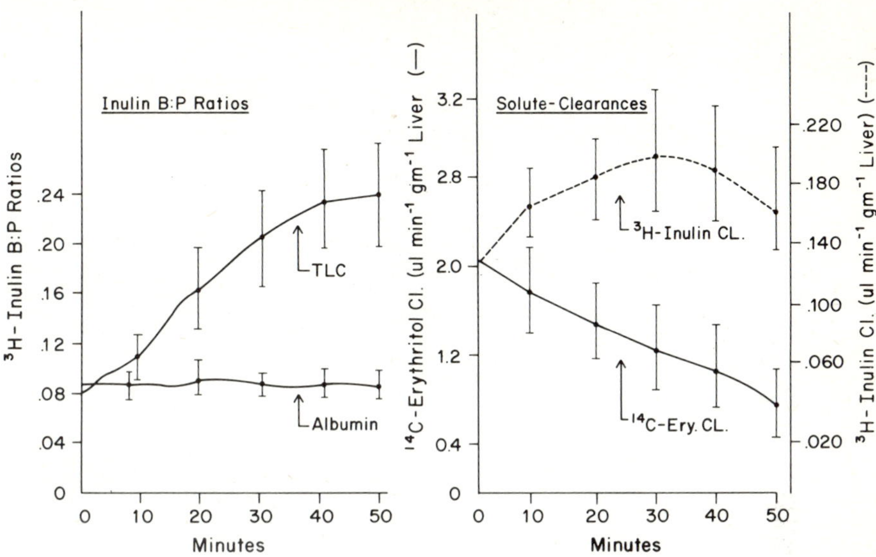

Figure 9 The effect of taurolithocholate infusions on the bile–plasma ratio (B : P) and biliary clearance of the solutes [^{3}H]inulin and [^{14}C]erythritol. Five μCi of [^{3}H]inulin and 2.5 μCi of [^{14}C]erythritol were administered intravenously following bile duct cannulation of male Sprague–Dawley rats. An hour was allowed for equilibration of the isotopes in these experiments before infusion of taurolithocholate (0.2 μm/min per 100 g). Control studies were performed with infusions of 10% albumin–saline which was used as the lithocholate carrier. Bile–plasma ratios of the isotopes were corrected for biliary dead space and changing transit times as cholestasis progressed[27]. No effects were noted following albumin infusions alone on either bile flow or [^{3}H]inulin or [^{14}C]erythritol B : P ratios, shown for the former in the left-hand panel of this figure. The clearance of these solutes into bile (corrected B : P ratios × bile flow) are illustrated in the right-hand panel. Data is expressed as the mean and standard deviation of nine experiments and are significantly different at all time periods after 10 min of TLC infusion

canaliculi (Figure 10)[24]. These findings thus lead to the conclusion that cholestasis may develop because secreted bile diffuses back across the leaky biliary membranes as a consequence of an increase in membrane permeability.

DISCUSSION AND CONCLUSIONS

Although it will be necessary to continue to examine the molecular events which underly the mechanism of TLC cholestasis, the accumulated evidence suggests that there is a direct interaction of this bile acid with components of the bile canalicular membrane. Javitt and Emerman previously suggested that lithocholate produced cholestasis because it exceeded its aqueous solubility in

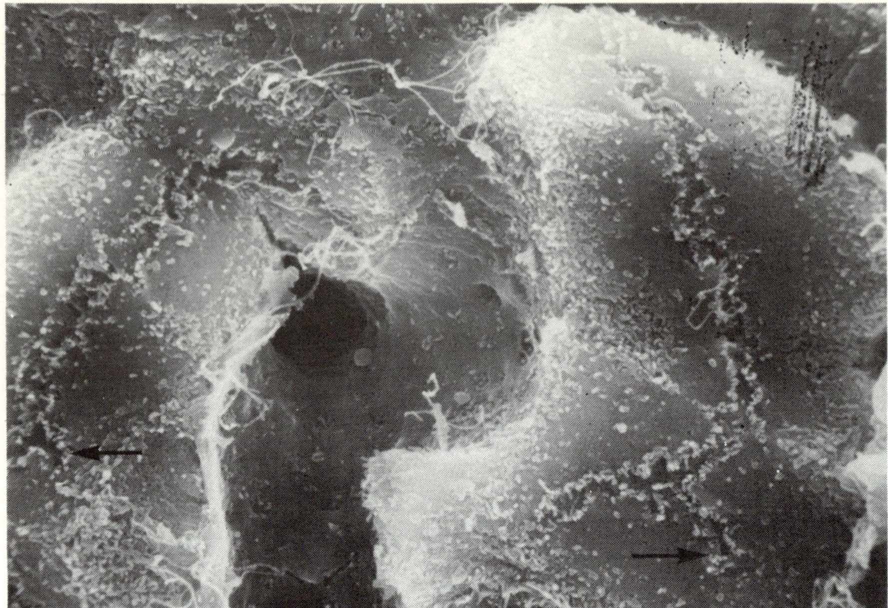

Figure 10 A scanning electron micrograph of rat liver demonstrating two hepatocytes with terminal canaliculi (arrows). Simultaneous infusions of equimolar sodium taurocholate (0.23 μm/min per 100 g rat) prevent disruption of the membrane and gross distortions of the terminal portion of the canaliculi. However, irregularities in the wall of the canaliculus still remain (× 5000)

bile and formed precipitates in bile which obstructed flow[3]. In support of this view, biliary precipitates can be demonstrated after infusions of lithocholate, and Miyai *et al.* have illustrated electron lucent, crystalline material in the lumen of bile canaliculi[8]. However the finding that lithocholate accumulates in the membrane of the canaliculus and increases membrane cholesterol, despite infusion rates which are well below the transport maximum for taurocholate, suggests that there are fundamental changes in the molecular organization of the membrane which are independent of biliary obstruction, even though obstruction from the lamellar projections of canalicular membranes and biliary precipitates may contribute to the development of cholestasis. While it is possible that lithocholate inhibits bile secretion after it has already been secreted into bile, other experimental evidence suggests that the toxicity of lithocholate may be related directly to the binding of the bile acid to components of the membrane. In the first place, the membrane abnormalities are distinctive, and are not commonly observed in animals or patient material obtained following biliary obstruction. Secondly, lithocholate is a haemolysin[29,29] and more toxic to the red cell membrane than other primary secondary bile acids. Since the relative haemolytic effect of primary and secondary bile acids is paralleled by their binding affinity to proteins such as albumin[30], lithocholate should bind more avidly than other bile acids to proteins in the membrane of the canaliculus or adjacent structures such as the microfilaments. The demonstration of high concentrations of lithocholate in the isolated bile canalicular membrane preparations supports this view[25]. Thirdly, when lithocholate is fed to rhesus monkeys, spherical or spiculated erythrocytes are formed simultaneously with increases in plasma levels of lithocholate[31]. This effect on the red cell membrane is consistent with the known effects of a variety of anionic amphipathic molecules which intercalate into the outer layer of membrane bilayers and which in the red blood cell produce crenated cells by expanding the outer layer relative to the cytoplasmic side[32,33]. Furthermore, lithocholate, unlike taurocholate, does not readily solubilize phosphatidyl choline into mixed micelles at physiological pH (M. C. Carey — personal communication) and it is for this reason that lithocholate may become lodged in the interstices of the membrane while taurocholate does not. Thus molecular events which are analogous to lithocholate's effects on the red cell membrane may be responsible for the morphological alterations in the bile canaliculus following administration of lithocholate. These modifications in membrane structure may increase membrane permeability and contribute to the reduction in bile flow.

Parenthetically, the distinctive morphological abnormalities in bile canalicular membranes observed in children with progressive cholestasis (Byler's disease[15]) are quite similar to those observed in the animal model suggesting that there may be a familial defect in bile acid metabolism which results in the accumulation of monohydroxy bile acids in the vicinity of the bile canaliculus in these patients.

SUMMARY

Lithocholate (3-hydroxycholanoic acid) or its taurine conjugate reproducibly inhibit the secretion of bile when infused into experimental animals. This bile acid-induced impairment of bile secretory function is associated with distinctive alterations in both the plasma membrane of the hepatocyte which forms the bile canaliculus and the surrounding microfilaments within the pericanalicular ectoplasm and junctional complex. The lumen of the canaliculus is widened, microvilli are lost and there is a striking thickening of the pericanalicular microfilaments which protrude into the lumen in the form of ectoplasmic lamellae. In focal areas, the integrity of the membrane is disrupted. These structural alterations in the canaliculus occur simultaneously with an increase in permeability of the biliary tree to solutes such as inulin which are normally restricted in their entry into bile.

Bile secretion, membrane morphology and biliary permeability to inulin are all restored to normal with simultaneous infusions of taurocholate, a micelle-forming bile acid, but remain altered when the non-micelle-forming triketo bile acid, dehydrocholate, is infused. These studies suggest that monohydroxy bile acids produce cholestasis through a direct toxic effect on the canalicular membrane and its associated actin-containing microfilaments, which leads to an increase in permeability to bile.

Acknowledgements

Portions of this work were supported by US Public Health Service grant No. AM-17153 and the L. L. Sinton Trust Fund. Dr Boyer is a recipient of an Academic Career Development Award No. AM-70218 from the Institute of Arthritis, Metabolism and Digestive Diseases.

References

1. Ivy, A. C. (1941). The applied physiology of bile secretion and bile salt therapy. *J. Am. Med. Assoc.*, **117,** 1151
2. Javitt, N. B. (1966). Cholestasis in rats induced by taurolithocholate. *Nature* (*London*), **210,** 1262
3. Javitt, N. B. and Emerman, S. (1968). Effect of sodium taurolithocholate on bile flow and bile acid excretion. *J. Clin. Invest.*, **47,** 1002
4. Schaffner, F. and Javitt, N. B. (1966). Morphologic changes in hamster liver during intrahepatic cholestasis induced by taurolithocholate. *Lab. Invest.*, **15,** 1783
5. Miyai, K., Price, V. M. and Fisher, M. M. (1971). Bile acid metabolism in mammals. Ultrastructural studies on the intrahepatic cholestasis induced by lithocholic and chenodeoxycholic acids in the rat. *Lab. Invest.*, **24,** 292
6. King, J. E. and Schoenfield, L. J. (1971). Cholestasis induced by sodium taurolithocholate in isolated hamster liver. *J. Clin. Invest.*, **50,** 2305
7. Priestly, B. G., Coté, M. G. and Plaa, G. L. (1971). Biochemical and morphological parameters of taurolithocholate-induced cholestasis. *Can. J. Physiol. Pharmacol.*, **49,** 1078
8. Miyai, K., Mayr, W. W. and Richardson, A. L. (1975). Acute cholestasis induced by lithocholic acid in the rat — a freeze-fracture replica and thin section study. *Lab. Invest.*, **32,** 527
9. Layden, T. J., Schwarz, J. and Boyer, J. L. (1975). Scanning electron microscopy of the rat liver — studies of the effect of taurolithocholate and other models of cholestasis. *Gastroenterology*, **69,** 724
10. Palmer, R. H. (1976). Toxic effects of lithocholate liver and biliary tree. In: W. Taylor (ed.). *The Hepatobiliary System — Fundamental and Pathological Mechanisms*, pp. 227–240, NATO Advanced Study Institute Series. (New York and London: Plenum Press)
11. Palmer, R. H. (1972). Bile acids, liver injury, and liver disease. *Arch. Int. Med.*, **130,** 606
12. Cowen, A. E., Korman, M. G., Hofmann, A. F. and Cass, O. W. (1975). Metabolism of lithocholate in healthy man. I. Biotransformation and biliary excretion of intravenously administered lithocholate, lithocholylglycine and their sulfates. *Gastroenterology*, **69,** 59
13. Cowen, A. E., Korman, M. G., Hofmann, A. F., Cass, O. W. and Coffin, S. B. (1975). Metabolism of lithocholate in healthy man. II. Enterohepatic circulation. *Gastroenterology*, **69,** 67
14. Cowen, A. E., Korman, M. G., Hofmann, A. F. and Thomas, P. J. (1975). Metabolism of lithocholate in healthy man. III. Plasma disappearance of radioactivity after intravenous injection of labeled lithocholate and its derivatives. *Gastroenterology*, **69,** 77

15. Linarelli, L. G., Williams, C. N. and Phillips, M. J. (1972). Byler's disease: fatal intrahepatic cholestasis. *J. Pediatr.*, **81**, 484
16. Williams, C. N., Kaye, R., Baker, L., Hurwitz, R. and Senior, J. R. (1972). Progressive familial cholestatic cirrhosis and bile acid metabolism. *J. Pediatr.*, **81**, 493
17. Banfield, W. J., Thaler, M. M., Alagille, D. and Admirand, W. H. (1974). Bile acid concentrations in Byler's disease. *Gastroenterology*, **67**, A-2/779
18. DeVos, R., DeWolf-Peeters, C., Desmet, V., Eggermont, E. and VanAcker, K. (1975). Progressive intrahepatic cholestasis (Byler's disease): case report. *Gut*, **16**, 943
19. Clayton, R. J., Iber, F. L., Reubner, B. H. and McKusick, V. A. (1969). Byler's disease: fatal familial intrahepatic cholestasis in an Amish kindred. *Am. J. Dis. Child.*, **117**, 112
20. Erlinger, S., Dhumeaux, D., Berthelot, P. and Dumont, M. (1970). Effects of inhibitors of sodium transport on bile formation in the rabbit. *Am. J. Physiol.*, **219**, 416
21. Boyer, J. L. and Reno, D. (1975). Properties of ($Na^+ + K^+$)-activated ATPase in rat liver plasma membranes enriched with bile canaliculi. *Biochem. Biophys. Acta*, **401**, 59
22. Popper, H. and Schaffner, F. (1970). The pathophysiology of cholestasis. *Human Pathol.*, **1**, 1
23. Phillips, M. J., Oda, M., Mak, E., Fisher, M. M. and Jeejeebhoy, K. N. (1975). Microfilament dysfunction as a possible cause of intrahepatic cholestasis. *Gastroenterology*, **69**, 48
24. Layden, T. J. and Boyer, J. L. (1977). Taurolithocholate-induced cholestasis: taurocholate, but not dehydrocholate, reverses cholestasis and bile canalicular membrane injury. *Gastroenterology*,
25. Yousef, I. M., Kakis, G. and Fisher, M. M. (1976). Lithocholate-induced intrahepatic cholestasis. *Gastroenterology*, **70**, 996
26. Boyer, J. L. (1975). Canalicular bile formation in the isolated perfused rat liver. *Am. J. Physiol.*, **221**, 1156
27. Hacki, W. and Paumgartner, G. (1973). Assessment of bile salt independent bile formation by injection of taurocholate. In: *The Liver, Quantitative Aspects of Structure and Function*, pp. 360–367. (Basel: Karger)
28. Palmer, R. H. (1964). Hemolytic effects of steroids. *Nature* (*London*), **201**, 1134
29. Kappas, A. and Palmer, R. H. (1963). Selected aspects of steroid pharmacology. *Pharmacol. Rev.*, **15**, 123
30. Rudman, D. and Kendall, F. E. (1959). Bile acid content of human serum. II. The binding of cholanic acids by human plasma proteins. *J. Clin. Invest.*, **36**, 538
31. Cooper, R. A., Garcia, F. A. and Trey, C. (1972). The effect of lithocholic acid on red cell membranes *in vivo*. *J. Lab. Clin. Med.*, **79**, 7
32. Deuticke, B. (1968). Transformation and restoration of biconcave shape of erythrocytes induced by amphiphilic agents and changes of ionic environment. *Biochem. Biophys. Acta*, **163**
33. Sheetz, M. P. and Singer, S. J. (1974). Biological membranes as bilayer couples. A molecular mechanism of drug–erythrocyte interactions. *Proc. Natl. Acad. Sci. USA*, **71**, 4457

Summary

HANS POPPER

This summary attempts to extract from the data presented at this symposium the points which help in understanding abnormal function and structure of the hepatocytes, and specifically explore what has been learned about the mechanism of hepatocellular necrosis. Some observations will have to be repeated, but from a different viewpoint.

We heard much about the complex structure of the liver cell membrane, concluded also from the study of the plasma and organelle membranes of cells other than hepatocytes. The key to the understanding is the *fractionation of membrane components*, the purity of which is being controlled by (a) electron microscopy; (b) enzyme histochemical demonstration of the reaction product, particularly of the enzymes on the outer surface, the ectoenzymes, most of which are glycoproteins; (c) chemical determination with emphasis on the activities of enzymes characteristic for the membranes, for instance, of glucose phosphatase for the endoplasmic reticulum and of 5′-nucleotidase for the cell membrane; and (d) specific binding of substances to the membrane including the use of probes with specific agents.

Several *characteristics of the hepatocellular membrane* have been recognized, less from studies on the liver than by extrapolation from investigations of the plasma membranes of other cells, of organelle membranes, and also of cell cultures in which, however, the separation of the various domains is not necessarily maintained. All studies indicate the dynamic state of the membranes, which changes under many circumstances.

First, transmembrane control exists, presumably also through the gap junctions between adjacent hepatocytes.

Secondly, the membrane helps to stabilize and compartmentalize the hepatocyte, although the cytosol, in which membranes are not demonstrable, presumably has an organization also and does not represent a 'lake' of equal constitution throughout the cell. Much emphasis has been placed upon movement within the membranes, i.e. flow of membrane components and of substances passing through the membranes. Both lateral and a transverse flow have been stressed, the latter giving the membrane an 'oily' character.

The third problem is the specialization of the heterogeneous membrane, even in the same domain. Constraints as well as translocations operate to produce a 'mosaic'. Despite the transverse asymmetry of phospholipids and proteins[1], a continuous hydrophobic core is maintained but transmembranous transport of hydrophilic substances is allowed. Glycoproteins and other carbohydrate-containing compounds project beyond the surface of the membrane. Lateral asymmetry or heterogeneity is maintained for the macromolecules and particularly the receptors, in part by fixation by microskeletal insertion. This is disturbed, for instance, in immunological reactions which produce differentiated plaques (capping) demonstrated in lymphocytes.

The formation and degradation of the plasma membrane components in their renewal is a fourth problem also concerning abnormalities of the membrane. The proteins of the membrane and those secreted by the liver cells are formed within the cells and undergo postsynthetic modifications, particularly glycosylation in the endoplasmic reticulum and in the Golgi zone, and some are linked with lipid moieties. The mode of their transport to the cell membrane as molecules, as bulk (micelles), and some in vesicles is not established for many of them. This includes their mode of insertion in the membrane and the possibility of subsequent lateral diffusion or distribution. These processes must be modified during the cycle of cell division. Alterations of the membrane induced by pathological processes or by drugs may thus result in (a) variations in synthesis and degradation, changing the quantity of their components, (b) *de novo* appearance and loss of membrane constituents, and (c) changes in cell membrane specialization.

Membrane components may be internalized into the cells during endocytosis with formation of vesicles with recycling (reuse) of the membrane material. Finally, exocytosis results in shedding of membrane material into the perisinusoidal and canalicular space.

This dynamic state offers several possibilities for *membrane injury as the potential basis of hepatocellular injury*. These include

(a) faulty assembly of the membrane components in general;
(b) failure of glycoprotein formation, for instance in galactosamine intoxication;
(c) alteration of phospholipids as key components of the membranes;
(d) reduced pool of sugar donor nucleotides, as for instance in galactosamine intoxication;
(e) alteration of the receptor activity, for instance by removal of the sialic acid extensions, permitting some and preventing other forms of binding;
(f) detergent action of bile acids;
(g) change of the amino acid metabolism;
(h) disturbance of the ion pumps, primarily interference with its Na^+–K^+-ATPase; and
(j) alteration of the fibrous glycolipids, with the dolichols being an important component. The latter may be related to alterations of the cellular vitamin A metabolism, which has been considered, at least in the

lipocytes, the presumed precursors of hepatic fibroblasts in the parenchyma[2].

Turning to *lesions of specific structures of the membrane*, the possibility of alterations of the receptors which bind peptide hormones, like insulin or glucagon, neurotransmitters and drugs, deserves consideration though only little information proves a relation to hepatic injury. Immunological interference with insulin receptors causes diabetes mellitus[3]. Similar processes have been invoked in Graves' disease and myasthenia gravis[4]. Uncontrolled proliferation as in cirrhosis or even in hepatic tumours might possibly be the result of receptor failure. Receptor alterations are associated with immunological injurious processes; evidence was presented here that attack of antigens of viral origin in the membrane by killer lymphocytes (under the possible influence of antibodies) is the basis of viral hepatitis B. During acute hepatitis these hepatocytes are eliminated, while, presumably, chronic active hepatitis is the result of a suppressed immune reaction. The role of other antigens including exposed or altered membrane lipoproteins is probably similar, while the pathogenetic role of the immune markers, like antibodies against smooth muscle, is today doubtful. Although the importance of hepatic receptor failure in extrahepatic manifestations such as in obesity is well established, the hypothesis that hepatocellular injury may result from receptor failure and faulty recognition is attractive but not yet sufficiently supported by facts.

Disturbance of the cytoskeleton is better substantiated in that a complex process of polymerization and depolymerization is disturbed although its regulation is not well known. Lesions can be produced by various agents; the contractile actin–myosin filaments are altered by cytochalasin B, Concanavalin A, and phalloidin, with different actions of these compounds. The rigid microtubules are changed by colchicine, griseofulvin, and vinblastine. Several consequences have to be considered. The first is secretory failure on the perisinusoidal membrane involving albumin, lipoproteins, globulins, fibrinogen, transferrin, and triglycerides. Secretion of biliary substances at the canalicular membrane is possibly disturbed in cytoskeletal failure, more probably of microfilaments. The second alteration is increased endocytosis demonstrated on the cell surface and accounting for uptake of some drugs but also for cytoplasmic inclusions in the nuclei. Pinocytosis is increased after hepatectomy[6]. Excess pigment formation in lysosomes may be the result of similar processes. Finally, the poorly understood abnormal appearance of bile ductules in various liver diseases may be related to changes in the cytoskeleton; increased filaments have been described under these circumstances[7], as well as abnormal accumulation of bile pigments. Microskeletal failure may cause abnormal exocytosis with resulting ion shifts, a disturbance in membrane tone can readily be accepted, and an antitubular action, for instance by griseofulvin, may produce Mallory's hyalin which, however, consists of excess microfilaments.

If one then attempts to explain the conventional pathological features of the hepatocytes by membrane alterations, several possibilities can be entertained. In *hepatocellular degeneration*, excess endocytosis may result in autophagic vacuoles, which otherwise are derived from focal cytoplasmic degeneration. Steatosis may also be the consequence of faulty secretion of triglycerides related to tubulin failure. Plasma membrane receptor failure may account for other metabolic alterations, and destruction of membranous antigen by immunological processes has already been mentioned. In alcoholic liver injury, tubulin failure accounts not only for hepatic accumulation of lipids, but apparently even to a greater degree of proteins normally secreted by hepatocytes into the serum, like albumin, and reference was made to the formation of hyalin.

In *cholestasis* several processes have to be listed, such as (a) loss of tone of the canalicular membrane from microfilament failure, (b) increased permeability of the canalicular membrane, (c) disturbed secretion from microskeletal failure, and (d) altered bile acid metabolism which may affect the membrane. Whether micelles play a role in transport from the hepatocyte, in addition to micelle formation in the bile, is not established. Bile acids, however, act in the normal release of the ecto-enzymes on the luminal surface of the canaliculi, such as ATPase, as well as of membrane phospholipids and possibly also cholesterol under normal conditions and this is modified in cholestasis. The detergent function of the bile acids is better established in the sequelae of cholestasis[8], possibly in causing membrane receptor failure and hyalin identical to Mallory's hyalin of alcoholics which has been demonstrated in chronic cholestasis[9].

Changes of the plasma membrane have been particularly investigated in *carcinomatous and transformed cells* and these studies have provided most of the information available today about lesions of the cell membrane. This may also apply to abnormal hepatocellular proliferation including hepatocellular cancer. An excess of fucose and of transferases has been demonstrated; membrane proteins and plasma membrane transport are increased; hormonal stimulation, presumably by receptors, is enhanced, and there are indications of greater membrane fluidity and reduced density inhibition.

One of the key problems of this symposium, the role of membrane failure in *the pathogenesis of hepatocellular necrosis*, remains poorly understood. The initial target is not established except in acute cyanide intoxication when cellular respiration and energy supply stop suddenly; no morphological changes are recognized since organelle structure is not altered. This already indicates that hepatocellular death is usually a multistep process[10] in which the separation of the last reversible from the first irreversible step is difficult. The point of no return[11], for instance, has not been recognized in experimental hypoxic injury when reduction of glycolysis, of the NAD/NADH and ATP/ADP ratio, disintegration of polysomes with reduction of ribosomes, and loss of effectiveness of tRNA were recognized[12] and disturbance of the gap functions was observed[13]. Inhibition of macromolecular synthesis does not cause hepatocytic necrosis[14], nor is impaired DNA synthesis fatal for the hepatocytes because of their long lifespan. Inhibition of protein synthesis, for instance, after administration of ethionine and antibiotics such as puromycin, does not necessarily cause death of the hepatocytes. Indeed, such inhibition of protein synthesis in CCl_4 intoxication by cycloheximide may protect, and it has been suggested that this is the result of inhibition of the synthesis of crucial hydrolytic enzymes or of 'lethal' proteins[14]. This again illustrates that hepatocellular death is an active, sequential process[10]. Damage to macromolecules may play a role, as exemplified by the inhibition of RNA polymerase B in alpha-amanitine intoxication[15]. Covalent binding to macromolecules does not necessarily lead to necrosis, though it may produce an immunogen causing liver cell necrosis, and DNA binding plays a role in carcinogenesis. Reduction of energy supply up to a critical level does not produce necrosis nor is lysosomal activation a primary process in hepatocellular death, but rather a secondary scavenger action.

The following processes associated with cell membrane lesions, however, may operate in causing necrosis of hepatocytes

(a) faulty formation of the membranes, particularly their glycoproteins, as seen in galactosamine intoxication;
(b) cytoskeleton failure, particularly of the actin–myosin microfilaments, exemplified by effects of phalloidin;
(c) immunological attack as shown in hepatitis B and by the possibility of abnormal attachment of sensitized lymphocytes when the receptor conditions have been altered by enzymatic removal of parts of the oligosaccharide extensions[16];
(d) cholestasis may raise bile acids to a level that their detergent activity depletes canalicular and possibly plasma membranes of essential components which are not adequately replaced;
(e) ion pump failure may result in an influx of calcium into the cell, associated with depolarization of the membrane and a rise in the intracellular calcium concentration above a tolerable level[17]. However, calcium pumps are not established in the plasma membrane, only in the membrane of the organelles, and the relation between calcium influx and membrane injury remains poorly understood[18] despite the experimental indication that this calcium influx seems to be a critical event.

To answer finally the last question in the introduction, the following seem, despite many common features like the ion pumps, to distinguish the perisinusoidal from the bile canalicular membranes: (a) different ecto-enzymes, used also as marker enzymes, (b) absence of established receptors from the bile canalicular membranes (although a recent study claims that microtubules are involved in canalicular bile salt secretion[19]), (c) participation of microtubules only in secretion through the perisinusoidal membrane, (d) different mechanisms of facilitated transport of bile acids and possibility of other anions; the solubilization of the phospholipids by bile acids *in vitro* from the bile canalicular membrane is, in contrast to the plasma membrane, distinctly selective[20].

The review of the material presented here, supported by the literature, creates a strong feeling that less is established and more is hypothetical as to the role of the alterations of the hepatocellular plasma membrane in disease than in other organ systems. This, however, stimulates rather than discourages future research because of the potential clinical significance of forthcoming information.

We might conclude that alterations of the structural membrane carbohydrates, in addition to secreted glycoproteins, play an important role in liver injury whether they are glycoproteins, such as tubulin and ecto-enzymes, glycolipids, glycolipoproteins, or the oligosaccharide extensions of the receptors. On a personal note, it is now 50 years since Dische and I[21] described the first method to measure total carbohydrates not based on reduction techniques and this gives me the courage to propose another symposium, on the role of structural carbohydrates in liver injury.

References

1. Rothman, J. E. and Lenard, J. (1977). Membrane asymmetry. The nature of membrane asymmetry provides clues to the puzzle of how membranes are assembled. *Science,* **195,** 743
2. Kent, G., Gay, S., Inouye, T., Minick, O.T. and Popper, H. (1976). Vitamin A-containing lipocytes and formation of type III collagen in liver injury. *Proc. Natl. Acad. Sci. USA,* **73,** 3719
3. Kahn, C. R., Flier, J. S., Bar, R. S., Archer, J. A., Gorden, P., Martin, M. M. and Roth, J. (1976). The syndromes of insulin resistance and acanthosis nigricans. Insulin-receptor disorders in man. *N. Engl. J. Med.,* **294,** 739
4. Kahn, C. R., Megyesi, K., Bar, R. S., Eastman, R. C. and Flier, J. S. (1977). Receptors for peptide hormones. New insights into the pathophysiology of disease states in man. *Ann. Intern. Med.,* **86,** 205
5. Hopf, U., Meyer zum Büschenfelde, K.-H. and Arnold, W. (1976). Detection of a liver–membrane autoantibody in HBsAg-negative chronic active hepatitis. *N. Engl. J. Med.,* **294,** 578
6. Mori, M. and Novikoff, A. B. (1977). Induction of pinocytosis in rat hepatocytes by partial hepatectomy. *J. Cell Biol.,* **72,** 695
7. Chedid, A., Spellberg, M. A. and DeBeer, R. A. (1974). Ultrastructural aspects of primary biliary cirrhosis and other types of cholestatic liver disease. *Gastroenterology,* **67,** 858
8. Popper, H., Schaffner, F. and Denk, H. (1976). Molecular pathology of cholestasis. In: W. Taylor (ed.). *The Hepatobiliary System — Fundamental and Pathological Mechanisms.* (New York: Plenum Press)
9. Gerber, M. A., Orr, W., Denk, H., Schaffner, F. and Popper, H. (1973). Hepatocellular hyalin in cholestasis and cirrhosis: its diagnostic significance. *Gastroenterology,* **64,** 89
10. Farber, E. (1971). Biochemical pathology. *Ann. Rev. Pharmacol.,* **11,** 71
11. Van Lancker, J. L. (1976). *Molecular and Cellular Mechanisms in Disease, Vol. 2.,* p. 627. (Heidelberg: Springer-Verlag)
12. Cajone, F., Schiaffonati, L. and Bernelli-Zazzera, A. (1976). Protein synthesis in liver injury. Soluble factors of protein synthesis in the cytosol from ischemic rat liver. *Lab. Invest.,* **34,** 387

13. Peracchia, C. (1977). Gap junctions. Structural changes after uncoupling procedures. *J. Cell Biol.*, **72,** 628
14. Lieberman, M. W. (1972). DNA metabolism, cell death, and cancer chemotherapy In: E. Farber (ed.). *The Pathology of Transcription and Translation.* (New York: Marcel Dekker)
15. Stripe, F. and Fiume, L. (1967). Studies on the pathogenesis of liver necrosis by α-amanitin. Effect of α-amanitin on ribonucleic acid synthesis and on ribonucleic acid polymerase in mouse liver nuclei. *Biochem. J.*, **105,** 779
16. Novogrodsky, A. and Ashwell, G. (1977). Lymphocyte mitogenesis induced by a mammalian liver protein that specifically binds desialylated glycoproteins. *Proc. Natl. Acad. Sci. USA*, **74,** 676
17. Farber, J. L. and El-Mofty, S. K. (1975). The biochemical pathology of liver cell necrosis. *Am. J. Pathol.*, **81,** 237
18. Farber, J. L., El-Mofty, S. K., Schanne, F. A. X., Aleo, J. J. and Serroni, A. (1977). Intracellular calcium homeostasis in galactosamine-intoxicated rat liver cells. Active sequestration of calcium by microsomes and mitochondria. *Arch. Biochem. Biophys.*, **178,** 617
19. Gelrud, L. G., Gregory, D. H., Vlahcevic, Z. R. and Swell, L. (1977). The interrelationship of microtubules, bile acids, bile flow and canalicular ($Na^{+}K^{+}$) ATPase. A unifying concept. *Gastroenterology*, **72,** A-126/1185
20. Yousef, I. M. and Fisher, M. M. (1976). *In vitro* effect of free bile acids on the bile canalicular membrane phospholipids in the rat. *Can. J. Biochem.*, **54,** 1040
21. Dische, Z. and Popper, H. (1926). Uber eine neue Mikrobestimmungsmethode der Kohlehydrate in Organen und Körpersäften. *Biochem. Z.*, **173,** 371

Index